ORGANIC SYNTHESES

ORGANIC SYNTHESES

AN ANNUAL PUBLICATION OF SATISFACTORY
METHODS FOR THE PREPARATION
OF ORGANIC CHEMICALS

VOLUME 86
2009

JOHN A. RAGAN
VOLUME EDITOR

The procedures in this text are intended for use only by persons with prior training in the field of organic chemistry. In the checking and editing of these procedures, every effort has been made to identify potentially hazardous steps and to eliminate as much as possible the handling of potentially dangerous materials; safety precautions have been inserted where appropriate. If performed with the materials and equipment specified, in careful accordance with the instructions and methods in this text, the Editors believe the procedures to be very useful tools. However, these procedures must be conducted at one's own risk. Organic Syntheses, Inc., its Editors, who act as checkers, and its Board of Directors do not warrant or guarantee the safety of individuals using these procedures and hereby disclaim any liability for any injuries or damages claimed to have resulted from or related in any way to the procedures herein.

For general information on our other products and services or for technical support, please contact our Customer Care Department within the United States at (800) 762-2974, outside the United States at (317) 572-3993 or fax (317) 572-4002.

Wiley also publishes its books in a variety of electronic formats. Some content that appears in print may not be available in electronic formats. For more information about Wiley products, visit our web site at www.wiley.com.

Library of Congress Catalog Card Number: 21-17747
ISBN 978-0-470-55614-6

Printed in the United States of America
10 9 8 7 6 5 4 3 2 1

ORGANIC SYNTHESES

Out of print.
†*Deceased.*

Out of print.
†*Deceased.*

Out of print.
†*Deceased.*

*Out of print.
†*Deceased.*

NOTICE

Beginning with Volume 84, the Editors of *Organic Syntheses* initiated a new publication protocol, which is intended to shorten the time between submission of a procedure and its appearance as a publication. Immediately upon completion of the successful checking process, procedures are assigned volume and page numbers and are then posted on the Organic Syntheses website (www.orgsyn.org). The accumulated procedures from a single volume are assembled once a year and submitted for publication in both hard cover and soft cover editions. The soft cover edition of this volume is produced by a rapid and inexpensive process, and is sent at no charge to members of the Organic Division of the American Chemical Society and The Society of Synthetic Organic Chemistry, Japan. The soft cover edition is intended as the personal copy of the owner and is not for library use. The hard cover edition is published by John Wiley and Sons, Inc., in the traditional format, and it differs in content primarily by the inclusion of an index. The hard cover edition is intended primarily for library collections and is available for purchase through the publisher. Incorporation of graphical abstracts into the Table of Contents began with Volume 77. Annual volumes 70–74, 75–79 and 80–84 have been incorporated into five-year versions of the collective volumes of *Organic Syntheses*. Collective Volumes IX, X and XI are available for purchase in the traditional hard cover format from the publishers. The Editors hope that the new Collective Volume series, appearing twice as frequently as the previous decennial volumes, will provide a permanent and timely edition of the procedures for personal and institutional libraries. The Editors welcome comments and suggestions from users concerning the new editions.

Organic Syntheses, Inc., joined the age of electronic publication in 2001 with the release of its free web site (www.orgsyn.org). Organic Syntheses, Inc., fully funded the creation of the free website at www.orgsyn.org in a partnership with CambridgeSoft Corporation and Data-Trace Publishing Company. The site is accessible to most internet browsers using Macintosh and Windows operating systems and may be used with or without a ChemDraw plugin. Because of continually evolving system requirements, users should review software compatibility at the website prior to use. John Wiley & Sons, Inc., and Accelrys, Inc., partnered with Organic Syntheses, Inc., to develop the new database (www.mrw.interscience.wiley.com/osdb) that is available for license with internet solutions from John Wiley & Sons, Inc. and intranet solutions from Accelrys, Inc.

Both the commercial database and the free website contain all annual and collective volumes and indices of *Organic Syntheses*. Chemists can draw structural queries and combine structural or reaction transformation queries with full-text and bibliographic search terms, such as chemical name, reagents, molecular formula, apparatus, or even hazard warnings or phrases. The preparations are categorized into reaction types, allowing search by category. The contents of individual or collective volumes can be browsed by lists of titles, submitters' names, and volume and page references, with or without reaction equations.

The commercial database at www.mrw.interscience.wiley.com/osdb also enables the user to choose his/her preferred chemical drawing package, or to utilize several freely available plug-ins for entering queries. The user is also able to cut and paste existing structures and reactions directly into the structure search query or their preferred chemistry editor, streamlining workflow. Additionally, this database contains links to the full text of primary literature references via CrossRef, ChemPort, Medline, and ISI Web of Science. Links to local holdings for institutions using open url technology can also be enabled. The database user can limit his/her search to, or order the search results by, such factors as reaction type, percentage yield, temperature, and publication date, and can create a customized table of reactions for comparison. Connections to other Wiley references are currently made via text search, with cross-product structure and reaction searching to be added in the near future. Incorporations of new preparations will occur as new material becomes available.

INFORMATION FOR AUTHORS OF PROCEDURES

Organic Syntheses welcomes and encourages submissions of experimental procedures that lead to compounds of wide interest or that illustrate important new developments in methodology. Proposals for *Organic Syntheses* procedures will be considered by the Editorial Board upon receipt of an outline proposal as described below. A full procedure will then be invited for those proposals determined to be of sufficient interest. These full procedures will be evaluated by the Editorial Board, and if approved, assigned to a member of the Board for checking. In order for a procedure to be accepted for publication, each reaction must be successfully repeated in the laboratory of a member of the Editorial Board at least twice, with similar yields (generally ±5%) and selectivity to that reported by the submitters.

Organic Syntheses Proposals

A cover sheet should be included providing full contact information for the principal author and including a scheme outlining the proposed reactions (an *Organic Syntheses* Proposal Cover Sheet can be down-loaded at orgsyn.org). Attach an outline proposal describing the utility of the methodology and/or the usefulness of the product. Identify and reference the best current alternatives. For each step, indicate the proposed scale, yield, method of isolation and purification, and how the purity of the product is determined. Describe any unusual apparatus or techniques required, and any special hazards associated with the procedure. Identify the source of starting materials. Enclose copies of relevant publications (attach pdf files if an electronic submission is used).

Submit proposals by mail or as e-mail attachments to:

Professor Charles K. Zercher
Associate Editor, Organic Syntheses
Department of Chemistry
University of New Hampshire
23 Academic Way, Parsons Hall
Durham, NH 03824

For electronic submissions: *org.syn@unh.edu*

Submission of Procedures

Authors invited by the Editorial Board to submit full procedures should prepare their manuscripts in accord with the Instructions to Authors which are described below or may be downloaded at orgsyn.org. Submitters are also encouraged to consult this volume of *Organic Syntheses* for models with regard to style, format, and the level of experimental detail expected in *Organic Syntheses* procedures. Manuscripts should be submitted to the Associate Editor. Electronic submissions are encouraged; procedures will be accepted as e-mail attachments in the form of Microsoft Word files with all schemes and graphics also sent separately as ChemDraw files.

Procedures that do not conform to the Instructions to Authors with regard to experimental style and detail will be returned to authors for correction. Authors will be notified when their manuscript is approved for checking by the Editorial Board, and it is the goal of the Board to complete the checking of procedures within a period of no more than six months.

Additions, corrections, and improvements to the preparations previously published are welcomed; these should be directed to the Associate Editor. However, checking of such improvements will only be undertaken when new methodology is involved.

NOMENCLATURE

Both common and systematic names of compounds are used throughout this volume, depending on which the Volume Editor felt was more appropriate. The Chemical Abstracts indexing name for each title compound, if it differs from the title name, is given as a subtitle. Systematic

Chemical Abstracts nomenclature, used in the Collective Indexes for the title compound and a selection of other compounds mentioned in the procedure, is provided in an appendix at the end of each preparation. Chemical Abstracts Registry numbers, which are useful in computer searching and identification, are also provided in these appendices. Whenever two names are concurrently in use and one name is the correct Chemical Abstracts name, that name is preferred.

ACKNOWLEDGMENT

Organic Syntheses wishes to acknowledge the contributions of Merck & Co. and Pfizer, Inc. to the success of this enterprise through their support, in the form of time and expenses, of members of the Board of Editors.

INSTRUCTIONS TO AUTHORS

All organic chemists have experienced frustration at one time or another when attempting to repeat reactions based on experimental procedures found in journal articles. To ensure reproducibility, *Organic Syntheses* requires experimental procedures written with considerably more detail as compared to the typical procedures found in other journals and in the "Supporting Information" sections of papers. In addition, each *Organic Syntheses* procedure is carefully "checked" for reproducibility in the laboratory of a member of the Board of Editors.

Even with these more detailed procedures, the experience of *Organic Syntheses* editors is that difficulties often arise in obtaining the results and yields reported by the submitters of procedures. To expedite the checking process and ensure success, we have prepared the following "Instructions for Authors" as well as a *Checklist for Authors* and *Characterization Checklist* to assist you in confirming that your procedure conforms to these requirements. These checklists, which are available at *www.orgsyn.org,* should be completed and submitted together with your procedure. Procedures submitted to *Organic Syntheses* will be carefully reviewed upon receipt. Procedures lacking any of the required information will be returned to the submitters for revision.

Scale and Optimization

The appropriate scale for procedures will vary widely depending on the nature of the chemistry and the compounds synthesized in the procedure. However, some general guidelines are possible. For procedures in which the principal goal is to illustrate a synthetic method or strategy, it is expected, in general, that the procedure should result in at least 5 g and no more than 50 g of the final product. In cases where the point of the procedure is to provide an efficient method for the preparation of a useful reagent or synthetic building block, the appropriate scale may be larger, but in general should not exceed 100 g of final product. Exceptions to these guidelines may be granted in special circumstances. For example, procedures describing the preparation of reagents employed as catalysts will often be acceptable on a scale of less than 5 g.

In considering the scale for an *Organic Syntheses* procedure, authors should also take into account the cost of reagents and starting materials. In general, the Editors will not accept procedures for checking in which the cost of any one of the reactants exceeds $500 for a single full-scale run. Authors are requested to identify the most expensive reagent or starting material on the procedure submission checklist and to estimate its cost per run of the procedure.

It is expected that all aspects of the procedure will have been optimized by the authors prior to submission, and that each reaction will have been carried out at least twice on exactly the scale described in the procedure. It is appropriate to report the weight, yield, and purity of the product of each step in the procedure as a range. In any case where a reagent is employed in significant excess, a Note should be included explaining why an excess of that reagent is necessary. If possible, the Note should indicate the effect of using amounts of reagent less than that specified in the procedure.

Reaction Apparatus

Describe the size and type of flask (number of necks) and indicate how *every* neck is equipped.

"A 500-mL, three-necked, round-bottomed flask equipped with an overhead mechanical stirrer, 250-mL pressure-equalizing addition funnel fitted with an argon inlet, and a rubber septum is charged with...."

Indicate how the reaction apparatus is dried and whether the reaction is conducted under an inert atmosphere. This can be incorporated in the text of the procedure or included in a Note.

"The apparatus is flame-dried and maintained under an atmosphere of argon during the course of the reaction."

In the case of procedures involving unusual glassware or especially complicated reaction setups, authors are encouraged to include a photograph or drawing of the apparatus in the text or in a Note (for examples, see *Org. Syn.*, Vol. 82, 99 and Coll. Vol. X, pp 2, 3, 136, 201, 208, and 669).

Reagents and Starting Materials

All chemicals employed in the procedure must be commercially available or described in an earlier *Organic Syntheses* or *Inorganic Syntheses* procedure. For other compounds, a procedure should be included either as one or more steps in the text or, in the case of relatively straightforward preparations of reagents, as a Note. In the latter case, all requirements with regard to characterization, style, and detail also apply.

In one or more Notes, indicate the purity or grade of each reagent, solvent, etc. It is desirable to also indicate the source (company the chemical was purchased from), particularly in the case of chemicals where it is suspected that the composition (trace impurities, etc.) may vary from one supplier to another. In cases where reagents are purified, dried, "activated" (e.g., Zn dust), etc., a detailed description of the procedure used should be included in a Note. In other cases, indicate that the chemical was "used as received".

> "Diisopropylamine (99.5%) was obtained from Aldrich Chemical Co., Inc. and distilled under argon from calcium hydride before use. THF (99+%) was obtained from Mallinckrodt, Inc. and distilled from sodium benzophenone ketyl. Diethyl ether (99.9%) was purchased from Aldrich Chemical Co., Inc. and purified by pressure filtration under argon through activated alumina. Methyl iodide (99%) was obtained from Aldrich Chemical Co., Inc. and used as received."

The amount of each reactant should be provided in parentheses in the order mL, g, mmol, and equivalents with careful consideration to the correct number of significant figures.

The concentration of solutions should be expressed in terms of molarity or normality, and not percent (e.g., 1 N HCl, 6 M NaOH, not "10% HCl").

Reaction Procedure

Describe every aspect of the procedure clearly and explicitly. Indicate the order of addition and time for addition of all reagents and how each is added (via syringe, addition funnel, etc.).

Indicate the temperature of the reaction mixture (preferably internal temperature). Describe the type of cooling (e.g., "dry ice-acetone bath") and heating (e.g., oil bath, heating mantle) methods employed. Be careful to describe clearly all cooling and warming cycles, including initial and final temperatures and the time interval involved.

Describe the appearance of the reaction mixture (color, homogeneous or not, etc.) and describe all significant changes in appearance during the course of the reaction (color changes, gas evolution, appearance of solids, exotherms, etc.).

Indicate how the reaction can be monitored to determine the extent of conversion of reactants to products. In the case of reactions monitored by TLC, provide details in a Note, including eluent, R_f values, and method of visualization. For reactions followed by GC, HPLC, or NMR analysis, provide details on analysis conditions and relevant diagnostic peaks.

"The progress of the reaction was followed by TLC analysis on silica gel with 20% EtOAc-hexane as eluent and visualization with *p*-anisaldehyde. The ketone starting material has $R_f = 0.40$ (green) and the alcohol product has $R_f = 0.25$ (blue)."

Reaction Workup

Details should be provided for reactions in which a "quenching" process is involved. Describe the composition and volume of quenching agent, and time and temperature for addition. In cases where reaction mixtures are added to a quenching solution, be sure to also describe the setup employed.

"The resulting mixture was stirred at room temperature for 15 h, and then carefully poured over 10 min into a rapidly stirred, ice-cold aqueous solution of 1 N HCl in a 500-mL Erlenmeyer flask equipped with a magnetic stirbar."

For extractions, the number of washes and the volume of each should be indicated.

For concentration of solutions after workup, indicate the method and pressure and temperature used.

"The reaction mixture is diluted with 200 mL of water and transferred to a 500-mL separatory funnel, and the aqueous phase is separated and extracted with three 100-mL portions of ether. The combined organic layers are washed with 75 mL of water and 75 mL of saturated NaCl solution, dried over MgSO$_4$, filtered, and concentrated by rotary evaporation (25°C, 20 mmHg) to afford 3.25 g of a yellow oil."

"The solution is transferred to a 250-mL, round-bottomed flask equipped with a magnetic stirbar and a 15-cm Vigreux column fitted with a short path distillation head, and then concentrated by careful distillation at 50 mmHg (bath temperature gradually increased from 25 to 75°C)."

In cases where solid products are filtered, describe the type of filter funnel used and the amount and composition of solvents used for washes.

"... and the resulting pale yellow solid is collected by filtration on a Büchner funnel and washed with 100 mL of cold (0°C) hexane."

When solid or liquid compounds are dried under vacuum, indicate the pressure employed (rather than stating "reduced pressure" or "dried *in vacuo*").

"... and concentrated at room temperature by rotary evaporation (20 mmHg) and then at 0.01 mmHg to provide. ..."

"The resulting colorless crystals are transferred to a 50-mL, round-bottomed flask and dried overnight in a 100°C oil bath at 0.01 mmHg."

Purification: Distillation

Describe distillation apparatus including the size and type of distillation column. Indicate temperature (and pressure) at which all significant fractions are collected.

"... and transferred to a 100-mL, round-bottomed flask equipped with a magnetic stirbar. The product is distilled under vacuum through a 12-cm, vacuum-jacketed column of glass helices (Note 16) topped with a Perkin triangle. A forerun (ca. 2 mL) is collected and discarded, and the desired product is then obtained, distilling at 50–55°C (0.04–0.07 mmHg). ..."

Purification: Column Chromatography

Provide information on TLC analysis in a Note, including eluent, R_f values, and method of visualization.

Provide dimensions of column and amount of silica gel used; in a Note indicate source and type of silica gel.

Provide details on eluents used, and number and size of fractions.

"The product is charged on a column (5 × 10 cm) of 200 g of silica gel (Note 15) and eluted with 250 mL of hexane. At that point, fraction collection (25-mL fractions) is begun, and elution is continued with 300 mL of 2% EtOAc-hexane (49:1 hexanes:EtOAc) and then 500 mL of 5% EtOAc-hexane (19:1 hexanes:EtOAc). The desired product is obtained in fractions 24–30, which are concentrated by rotary evaporation (25°C, 15 mmHg). ..."

Purification: Recrystallization

Describe procedure in detail. Indicate solvents used (and ratio of mixed solvent systems), amount of recrystallization solvents, and temperature protocol. Describe how crystals are isolated and what they are washed with.

"The solid is dissolved in 100 mL of hot diethyl ether (30°C) and filtered through a Büchner funnel. The filtrate is allowed to cool to room temperature, and 20 mL of hexane is added. The solution is cooled at −20°C overnight and the resulting crystals are collected by suction filtration on a Büchner funnel, washed with 50 mL of ice-cold hexane, and then transferred to a 50-mL, round-bottomed flask and dried overnight at 0.01 mmHg to provide. . . ."

Characterization

Physical properties of the product such as color, appearance, crystal forms, melting point, etc. should be included in the text of the procedure. Comments on the stability of the product to storage, etc. should be provided in a Note.

In a Note, provide data establishing the identity of the product. This will generally include IR, MS, ^{1}H-NMR, and ^{13}C-NMR data, and in some cases UV data. Copies of the proton and carbon NMR spectra for the products of each step in the procedure should be submitted showing integration for all resonances. Submission of copies of NMR spectra and other nuclei are encouraged as appropriate.

In the same Note, provide quantitative analytical data establishing the purity of the product. Elemental analysis for carbon and hydrogen (and nitrogen if present) agreeing with calculated values within 0.4% is preferred. However, GC data (for distilled or vacuum-transferred samples) and/or HPLC data (for material isolated by column chromatography) may be acceptable in some cases.

In procedures involving non-racemic, enantiomerically enriched products, optical rotations should generally be provided, but enantiomeric purity must be determined by another method such as chiral HPLC or GC analysis.

In cases where the product of one step is used without purification in the next step, a Note should be included describing how a sample of the product can be purified and providing characterization data for the pure material. Copies of the proton NMR spectra of both the product both *before* and *after* purification should be submitted.

Hazard Warnings

Any significant hazards should be indicated in a statement at the beginning of the procedure in italicized type. Efforts should be made to avoid the use of toxic and hazardous solvents and reagents when less hazardous alternatives are available.

Discussion Section

The style and content of the discussion section will depend on the nature of the procedure.

For procedures that provide an improved method for the preparation of an important reagent or synthetic building block, the discussion should focus on the advantages of the new approach and should describe and reference all of the earlier methods used to prepare the title compound.

In the case of procedures that illustrate an important synthetic method or strategy, the discussion section should provide a mini-review on the new methodology. The scope and limitations of the method should be discussed, and it is generally desirable to include a table of examples. Competing methods for accomplishing the same overall transformation should be described and referenced. A brief discussion of mechanism may be included if this is useful for understanding the scope and limitations of the method.

Style and Format

Articles should follow the style guidelines used for organic chemistry articles published in the ACS journals such as *J. Am. Chem. Soc., J. Org. Chem., Org. Lett.*, etc. as described in the the ACS Style Guide (3rd Ed.). The text of the procedure should be constructed using a standard word processing program, like MS Word, with 14-point Times New Roman font. Chemical structures and schemes should be drawn using the standard ACS drawing parameters (in ChemDraw, the parameters are found in the "ACS Document 1996" option) with a maximum width of 6 inches. The graphics files should be inserted into the document at the correct location and the graphics files should also be submitted separately. All Tables that include structures should be entirely prepared in the graphics (ChemDraw) program and inserted into the word processing file at the appropriate location. Tables that include multiple, separate graphics files prepared in the word processing program will require modification.

Biographies and Photographs of Authors

Photographs and 100-word biographies of all authors should be submitted as separate files at the time of the submission of the procedure. The format of the biographies should be similar to those in the Volume 84 procedures found at the orgsyn.org website. Photographs can be accepted in a number of electronic formats, including tiff and jpeg formats.

HANDLING HAZARDOUS CHEMICALS

A Brief Introduction

General Reference: *Prudent Practices in the Laboratory*; National Academy Press; Washington, DC, 1995.

Physical Hazards

Fire. Avoid open flames by use of electric heaters. Limit the quantity of flammable liquids stored in the laboratory. Motors should be of the nonsparking induction type.

Explosion. Use shielding when working with explosive classes such as acetylides, azides, ozonides, and peroxides. Peroxidizable substances such as ethers and alkenes, when stored for a long time, should be tested for peroxides before use. Only sparkless "flammable storage" refrigerators should be used in laboratories.

Electric Shock. Use 3-prong grounded electrical equipment if possible.

Chemical Hazards

Because all chemicals are toxic under some conditions, and relatively few have been thoroughly tested, it is good strategy to minimize exposure to all chemicals. In practice this means having a good, properly installed hood; checking its performance periodically; using it properly; carrying out all operations in the hood; protecting the eyes; and, since many chemicals can penetrate the skin, avoiding skin contact by use of gloves and other protective clothing at all times.

a. Acute Effects. These effects occur soon after exposure. The effects include burn, inflammation, allergic responses, damage to the eyes, lungs, or nervous system (e.g., dizziness), and unconsciousness or death (as from overexposure to HCN). The effect and its cause are usually obvious and so are the methods to prevent it. They generally arise from inhalation or skin contact, so should not be a problem if one follows the admonition "work in a hood and keep chemicals off your hands". Ingestion is a rare route, being generally the result of eating in the laboratory or not washing hands before eating.

b. Chronic Effects. These effects occur after a long period of exposure or after a long latency period and may show up in any of numerous organs. Of the chronic effects of chemicals, cancer has received the most attention lately. Several dozen chemicals have been demonstrated to be carcinogenic in man and hundreds to be carcinogenic to animals. Although there is no simple correlation between carcinogenicity in animals and in man, there is little doubt that a significant proportion of the chemicals used in laboratories have some potential for carcinogenicity in man. For this and other reasons, chemists should employ good practices at all times.

The key to safe handling of chemicals is a good, properly installed hood, and the referenced book devotes many pages to hoods and ventilation. It recommends that in a laboratory where people spend much of their time working with chemicals there should be a hood for each two people, and each should have at least 2.5 linear feet (0.75 meter) of working space at it. Hoods are more than just devices to keep undesirable vapors from the laboratory atmosphere. When closed they provide a protective barrier between chemists and chemical operations, and they are a good containment device for spills. Portable shields can be a useful supplement to hoods, or can be an alternative for hazards of limited severity, e.g., for small-scale operations with oxidizing or explosive chemicals.

Specialized equipment can minimize exposure to the hazards of laboratory operations. Impact resistant safety glasses are basic equipment and should be worn at all times. They may be supplemented by face shields or goggles for particular operations, such as pouring corrosive liquids. Because skin contact with chemicals can lead to skin irritation or sensitization or, through absorption, to effects on internal organs, protective gloves should be worn at all times.

Laboratories should have fire extinguishers and safety showers. Respirators should be available for emergencies. Emergency equipment should be kept in a central location and must be inspected periodically.

MSDS (Materials Safety Data Sheets) sheets are available from the suppliers of commercially available reagents, solvents, and other chemical materials; anyone performing an experiment should check these data sheets before initiating an experiment to learn of any specific hazards associated with the chemicals being used in that experiment.

DISPOSAL OF CHEMICAL WASTE

General Reference: *Prudent Practices in the Laboratory* National Academy Press, Washington, D.C. 1996

Effluents from synthetic organic chemistry fall into the following categories:

1. **Gases**

   1a. Gaseous materials either used or generated in an organic reaction.

   1b. Solvent vapors generated in reactions swept with an inert gas and during solvent stripping operations.

   1c. Vapors from volatile reagents, intermediates and products.

2. **Liquids**

   2a. Waste solvents and solvent solutions of organic solids (see item 3b).

   2b. Aqueous layers from reaction work-up containing volatile organic solvents.

   2c. Aqueous waste containing non-volatile organic materials.

   2d. Aqueous waste containing inorganic materials.

3. **Solids**

   3a. Metal salts and other inorganic materials.

   3b. Organic residues (tars) and other unwanted organic materials.

   3c. Used silica gel, charcoal, filter aids, spent catalysts and the like.

The operation of industrial scale synthetic organic chemistry in an environmentally acceptable manner* requires that all these effluent categories be dealt with properly. In small scale operations in a research or

---

*An environmentally acceptable manner may be defined as being both in compliance with all relevant state and federal environmental regulations *and* in accord with the common sense and good judgment of an environmentally aware professional.

academic setting, provision should be made for dealing with the more environmentally offensive categories.

1a. Gaseous materials that are toxic or noxious, e.g., halogens, hydrogen halides, hydrogen sulfide, ammonia, hydrogen cyanide, phosphine, nitrogen oxides, metal carbonyls, and the like.

1c. Vapors from noxious volatile organic compounds, e.g., mercaptans, sulfides, volatile amines, acrolein, acrylates, and the like.

2a. All waste solvents and solvent solutions of organic waste.

2c. Aqueous waste containing dissolved organic material known to be toxic.

2d. Aqueous waste containing dissolved inorganic material known to be toxic, particularly compounds of metals such as arsenic, beryllium, chromium, lead, manganese, mercury, nickel, and selenium.

3. All types of solid chemical waste.

Statutory procedures for waste and effluent management take precedence over any other methods. However, for operations in which compliance with statutory regulations is exempt or inapplicable because of scale or other circumstances, the following suggestions may be helpful.

Gases

Noxious gases and vapors from volatile compounds are best dealt with at the point of generation by "scrubbing" the effluent gas. The gas being swept from a reaction set-up is led through tubing to a large trap to prevent suck-back and into a sintered glass gas dispersion tube immersed in the scrubbing fluid. A bleach container can be conveniently used as a vessel for the scrubbing fluid. The nature of the effluent determines which of four common fluids should be used: dilute sulfuric acid, dilute alkali or sodium carbonate solution, laundry bleach when an oxidizing scrubber is needed, and sodium thiosulfate solution or diluted alkaline sodium borohydride when a reducing scrubber is needed. Ice should be added if an exotherm is anticipated.

Larger scale operations may require the use of a pH meter or starch/iodide test paper to ensure that the scrubbing capacity is not being exceeded.

When the operation is complete, the contents of the scrubber can be poured down the laboratory sink with a large excess (10–100 volumes) of water. If the solution is a large volume of dilute acid or base, it should be neutralized before being poured down the sink.

Liquids

Every laboratory should be equipped with a waste solvent container in which *all* waste organic solvents and solutions are collected. The contents of these containers should be periodically transferred to properly labeled waste solvent drums and arrangements made for contracted disposal in a regulated and licensed incineration facility.**

Aqueous waste containing dissolved toxic organic material should be decomposed *in situ*, when feasible, by adding acid, base, oxidant, or reductant. Otherwise, the material should be concentrated to a minimum volume and added to the contents of a waste solvent drum.

Aqueous waste containing dissolved toxic inorganic material should be evaporated to dryness and the residue handled as a solid chemical waste.

Solids

Soluble organic solid waste can usually be transferred into a waste solvent drum, provided near-term incineration of the contents is assured.

Inorganic solid wastes, particularly those containing toxic metals and toxic metal compounds, used Raney nickel, manganese dioxide, etc. should be placed in glass bottles or lined fiber drums, sealed, properly labeled, and arrangements made for disposal in a secure landfill.** Used mercury is particularly pernicious and small amounts should first be amalgamated with zinc or combined with excess sulfur to solidify the material.

Other types of solid laboratory waste including used silica gel and charcoal should also be packed, labeled, and sent for disposal in a secure landfill.

Special Note

Since local ordinances may vary widely from one locale to another, one should always check with appropriate authorities. Also, professional disposal services differ in their requirements for segregating and packaging waste.

**If arrangements for incineration of waste solvent and disposal of solid chemical waste by licensed contract disposal services are not in place, a list of providers of such services should be available from a state or local office of environmental protection.

PREFACE

One of the most tangible benefits I realized upon joining the Organic Division of the American Chemical Society was my annual receipt of the most recent volume of *Organic Syntheses*. Throughout my graduate career I made a practice of reading each procedure in some detail (a convenient way to dispose of five minutes while waiting for a TLC plate to develop in those ancient days before the internet). This provided me with exposure to the recent chemical literature from a practical, hands-on perspective, and frequently fostered ideas for experimental techniques that could apply to my own research.

Such experimental procedures are the bedrock of organic chemistry. While a detailed procedure is sometimes sufficient for the successful transfer of a synthetic procedure from one lab to another, this is not always the case. It is likely that all chemists have encountered a problem in reproducing an experimental procedure from the literature; when this occurs, it is inevitable that you question whether the cause for this failure is a shortcoming in your own experimental technique, or a key element of the procedure that is not adequately described. As a graduate student I experienced a dramatic example of this type of challenge when a late-stage coupling in our synthesis of FK506 failed to transfer from a co-worker's lab to my lab one floor down in the same building. Following a concerted and successful effort to transfer this procedure (involving a colleague's weekend return to Cambridge from North Carolina), the most likely culprit was identified as the source of inert atmosphere. Early, successful runs were performed under house nitrogen, whereas several failed experiments were run under argon from a cylinder. Upon returning to the use of house nitrogen, the coupling was successfully reproduced (see *Strategies and Tactics in Organic Synthesis*, vol. 3, pp. 440–441 for a full description of this story).

While such transfers of procedures were rare in my graduate and postdoctoral career, they are routine in the pharmaceutical industry, and occur early and frequently in a project's lifecycle. The chemist who develops a procedure on laboratory scale may participate in the scale-up in a consulting role, but the direct responsibility for "on the floor" operations will lie elsewhere during scale-up to kilo lab or pilot plant. Based on my 14 years in Chemical Research and Development at Pfizer, I can comfortably say that successful technology transfers from lab to manufacturing represent our most critical responsibility as synthetic organic chemists. Discovery of a new chemical transformation to efficiently prepare a key intermediate or drug candidate on lab scale provides great substance for a staff report; but if it cannot be successfully executed on multi-kilogram scale, its practical value is negligible.

Organic Syntheses helps provide solutions to these challenges. All of the procedures in this volume have been performed successfully in both the submitters' and the checkers' labs, and thus meet a significantly higher standard of reliability and reproducibility than an unchecked procedure. The 35 procedures in Volume 86 represent a wide variety of interesting compounds and synthetic methods. As with many previous volumes, methods for the enantioselective synthesis of chiral molecules are well represented. The burgeoning field of organocatalysis is represented by List's proline-mediated Mannich reaction to form **TERT-BUTYL (1S,2S)-2-METHYL-3-OXO-1-PHENYLPROPYL-CARBAMATE** (p. 11). Ellman and co-workers provide a useful method for the preparation of chiral diaryl methanamines by metal-catalyzed addition of a boronic acid to an *in situ* generated imine in their preparation of **(S)-TERT-BUTYL (4-CHLOROPHENYL) (THIOPHEN-2-YL)METHYLCARBAMATE** (p. 360). Two catalytic asymmetric allylations are represented in this volume: Braun and co-workers' palladium-catalyzed synthesis of **(S)-(−)-2-ALLYLCYCLO-HEXANONE** (p. 47), and Stoltz and co-workers' decarboxylative allylation to prepare **(S)-2-ALLYL-2-METHYLCYCLOHEXANONE** (p. 194). A chiral auxiliary aldol-type method is represented by Urpi and co-workers' preparations of **(S)-4-ISOPROPYL-N-PROPANOYL-1, 3-THIAZOLIDINE-2-THIONE** (p. 70) and *anti* **α-METHYL-β-METHOXY CARBOXYLIC COMPOUNDS** (p. 81). A cinchonine-catalyzed Mannich addition of a β-keto ester to an *in situ* generated imine is described in Schaus and co-workers' **ENANTIOSELECTIVE PREPARATION OF DIHYDROPYRIMIDONES** (p. 236). Williams and co-workers describe the diastereoselective allylation of a useful proline-derived chiral template, **(3R,7aS)-3-(TRICHLOROMETHYL)-TETRAHYDROPYRROLO-[1,2-C]OXAZOL-1(3H)-ONE** (p. 262). Jackson and co-workers provide the preparation of a *TERT*-butylglycinol-derived chiral ligand and its use in a vanadium-catalyzed oxidation to prepare **(S)-(−)-METHYL p-BROMOPHENYL SULFOXIDE** (p. 121).

Cross-coupling methods continue to be an active area of research in both academic and industrial settings, and are represented by three procedures in this volume. Denmark and co-workers describe a cost-effective and practical polyvinylsiloxane cross-coupling method in the preparation of **3-VINYLQUINOLINE AND 4-VINYLBENZO-PHENONE** (p. 274). Daugulis and co-workers describe the direct arylation of an electron-rich heterocycle with an arylchloride in their preparation of **5-PHENYL 2 ISOBUTYLTHIAZOLE** (p. 105). Burke and co-workers describe the utility of a boron-protected bromoboronic acid in an iterative cross-coupling sequence to prepare **4-(p-TOLYL)-PHENYLBORONIC ACID** (p. 344).

In addition to providing examples of useful synthetic methods, *Organic Syntheses* has also provided preparations of many useful and interesting compounds; the contents of this volume provide several such examples. Stoltz and co-workers describe an improved synthesis of an oxazoline phosphine ligand (used in their decarboxylative allylation procedure, p. 194) in their preparation of **(S)-TERT-ButylPHOX** (p. 181). Synthesis of an unsymmetrically substituted triyne via a desilylative bromination/cross-coupling sequence is described by Kim and co-workers in the preparation of **(7-(BENZYLOXY)HEPTA-1,3,5-TRIYNYL) TRIISOPROPYLSILANE** (p. 225). Synthesis of the very interesting **TETRAKIS(DIMETHYLAMINO)ALLENE** is described by Fürstner and co-workers (p. 298). Olofsson and co-workers provide an efficient synthesis of **BIS(4-TERT-BUTYLPHENYL)IODONIUM TRIFLATE** (p. 308), and Donahue and co-workers provide a practical synthesis of **4,5-DIMETHYL-1,3-DITHIOL-2-ONE** (p. 333).

Heterocycles are also well represented in this volume. Stevens and co-workers describe a useful rearrangement of an azirine in their preparation of **2-(3-BROMOPHENYL)-6-(TRIFLUOROMETHYL) PYRAZOLO-[1,5-a]PYRIDINE** (p. 18). Lebel and co-workers apply a rhodium-catalyzed C-H bond insertion of a nitrene to prepare **4-SPIROCYCLOHEXYLOXAZOLIDINONE** (p. 59). Kuethe and co-workers apply a palladium-catalyzed reductive cyclization of a nitro-styrene to prepare **[2-(4-FLUOROPHENYL)-1H-INDOL-4-YL]-1-PYRROLIDINYLMETHANONE** (p. 92). Simanek and co-workers describe the large scale preparation of **2-[3,3'-DI-(TERT-BUTOXY-CARBONYL)-AMINODIPROPYLAMINE]-4,6,-DICHLORO-1,3,5-TRIAZINE** and **1,3,5-[TRIS-PIPERAZINE]-TRIAZINE** (p. 141), and their subsequent coupling to prepare a **MELAMINE (TRI-AZINE) DENDRIMER** (p. 151). Hossain and co-workers describe the single pot conversion of salicylaldehydes and ethyl diazoacetate to **3-ETHOXYCARBONYL BENZOFURANS** (p. 172). Kwon and co-workers describe a phosphine-catalyzed [4+2] annulation to prepare **ETHYL 6-PHENYL-1-TOSYL-1,2,5,6-TETRAHYDRO-PYRIDINE-3-CARBOXYLATE** (p. 212). Kappe and co-workers provide two interesting multi-component condensations of an aminopyrazole, 1,3-diketone and aromatic aldehyde to provide **TETRAHY-DROPYRAZOLOQUINOLINONES AND TETRA-HYDRO-PYRAZOLOQUINAZOLINONES** (p. 252). Shi and co-workers describe a stereospecific palladium-catalyzed diamination of an olefin with **DI-*TERT*-BUTYL-DIAZIRIDINONE AS NITROGEN SOURCE** (p. 315).

Several procedures in this volume demonstrate the application of generally useful synthetic methods that fall outside the umbrella of the preceding topics. Lautens and co-workers provide an improved method for the synthesis of gem-dibromovinyl anilines, useful compounds for the synthesis of substituted indoles via tandem cross-couplings, as exemplified in their preparation of **2-(2,2-DIBROMOETHENYL)-BENZENAMINE** (p. 36). Landais and co-workers describe the use of a Birch reductive alkylation to prepare **(3,5-DIMETHOXY-1-PHENYL-CYCLOHEXA-2,5-DIENYL)-ACETONITRILE** (p. 1). Williams and co-workers describe the use of benzyl alcohol as an alkylating agent via the **RUTHENIUM-CATALYZED BORROWING HYDROGEN STRATEGY** (p. 28). Lebel and co-workers describe a single pot Curtius rearrangement in their preparation of *N-TERT*-BUTYL ADAMANTANYL-1-YL-CARBAMATE** (p. 113). Fukuyama and co-workers demonstrate a 1,2-diol protecting group that can be selectively removed in the presence of other acid-labile functionality in their preparation (and deprotection) of **(4-((4*R*,5*R*)-4,5-DIPHENYL-1,3-DIOXOLAN-2-YL)PHENOXY)-(*TERT*- BUTYL)DI-METHYLSILANE** (p. 130). Stoltz and co-workers describe an interesting addition of a β-keto ester enolate to an *in situ* generated aryne in their preparation of **METHYL 2-(2-ACETYLPHENYL) ACETATE** (p. 161). Coates and co-workers describe a chromium/cobalt-catalyzed low pressure carbonylation for the conversion of **EPOXIDES TO β-LACTONES** (p. 287). Nakamura and co-workers utilize their indium-catalyzed addition of active methylene compounds to an unactivated alkyne to prepare **ETHYL 2-ETHANOYL-2-METHYL-3-PHENYLBUT-3-ENOATE** (p. 325). Knochel and co-workers demonstrate the ortho-metallation of a weakly activated arene with (tmp)$_2$ Mg2LiCl in their preparation of **_TERT_-BUTYL ETHYL PHTHALATE** (p. 374).

Finally, I would like to gratefully acknowledge the authors and checkers of the procedures in this volume. Preparing and checking a procedure for _Organic Syntheses_ is no small task, and without the commitment of these authors this volume would not be possible. I would also like to thank my fellow board members for the opportunity to serve with them over the past five years. It has been a genuine treat for a chemist from the pharmaceutical industry to rub elbows with the cream of academic chemistry, and the board meetings and Org Syn dinners have provided me with a host of pleasant memories (comparing iPhone apps with Clayton Heathcock and Rick Danheiser at the dinner table will always be one of my favorites). I would also like to thank Tamim Braish, Lynne Handanyan, and Charlie Santa Maria of Pfizer for their support of my editing and checking responsibilities, and the numerous checkers (and authors) of procedures here at Pfizer. I am particularly grateful to my colleague and long-time friend Sarah Kelly for her recommendation of me to replace her on the board in 2005. Special thanks to Editor-in-Chief Rick Danheiser and Associate Editor Chuck Zercher for their support and guidance in my volume editing responsibilities.

JOHN A. RAGAN
Groton, Connecticut

CONTENTS

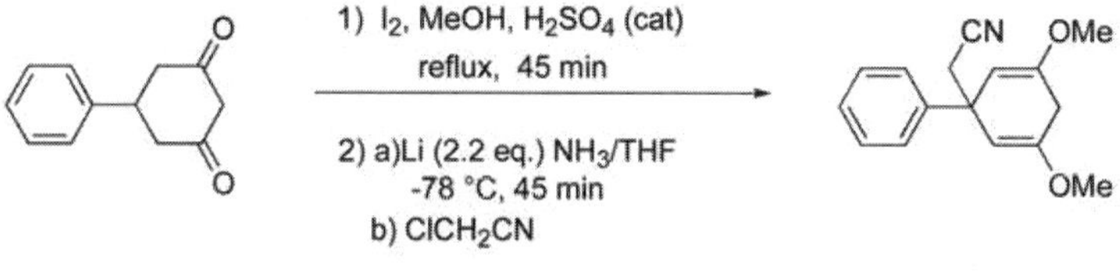
1) I2, MeOH, H2SO4 (cat)
reflux, 45 min
2) a)Li (2.2 eq.) NH3/THF
-78 °C, 45 min
b) ClCH2CN
CN
OMe
OMe
O
O

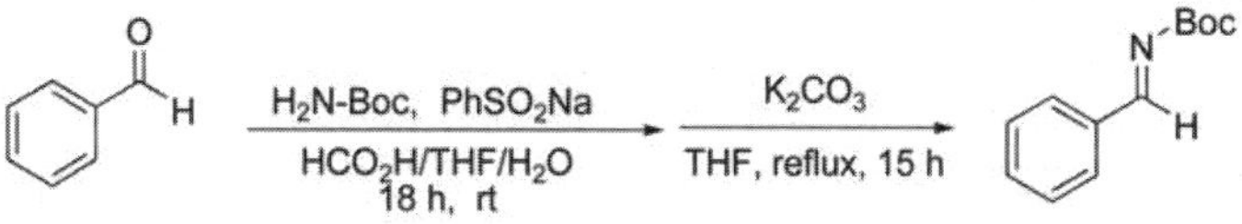
H2N-Boc, PhSO2Na
HCO2H/THF/H2O
18 h, rt
K2CO3
THF, reflux, 15 h
N
Boc
H
O

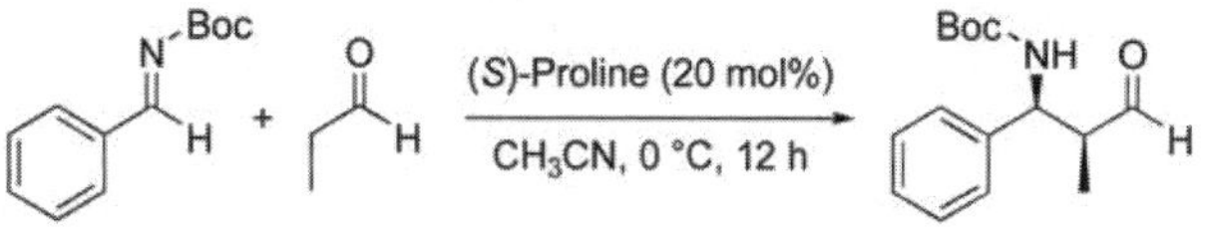
(S)-Proline (20 mol%)
CH3CN, 0 °C, 12 h
Boc
NH
N
H
O
+

SYNTHESIS OF PYRAZOLO[1,5-a]PYRIDINES VIA AZIRINES: PREPARATION OF 2-(3-BROMOPHENYL)-6-(TRIFLUOROMETHYL)PYRAZOLO[1,5-a]PYRIDINE

Stephen Greszler and Kirk L. Stevens

BENZYL ALCOHOL AS AN ALKYLATING AGENT USING THE RUTHENIUM-CATALYZED BORROWING HYDROGEN STRATEGY

Tracy D. Nixon, Paul A. Slatford, Michael K. Whittlesey and Jonathan M. J. Williams

2-(2,2-DIBROMOETHENYL)-BENZENAMINE

Christopher Bryan, Valentina Aurregi and Mark Lautens

(S)-(–)-2-ALLYLCYCLOHEXANONE

Manfred Braun, Panos Meletis, and Mesut Fidan

PREPARATION OF 4-SPIROCYCLOHEXYLOXAZOLIDINONE BY C-H BOND NITRENE INSERTION

Kim Huard and Hélène Lebel

PREPARATION OF (S)-4-ISOPROPYL-N-PROPANOYL-1,3-THIAZOLIDINE-2-THIONE

Erik Gálvez, Pedro Romea, and Fèlix Urpí

STEREOSELECTIVE SYNTHESIS OF *ANTI* α-METHYL- β-METHOXY CARBOXYLIC COMPOUNDS

Erik Gálvez, Pedro Romea, and Fèlix Urpi

SYNTHESIS OF 2-ARYLINDOLE-4-CARBOXYLIC AMIDES: [2-(4-FLUOROPHENYL)-1*H*-INDOL-4-YL]-1-PYRROLIDINYLMETHANONE

Jeffrey T. Kuethe and Gregory L. Beutner

PALLADIUM (II) ACETATE-BUTYLDI-1-ADAMANTYLPHOSPHINE CATALYZED ARYLATION OF ELECTRON-RICH HETEROCYCLES. PREPARATION OF 5-PHENYL-2-ISOBUTYLTHIAZOLE

Anna Lazareva, Hendrich A. Chiong, and Olafs Daugulis

MILD AND EFFICIENT ONE-POT CURTIUS REARRANGEMENT: PREPARATION OF *N-TERT*-BUTYL ADAMANTANYL-1-YL-CARBAMATE

Olivier Leogane and Hélène Lebel

ENANTIOSELECTIVE OXIDATION OF AN ALKYL ARYL SULFIDE: SYNTHESIS OF (*S*)-(–)-METHYL *p*-BROMOPHENYL SULFOXIDE

Carmelo Drago, Emma-Jane Walker, Lorenzo Caggiano and Richard F. W. Jackson

PROTECTION OF DIOLS WITH 4-(*tert*-BUTYLDIMETHYLSILYLOXY)BENZYLIDENE ACETAL AND ITS DEPROTECTION:(4-((4*R*,5*R*)-4,5-DIPHENYL-1,3-DIOXOLAN-2-YL)PHENOXY)(*TERT*-BUTYL)DIMETHYLSILANE

130

Hiroyuki Osajima, Hideto Fujiwara, Kentaro Okano, Hidetoshi Tokuyama, and Tohru Fukuyama

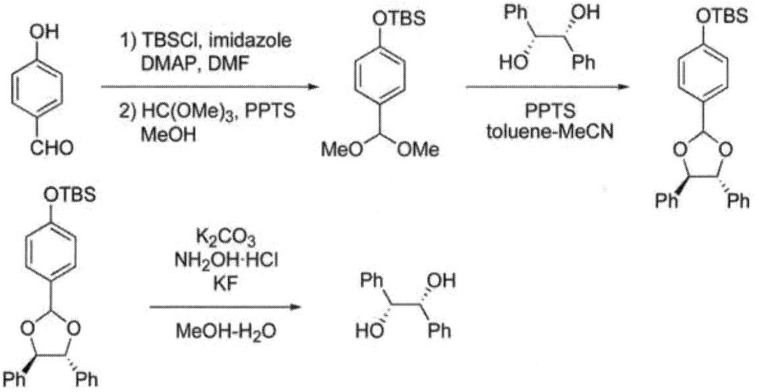

SYNTHESIS OF 2-[3,3'-DI-(*TERT*-BUTOXYCARBONYL)-AMINODIPROPYLAMINE]-4,6,-DICHLORO-1,3,5-TRIAZINE AS A MONOMER AND 1,3,5-[*TRIS*-PIPERAZINE]-TRIAZINE AS A CORE FOR THE LARGE SCALE SYNTHESIS OF MELAMINE (TRIAZINE) DENDRIMERS

141

Abdellatif Chouai, Vincent J. Venditto, and Eric E. Simanek

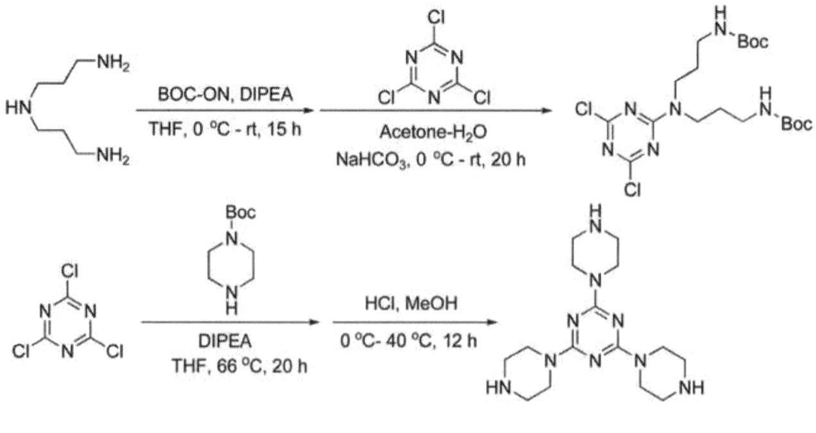

LARGE SCALE, GREEN SYNTHESIS OF A 151
GENERATION-1 MELAMINE (TRIAZINE) DENDRIMER
Abdellatif Chouai, Vincent J. Venditto, and Eric E. Simanek

THE DIRECT ACYL-ALKYLATION OF ARYNES. 161
PREPARATION OF METHYL 2-(2-ACETYLPHENYL)
ACETATE
David C. Ebner, Uttam K. Tambar, and Brian M. Stoltz

CONVENIENT PREPARATION OF 3-ETHOXYCARBONYL 172
BENZOFURANS FROM SALICYLALDEHYDES AND ETHYL
DIAZOACETATE
Matthew E. Dudley, M. Monzur Morshed, and M. Mahmun Hossain

PREPARATION OF (S)-*TERT*-BUTYLPHOX
Michael R. Krout, Justin T. Mohr, and Brian M. Stoltz

PREPARATION OF (S)-2-ALLYL-2-METHYLCYCLOHEXANONE

Justin T. Mohr, Michael R. Krout, and Brian M. Stoltz

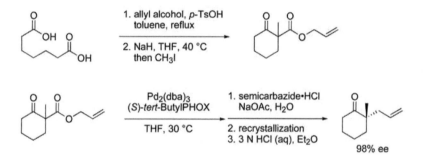

PHOSPHINE-CATALYZED [4+2] ANNULATION: SYNTHESIS OF ETHYL 6-PHENYL-1-TOSYL-1,2,5,6-TETRAHYDROPYRIDINE-3-CARBOXYLATE

Kui Lu and Ohyun Kwon

SYNTHESIS OF POLYYNES BY IN SITU DESILYLATIVE BROMINATION AND PALLADIUM-CATALYZED COUPLING: (7-(BENZYLOXY)HEPTA-1,3,5-TRIYNYL) TRIISOPROPYLSILANE

Soonho Hwang, Hee Ryong Kang, and Sanghee Kim

ENANTIOSELECTIVE PREPARATION OF DIHYDROPYRIMIDONES

Jennifer M. Goss, Peng Dai, Sha Lou, and Scott E. Schaus

ONE-POT MULTICOMPONENT PREPARATION OF TETRAHYDROPYRAZOLOQUINOLINONES AND TETRAHYDROPYRAZOLOQUINAZOLINONES

Toma N. Glasnov and C. Oliver Kappe

(3R,7aS)-3-(TRICHLOROMETHYL)TETRAHYDROPYRROLO [1,2-C]OXAZOL-1(3H)-ONE: AN AIR AND MOISTURE STABLE REAGENT FOR THE SYNTHESIS OF OPTICALLY ACTIVE α-BRANCHED PROLINES

Gerald D. Artman III, Ryan J. Rafferty, and Robert M. Williams

VINYLATION WITH INEXPENSIVE SILICON-BASED REAGENTS: PREPARATION OF 3-VINYLQUINOLINE AND 4-VINYLBENZOPHENONE

Scott E. Denmark and Christopher R. Butler

LOW PRESSURE CARBONYLATION OF EPOXIDES TO β-LACTONES

287

John W. Kramer, Daniel S. Treitler, and Geoffrey W. Coates

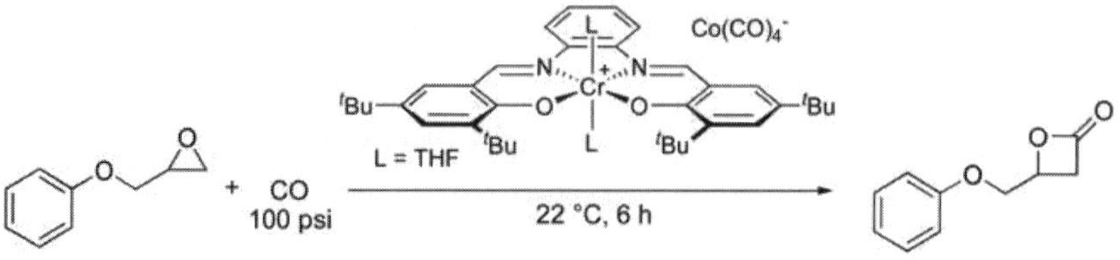

TETRAKIS(DIMETHYLAMINO)ALLENE

298

Alois Fürstner, Manuel Alcarazo, and Helga Krause

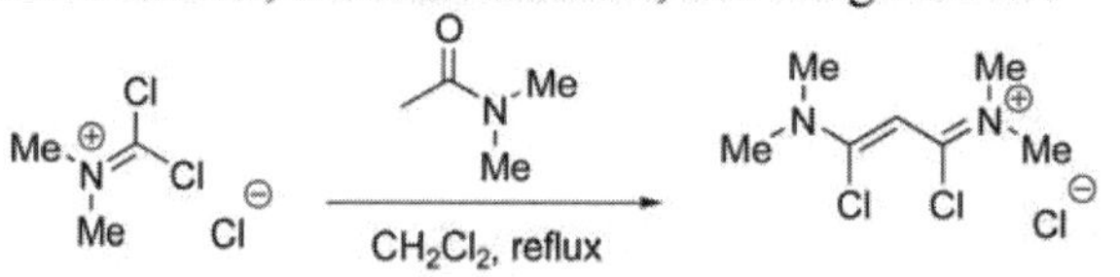

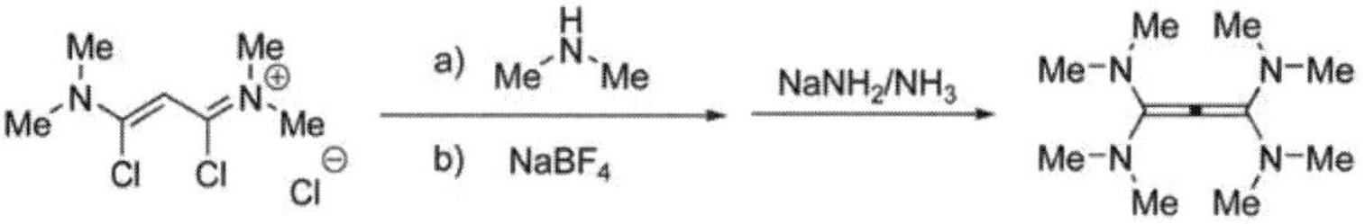

EFFICIENT ONE-POT SYNTHESIS OF BIS(4-*TERT*-BUTYLPHENYL)IODONIUM TRIFLATE

308

Marcin Bielawski and Berit Olofsson

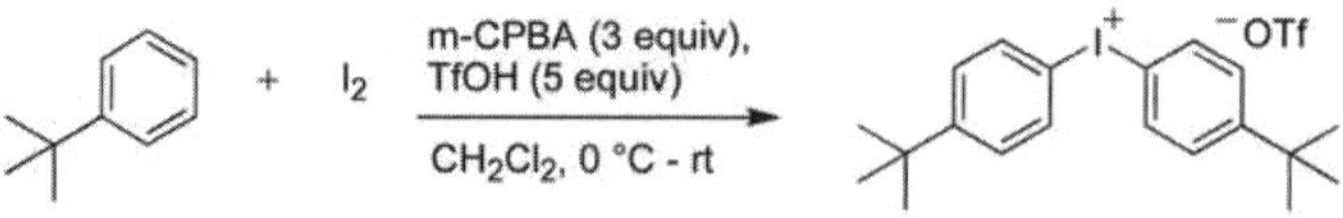

Pd(0)-CATALYZED DIAMINATION OF *TRANS*-1-PHENYL-
1,3-BUTADIENE WITH DI-*TERT*-BUTYLDIAZIRIDINONE
AS NITROGEN SOURCE
Haifeng Du, Baoguo Zhao, and Yian Shi

SYNTHESIS OF ETHYL 2-ETHANOYL-2-METHYL-3-
PHENYLBUT-3-ENOATE
Taisuke Fujimoto, Kohei Endo, Masaharu Nakamura, and
Eiichi Nakamura

SYNTHESIS OF 4,5-DIMETHYL-1,3-DITHIOL-2-ONE
Perumalreddy Chandrasekaran and James P. Donahue

B-PROTECTED HALOBORONIC ACIDS FOR ITERATIVE CROSS-COUPLING

Steven G. Ballmer, Eric P. Gillis, and Martin D. Burke

RHODIUM-CATALYZED ENANTIOSELECTIVE ADDITION OF ARYLBORONIC ACIDS TO *IN SITU* GENERATED *N*-BOC ARYLIMINES. PREPARATION OF (*S*)-*TERT*-BUTYL (4-CHLOROPHENYL)(THIOPHEN-2-YL)METHYLCARBAMATE

Morten Storgaard and Jonathan A. Ellman

MAGNESIATION OF WEAKLY ACTIVATED ARENES USING tmp₂Mg·2LiCl: SYNTHESIS OF *TERT*-BUTYL ETHYL PHTHALATE

Christoph J. Rohbogner, Andreas J. Wagner, Giuliano C. Clososki, and Paul Knochel

ERRATA

1,4-BIS(TRIMETHYLSILYL)BUTA-1,3-DIYNE   385
Graham E. Jones, David A. Kendrick, and Andrew B. Holmes

PREPARATION OF (3,5-DIMETHOXY-1-PHENYL-CYCLOHEXA-2,5-DIENYL)-ACETONITRILE THROUGH BIRCH REDUCTIVE ALKYLATION (BRA)

A.

B.

Submitted by Raphaël Lebeuf, Muriel Berlande, Frédéric Robert, and Yannick Landais.[1]

Checked by Marcus G. Schrems and Andreas Pfaltz.

1. Procedure

Caution! The Birch reductive alkylation reaction (part B) should be carried out in a well-ventilated hood to avoid exposure to ammonia.

A. 5-Phenyl-1,3-dimethoxybenzene. To a 250-mL, three-necked, round-bottomed flask, equipped with a magnetic stirring bar, a reflux condenser (central neck), an internal thermometer and a ground-glass stopper, 5-phenylcyclohexane-1,3-dione (7.98 g, 42.4 mmol) (Note 1) and methanol (80 mL) are added (Note 2). Concentrated sulfuric acid (1 mL, 95%) is added dropwise over 1 min and the suspension turns into a homogeneous solution. After 10 min iodine (23.7 g, 93.4 mmol, 2.20 equiv) is added as a solid in one portion. The mixture is refluxed for 45 min (oil bath; T = 85 °C) and then cooled down to room temperature. The solution is added to an aqueous sodium thiosulfate solution (0.5 M, 350 mL) and 150 mL of ether. Using a 1-L separatory funnel the aqueous phase is separated and the organic phase is washed with 2 × 100 mL of 0.5 M sodium thiosulfate solution, giving a yellowish organic phase. The aqueous phases are combined and extracted with 2 × 150 mL of ether. The combined organic phase was washed with brine (50 mL), dried over sodium sulfate (1.80 g) and concentrated on a rotary evaporator (600 mmHg, 40 °C, then 15 mmHg), (8.87 g, light brown oil). Silica gel (10 g) and ethyl acetate (50 mL) are added and agitated. The solvent is removed on a rotary evaporator (125 mmHg, 40 °C, then 15 mmHg). A sintered glass funnel (h × d: 9 cm × 5 cm) is charged with 20 g of silica gel (ca. 2.5 cm layer). A layer of sand (0.5 cm) is put on top followed by the crude material adsorbed on silica gel and another layer of sand (0.5 cm). A mixture of hexanes and ethyl acetate (50:1; 750 mL) is passed through the silica gel plug by applying vacuum from below and collected (Note 3). After evaporation of the solvent on a rotary evaporator and drying under vacuum (0.1 mmHg) biaryl **1** is obtained as a white solid (6.04-6.42 g, 66-71%, mp 61.0-61.9 °C) (Note 4).

B. (3,5-Dimethoxy-1-phenyl-cyclohexa-2,5-dienyl)-acetonitrile. A 500-mL, three-necked, round-bottomed flask is equipped with a magnetic stirring bar. The central neck of the reaction flask is fitted with a large cold-finger condenser. The flask with the condenser is heated in an oven overnight (120 °C). The two remaining necks are closed with ground-glass stoppers and the hot glassware is connected via the cold-finger condenser to a vacuum-nitrogen line. Vacuum is applied until the glassware is cooled down to room temperature and the glassware is flushed with nitrogen (Note 5). Biaryl **1** (6.08 g, 28.4 mmol) is introduced under nitrogen, and dissolved in anhydrous THF (100 mL) (Note 6). A T-joint with a bubbler for exhaust is introduced between the vacuum-nitrogen line and the cold-finger condenser. One ground-glass stopper is exchanged with a gas inlet valve connected to an ammonia tank (Note 7, 8) the other stopper with an internal thermometer. The flask and the cold-finger condenser are cooled to −78 °C (bath temperature) using acetone and dry-ice, then ammonia (200 mL approximately) is condensed (Note 9). The gas inlet is replaced with a stopper and lithium (435 mg, 62.7 mmol, 2.21 equiv) (Note 10) is added portion wise at −78 °C under a nitrogen flow during 10 min. During lithium addition the solution turns rapidly brown and finally brick red. After 45 minutes of stirring at this temperature, one of the glass stoppers is replaced by a septum and a cold solution of chloroacetonitrile (5.40 mL, 85.3 mmol, 3.00 equiv) in anhydrous THF (20 mL) is added dropwise over 10 min, keeping the temperature below −70 °C (Note 11). The mixture turns immediately brown. After 20 min of stirring, the reaction is quenched by addition of solid ammonium chloride (3.50 g). The cooling bath and condenser are removed and ammonia is allowed to evaporate under air (Note

2

12). When the reaction mixture has reached room temperature, a saturated ammonium chloride solution (150 mL) is added, followed by water (50 mL). The aqueous phase is extracted with ether (3 × 100 mL) in a 1-L separating funnel. The combined organic phases are washed with brine (150 mL), dried over sodium sulfate (2 g) and concentrated on a rotary evaporator (125 mmHg, 40 °C, then 15 mmHg). The resulting brown paste (8.20 g) is purified by flash chromatography (Note 13). After concentration of the product-containing fractions on a rotary evaporator and drying under vacuum (0.1 mmHg), the bis-enol ether is obtained in 2 fractions: white solid (5.11 g) with mp = 101.0-102.0 °C (Note 14), off-white fraction (1.17 g) with mp = 98.0-99.0 °C (Note 15) (total yield: 6.28-6.65 g, 87-88%).

2. Notes

1.   5-Phenylcyclohexa-1,3-dione (97%) was purchased from Alfa-Aesar and was used as received.

2.   Distillation over magnesium prior to use did not improve the yield. Sulfuric acid (ACS grade 95-98%) was purchased from Alfa-Aesar, sodium thiosulfate anhydrous was purchased from SDS, ether (99%) was purchased from SDS, ammonium chloride (99%) was purchased from Alfa-Aesar. The Checkers used methanol from J. T. BAKER (Baker analyzed), ammonium chloride from Fluka (≥99%), iodine from Riedel-de-Haën (purris. p.a.), sulfuric acid from VWR (AnalR NORMAPUR, 95%), sodium thiosulfate pentahydrate from Sigma-Aldrich (≥99.5%), ether form J. T. BAKER (Baker analyzed, distilled prior to use).

3.   The Submitters used column chromatography for purification of the product: silica gel column (25 x 4.5 cm charged with 100 g of silica gel Si60 43-60 μm, purchased from Merck), eluted with 1 L of a 95/5 petroleum/EtOAc mixture. The product has an R_f = 0.82 when eluting with a 90/10 petroleum ether/EtOAc mixture. Checkers: The product has a R_f = 0.42 when eluting with a 90/10 hexanes/EtOAc mixture. The purity (elemental analysis, ^{1}H NMR) of the material purified by column chromatography and by filtration over silica gel (Procedure A) was identical.

4.   Analytical data closely matched published data[2]. ^{1}H NMR (400.1 MHz, CDCl$_3$) δ: 3.86 ppm (s, 6 H), 6.49 (t, 1 H, J = 2.2 Hz), 6.75 (d, 2 H, J = 2.3 Hz), 7.34-7.38 (m, 1 H), 7.42-7.46 (m, 2 H), 7.58-7.60 (m, 2 H). ^{13}C

NMR (100.6 MHz, CDCl$_3$) δ: 55.4, 99.2, 105.4, 127.2, 127.5, 128.7, 141.2, 143.5, 161.0. IR (ATR) 3085, 3009, 2970, 2939, 2839, 1589, 1450, 1411, 1350, 1311, 1204, 1150, 1064, 1026, 926, 833, 764, 694, 656 cm^{-1}. MS (EI, 70 eV) m/z (%): 214 (M$^+$, 100), 185 (20), 128 (13). Anal Calcd for C$_{14}$H$_{14}$O$_2$: C, 78.48; H, 6.59. Found: C, 78.34; H, 6.66.

5.   Submitters: All connections and septa were secured with Parafilm-M®. The Checkers used GLINDEMANN®-sealing rings (PTFE) purchased from AMSI-Glas AG.

6.   The Checkers used THF (VWR, HPLC-grade) dried using a Pure-Solv™ system.[11]

7.   Anhydrous ammonia (N36) was purchased from Air Liquide in a steel cylinder (6 bar, 9.6 m^3). The Checkers used anhydrous ammonia form PanGas in a steel cylinder (10 L, 5.3 kg).

8.   A drying tower filled with potassium hydroxide pellets is placed between the cylinder and the flask in order to dry the ammonia.

9.   30-45 minutes are usually required to condense ammonia. An estimation of the volume of condensed ammonia was carried out by graduating the reaction flask with 300 mL of a solvent, prior to the experiment.

10.   Lithium wire, purchased from Aldrich (99.9%, 0.01% Na) is cut and washed in a beaker filled with dry pentane or cyclohexane. After weighing out the required amount, the wire is cut into 8-10 pieces in the beaker. Each piece is then hammered in an aluminum foil covered with mineral oil in order to protect the lithium surface from air and moisture. The pieces have to be flattened thoroughly to ensure rapid reaction. Each shiny flattened piece is finally washed quickly with dry pentane and introduced immediately into the reaction mixture under nitrogen atmosphere.

11.   Three equivalents of electrophile are required as lithium amide (LiNH$_2$) is formed during the reaction that can consume part of the electrophile. Chloroacetonitrile (>99% GC) was purchased from Fluka and used as received. The chloroacetonitrile solution is prepared in a dry 50-mL Schlenk tube (sealed with a septum) under nitrogen and is cooled down to −78 °C before addition via cannula. A nitrogen filled balloon is used for pressure regulation.

12.   The fume hood should be pulled down during evaporation of ammonia. A water bath may be used to accelerate the evaporation.

13.   The crude mixture was adsorbed on silica gel (10 g) before being loaded at the top of a silica gel column (h × d: 19 cm × 4 cm, 120 g

4

silica gel 43-60 μm, purchased from Merck). The fractions were collected in 50 mL tubes. The column is eluted with 800 mL of a 90/10 hexanes/EtOAc mixture, then 1 L of a 50/10 hexanes/EtOAc mixture and 1.2 L of a 30/10 hexanes/EtOAc mixture. A first fraction (tubes 32-43) contained 5.11 g of a colorless solid, mp 101-102 °C, a second fraction (tubes 44-60) contained 1.17 g of an off-white solid mp 98-99 °C (Note 15), the product has a $R_f = 0.13$ when eluting with a 90/10 hexanes/EtOAc mixture.

14. Analytical data of **2**: ^{1}H NMR (400.1 MHz, CDCl$_3$)δ: 2.87 (dt, 1 H, $J = 20.7$, 1.2 Hz), 2.93 (s, 2 H), 2.97 (dt, 1 H, $J = 20.7$, 1.2 Hz), 3.60 (s, 6 H), 4.70 (t, 2 H, $J = 1.2$ Hz), 7.22-7.27 (m, 1 H), 7.34-7.35 (m, 4 H). ^{13}C NMR (100.6 MHz, CDCl$_3$) δ: 31.1, 32.2, 44.9, 54.5, 54.5, 98.4, 118.0, 125.7, 126.8, 128.6, 146.3, 153.3. IR (ATR) 3057, 2996, 2960, 2929, 2899, 2828, 2252, 1694, 1406, 1209, 1196, 1164, 1146, 1022, 870, 809, 766, 703 cm^{-1}. MS (FAB) m/z (%) 256 ([M+H]$^+$, 26), 215 (100). HRMS [M+Na]$^+$ C$_{16}$H$_{17}$NNaO$_2$: Calcd 278.1157; found: 278.1160. Anal Calcd for C$_{16}$H$_{17}$NO$_2$: C, 75.27; H, 6.71; N, 5.49; found: C, 75.50; H, 6.82; N, 5.32.

15. The compound collected as a second fraction showed no impurities in the ^{1}H NMR spectrum. However, elemental analysis was outside the accepted range. Anal Calcd for C$_{16}$H$_{17}$NO$_2$: C, 75.27; H, 6.71; N, 5.49; found: C, 74.86; H, 6.74; N, 5.74.

Safety and Waste Disposal Information

All hazardous materials should be handled and disposed of in accordance with "Prudent Practices in the Laboratory"; National Academy Press; Washington, DC, 1995.

3. Discussion

Birch reductive alkylation (BRA)[3] has proven to be a valuable process for the synthesis of highly functionalized building blocks, including cyclohexadienes bearing a quaternary center. The strategy is well documented on aryls containing electron-withdrawing groups and this strategy was elegantly extended by Schultz[4] to diastereoselective reductive alkylation of chiral amides. More recently, Donohoe devised a Birch reductive alkylation of heteroarenes (furan, pyrrole, and pyridine) bearing electron-withdrawing groups using an ammonia-free protocol.[5] Pioneering studies[6] on Birch reductive alkylation of biaryls showed that mixture of compounds with modest level of regioselectivities and yields were obtained depending on the nature of the substituents on both aromatic rings. Recent investigations in our laboratory, however, demonstrated that BRA on biaryls bearing electron-donating groups (OMe) could be a very efficient process, affording good to excellent yields of alkylated products.[7] Minor variations on the nature and the number of substituents on the aromatic rings were effectively shown to strongly affect both the regiocontrol and the competition between protonation and alkylation. It was, however, possible to control the regioselectivity of the reduction with a careful choice of the nature of the substituents on the biaryls. Two methoxy groups on the first aromatic ring, *meta* to the biaryl linkage, were thus shown to activate the reduction of this ring. Substituents such as OH or NHR on the second ring, in any positions relative to the biaryl bond, were found to be compatible with this reduction, providing their acidic proton was removed with *n*-BuLi, prior to addition of Li and ammonia. In contrast, a methoxy group placed in *ortho* or *para* position was cleaved under these conditions. The resulting substituted cyclohexa-1,4-dienes are valuable intermediates for the synthesis of more elaborated targets and were used as starting materials for the synthesis of various alkaloids using desymmetrization processes.[8]

The above procedure is representative of the synthesis of diversely substituted arylcyclohexa-1,4-dienes. The strategy also features a convenient preparation of biaryls having a 3,5-dimethoxyphenyl moiety starting from a cyclohexa-1,3-dione[9], adapted from published protocols,[10] which has been found, on large scale, more straightforward and practical than palladium couplings. The Birch reductive alkylation follows a standard protocol and can be run on large scale without erosion of the yield, in general. This methodology has been successfully extended to others biaryls with minor modifications, as shown in Table 1.

Table 1

Entry	Biaryls	Conditions	Products	Yields[a]
1		1) Li, NH₃, THF -78 °C, 0.5 h 2) ClCH₂CN		88%
2		1) Li, NH₃, THF -78 °C, 0.5 h 2) ClCH₂CO₂Me		77%
3		1) Li, NH₃, THF -78 °C, 0.5 h 2) ClCH₂CN		91%
4		1) n-BuLi, THF, 15 min 2) Li/NH₃, THF -78 °C, 0.5 h 3) ClCH₂CN		86%
5		1) n-BuLi, THF, 15 min 2) Li/NH₃, THF -78 °C, 0.5 h 3) ClCH₂CN		50%
6		1) Li, NH₃, THF -33 °C, 1 h 2) ClCH₂CN		88%
7		1) n-BuLi, THF, 15 min 2) Li/NH₃, THF -33 °C, 1 h 3) ClCH₂CN		68%

[a] Yields given on a 1 g scale. With biaryls bearing a phenol (or an amine function) in *ortho* or *para* position (entry 4, 5, 7), yields tend to decrease (50% for entry 4 for example) upon scaling up (5 g) due to the higher basicity of the final anion, which may react with both ammonia and the electrophile.

1. Université Bordeaux 1; Institut des Sciences Moléculaires; UMR-CNRS 5255; 351, cours de la Libération; 33405 Talence Cedex, France.
2. Dol, G. C.; Kamer, P. C. J.; van Leeuwen, P. W. N. M. *Eur. J. Org. Chem.* **1998**, 359-364.
3. Rabideau, P. W.; Marcinow, Z. *Org. React.* **1992**, *42*, 1-334 and references therein.
4. Schultz, A. G. *Chem. Commun.* **1999**, 1263-1271 and references therein.
5. Donohoe, T. J.; House, D. *J. Org. Chem.* **2002**, *67*, 5015-5018 and references therein.
6. (a) Müller, P. M.; Pfister, R. *Helv. Chim. Acta* **1983**, *66*, 771-779. (b) Rabideau, P. W.; Peters, N. K.; Huser, D. L. *J. Org. Chem.* **1981**, *46*, 1593-1597.
7. Lebeuf, R.; Robert, F.; Landais, Y. *Org. Lett.* **2005**, *7*, 4557-4560.
8. For an example, see: Lebeuf, R.; Robert, F.; Schenk, K.; Landais, Y. *Org. Lett.* **2006**, *8*, 4755-4758.
9. For the synthesis of 5-substituted cyclohexanediones, see: Tamura, Y.; Yoshimoto, Y.; Kunimoto, K.; Tada, S.; Tomita, T.; Wada, T.; Seto, E.; Murayama, M.; Shibata, Y.; Nomura, A.; Ohata, K. *J. Med. Chem.* **1977**, *20*, 709-714.
10. Kotnis, A. S. *Tetrahedron Lett.* **1991**, *32*, 3441-3444 (trusting the text but not the misdrawn structures shown in table 1).
11. (a) Pangborn, A. B.; Giardello, M. A.; Grubbs, R. H.; Rosen, R. K.; Timmers, F. J. *Organometallics* **1996**, *15*, 1518-1520. (b) Alaimo, P. J.; Peters, D. W.; Arnold J.; Bergmann, R. G. *J. Chem. Educ.* **2001**, *78*, 64.

Appendix
Chemical Abstracts Nomenclature (Registry Number)

5-Phenylcyclohexane-1,3-dione (493-72-1)
Iodine (7553-56-2)
Sodium thiosulfate (7772-98-7)
Lithium (7439-93-2)
Chloroacetonitrile (107-14-2)

Yannick Landais was born in Angers (France) in 1962. He received his Ph.D. in chemistry from the University of Orsay (Paris XI) under the supervision of Dr. J.-P. Robin. After carrying out postdoctoral work with Prof. Ian Fleming at Cambridge University (1988-1990), he took a position of Assistant Professor at the University of Lausanne (1990-1997). He was then appointed at the University Bordeaux 1 where he is currently Professor of Organic Chemistry. His research interests are in asymmetric synthesis, radical chemistry, with a special emphasis on organosilicon chemistry and its applications in total synthesis of natural products. In 1997, he was awarded the Werner prize by the New Swiss Chemical Society.

Raphaël Lebeuf was born in 1980 in Carcassonne. He did all his undergraduate studies at the University Bordeaux 1 and started research in 2002 under the supervision of Prof. Y. Landais and Dr. F. Robert. During his Ph.D. studies, he worked on Birch reductive alkylations of biaryls and desymmetrization including hydroamination processes. After graduating from the University Bordeaux 1 in 2006, he accepted a postdoctoral position in Prof. Glorius group at Münster University (Germany).

Muriel Berlande was born in Saint-Etienne (France) in 1969. She studied chemistry at the University of Clermont-Ferrand where she did her Master of chemistry in 1994 under the supervision of Prof J. C. Gramain. She was then appointed as a laboratory assistant at the University Bordeaux 1 in 1997 and finally laboratory technician at the CNRS in 2000. She is currently involved in different multi-step organic syntheses and specialized in HPLC analysis.

Frédéric Robert studied in Grenoble where he received his Ph.D. in 1999 under the supervision of Dr. Andrew E. Greene and Dr. Yves Gimbert. He moved to Dartmouth College (New Hampshire, USA) for a postdoctoral study with Prof. Peter A. Jacobi (2000-2001), followed by a second postdoctorate at the University of Geneva with Prof. Peter Kündig (2001-2002). He was appointed Chargé de Recherches CNRS at the University Bordeaux 1 (2002). His current research focuses on the development of new desymmetrization processes for alkaloid synthesis.

Marcus G. Schrems was born in Groß-Umstadt (Germany) in 1979. He studied chemistry at the Technische Universität München (TUM), National University of Singapore (NUS) and Universtiy of Bergen (Norway) and graduated from TUM in 2005 after completing his Diplom-Thesis under the direction of R. Anwander and W. A. Herrmann. In 2006 he joined the lab of Andreas Pfaltz at University of Basel (Switzerland). He is currently working on Ir-catalyzed enantioselective hydrogenation of unfunctionalized olefins, focusing on tetrasubstituted olefins.

SYNTHESIS OF *TERT*-BUTYL (1*S*,2*S*)-2-METHYL-3-OXO-1-PHENYLPROPYLCARBAMATE BY ASYMMETRIC MANNICH REACTION

Submitted by Jung Woon Yang, Subhas Chandra Pan, and Benjamin List.[1]
Checked by Hisashi Mihara, and Masakatsu Shibasaki.

1. Procedure

A. tert-Butyl phenyl(phenylsulfonyl)methylcarbamate (1).[2] A 500-mL, two-necked, round-bottomed flask is equipped with a magnetic stirring bar and fitted with a glass stopper and an argon inlet. The flask is flushed with argon and charged with *tert*-butyl carbamate (13.00 g, 110.9 mmol, 1.00 equiv) (Note 1) and 40 mL of tetrahydrofuran (Note 2). Water (100 mL), sodium benzenesulfinate (18.21 g, 110.9 mmol, 1.00 equiv) (Note 3), and freshly distilled benzaldehyde (11.48 mL, 113.1 mmol, 1.02 equiv) (Note 4) are sequentially added in single portions, followed by formic acid (99%, 24.3 mL, 643 mmol) (Note 5). The reaction mixture is stirred for 18 h at room temperature under an argon atmosphere, during which time the desired product precipitates. The resulting white solid is filtered through a Büchner funnel (diameter 100 mm) and washed with distilled water (200 mL). The solid is transferred to a 500-mL single-necked, round-bottomed flask and is slurried in a mixture of hexane/dichloromethane (150/15 mL). The mixture is stirred for 2 h at room temperature, after which time it is collected by filtration in a Büchner funnel (diameter 100 mm) and washed with hexane/dichloromethane (91/9 mL). The solid is transferred to a 300-mL round-bottomed flask and dried under reduced pressure (0.5 mmHg) at 25 °C for 6 h to afford 30.79 g (80%) (Note 6) of the title compound **1** as a white solid.

B. *(E)-tert-Butyl benzylidenecarbamate* (**2**).[3] A 1-L, two-necked, round-bottomed flask equipped with a magnetic stirring bar, a rubber septum, and a reflux condenser capped with an inlet adaptor connected to an argon-vaccum manifold, is charged with anhydrous potassium carbonate (71.48 g, 517.2 mmol, 6.0 equiv) (Note 7). The solid is placed under vacuum (0.5 mmHg) and flame-dried. The flask is purged with argon and anhydrous tetrahydrofuran (600 mL) (Note 2) is added via cannula under argon at 25 °C. Then, the septum is removed and compound **1** (29.95 g, 86.20 mmol, 1.0 equiv) is added into the flask. The septum is exchanged for a glass stopper, and the resulting suspension is heated to reflux at 80–85 °C (oil bath temp) with vigorous stirring under argon. After 15 h, the reaction is cooled to 25 °C and the solid is filtered off through alternating layers of Celite (1 cm thick)/Na$_2$SO$_4$ (1 cm thick)/Celite (1 cm thick) using a Büchner funnel (diameter 100 mm), and washed with anhydrous tetrahydrofuran (100 mL) (Note 2). The filtrate is concentrated at 30 °C (water bath temperature) by rotary evaporation (20 mmHg) and dried under vacuum (0.5 mmHg) to give 17.68 g (>99%) (Note 8) of the corresponding *N*-Boc-imine **2** as colorless oil.

C. *(1S,2S)-2-Methyl-3-oxo-1-phenylpropylcarbamate (3)*. A 500-mL, three-necked, round-bottomed flask equipped with a magnetic stirring bar, thermometer, rubber septum and argon inlet is charged with *N*-Boc-imine **2** (5.00 g, 24.4 mmol, 1.0 equiv). The flask is flushed with dry argon and anhydrous acetonitrile (240 mL) (Note 9) is added via cannula under argon. Freshly distilled propionaldehyde (3.52 mL, 48.7 mmol, 2.0 equiv) (Note 10) is added quickly in one portion by syringe and the flask is sealed and purged with argon. The resulting colorless solution is cooled to 0 °C (internal temperature) with a cryostat. The septum is removed temporarily and (*S*)-proline (561 mg, 4.87 mmol, 20 mol %) (Note 11) is added. After the reaction mixture is stirred at 0 °C (internal temperature) for 12 h, it is treated with 80 mL of distilled water and warmed to 25 °C. The solution is stirred vigorously for 20 min at 25 °C, diluted with diethyl ether (100 mL), and transferred to a 500-mL separatory funnel. The layers are separated, and the aqueous layer is extracted with diethyl ether (100 mL). The combined organic phases are washed with brine (60 mL), dried over MgSO$_4$ (20 g), filtered through a Büchner funnel, and concentrated by rotary evaporator (30 °C, water bath temp, 20 mmHg). The resulting solid is triturated with cool *iso*-hexane (4 °C, 70 mL), filtered, washed with an additional portion of cool *iso*-hexane (4 °C, 15 mL), and dried under vacuum (0.5 mmHg, 25 °C) to afford 5.61 g (87%) (Notes 12, 13) of Mannich product **3**.

2. Notes

1.   *tert*-Butyl carbamate (≥98% purity) was purchased from Fluka and used as received.

2.   Anhydrous tetrahydrofuran was purchased from Kanto Chemicals and used as received by the Checkers. (Tetrahydrofuran dried by distillation from sodium and benzophenone under an argon atmosphere was used by the Submitters.)

3.   Benzenesulfinic acid sodium salt (98% purity) was purchased from Alfa Aesar and used as received.

4.   Benzaldehyde was purchased from Alfa Aesar and distilled before use (5 mmHg, bp 40 °C).

5.   Formic acid (98%) was purchased from Acros and used as received.

6.   Compound **1** exhibits the following physical and spectroscopic properties: mp: 153-154 °C; IR (KBr) ν 3362, 2975, 2495, 1713, 1311, 1146 cm^{-1}; ^{1}H NMR (500 MHz, (CD$_3$)$_2$CO) δ: 1.26 (s, 9 H), 6.08 (d, J = 10.9 Hz, 1 H), 7.45-7.47 (m, 3 H), 7.65-7.75 (m, 6 H), 7.95-7.97 (m, 2 H); ^{13}C NMR (125 MHz, (CD$_3$)$_2$CO) δ: 28.3, 75.3, 80.5, 129.2, 129.9, 130.3, 130.4, 130.6, 131.9, 134.7, 138.8, 154.9; Anal. Calcd for C$_{18}$H$_{21}$NO$_4$S: C, 62.23; H, 6.09; N, 4.03. Found: C, 62.29; H, 6.05; N, 4.07.

7.   Anhydrous potassium carbonate as granular form was purchased from Merck.

8.   Compound **2** exhibits the following physical and spectroscopic properties: IR (KBr) ν 3382, 2979, 1717, 1636, 1152 cm^{-1}; ^{1}H NMR (500 MHz, (CD$_3$)$_2$CO) δ: 1.58 (s, 9 H), 7.57-7.66 (m, 3 H), 7.98-8.00 (m, 2 H), 8.84 (s, 1 H); ^{13}C NMR (125 MHz, (CD$_3$)$_2$CO) δ: 28.1, 82.1, 129.9, 130.5, 134.2, 135.6, 163.5, 168.5; Anal. Calcd for C$_{12}$H$_{15}$NO$_2$: C, 70.22; H, 7.37; N, 6.82. Found: C, 69.82; H, 7.30; N, 6.71.

9.   Anhydrous acetonitrile was purchased from Fluka and used without further purification.

10. Propionaldehyde was purchased from Alfa Aesar and distilled before use (bp 50 °C).

11. (S)-Proline was purchased from Fluka and used as received.

12. TLC analysis was performed on Merck silica gel 60 F-254 plates using 15% ethyl acetate in hexanes as eluent. The product has an $R_f = 0.33$ (visualized with 254 nm UV lamp and stained with p-anisaldehyde reagent).

13. The enantiomeric ratio was determined to be >99:1 by chiral HPLC (CHIRALPAK AS-H column, 2% i-PrOH/heptane, 0.50 mL/min, 220 nm, t_R (minor: 41.2 min), t_R (major: 53.8 min). Compound **3** exhibits the following physical and spectroscopic properties: mp: 127-128 °C; $[\alpha]^{26.4}_D$ +14.6 (c 1.0, CHCl$_3$); IR (KBr) v 3379, 2977, 1721, 1683, 1524, 1174 cm^{-1}; ^{1}H NMR (500 MHz, CDCl$_3$) δ: 1.07 (d, $J = 7.0$ Hz, 3 H), 1.42 (s, 9 H), 2.87 (m, 1 H), 5.12 (br s, 1 H), 5.19 (br s, 1 H), 7.24-7.37 (m, 5 H), 9.72 (s, 1 H); ^{13}C NMR (125 MHz, CDCl$_3$) δ: 9.5, 28.5, 51.8, 54.9, 80.3, 126.9, 127.9, 129.0, 139.8, 155.3, 203.3; Anal. Calcd for C$_{15}$H$_{21}$NO$_3$: C, 68.42; H, 8.04; N, 5.32. Found: C, 68.30; H, 7.99; N, 5.40.

Safety and Waste Disposal Information

All hazardous materials should be handled and disposed of in accordance with "Prudent Practices in the Laboratory"; National Academy Press: Washington, DC, 1995.

3. Discussion

The catalytic asymmetric Mannich reaction is arguably the most useful approach to synthesize chiral β-amino carbonyl compounds.[4] In 2000, we discovered a proline-catalyzed direct three-component asymmetric Mannich reaction between aldehydes, 4-methoxyaniline, and ketones. This Mannich reaction required the use of an aniline as the amine component.[5] Since the N-substituent is usually employed as protecting group, it should be easily removable after the reaction has taken place. However, the removal of the commonly used p-methoxyphenyl (PMP) group from nitrogen often requires drastic oxidative conditions involving harmful reagents such as ceric ammonium nitrate (CAN) that are not compatible with all substrates. We have employed the *tert*-butoxycarbonyl (Boc) group as an easily removable protecting group to overcome this drawback.[6] For example, we found the reaction of propionaldehyde with benzaldehyde-derived N-Boc-imine[7] to

14

give the corresponding chiral $\beta^{2,3}$-amino aldehyde in high levels of diastereo- and enantioselectivities. Moreover, the product of the reaction was obtained by an aqueous workup/organic extraction process as stable, crystalline solid. Purification can be achieved by trituration with cool *iso*-hexane.

1. Max-Planck-Institut für Kohlenforschung, Kaiser-Wilhelm-Platz 1, D-45470 Mülheim an der Ruhr, Germany. E-mail: list@mpi-muelheim.mpg.de
2. Mccozzi, T., Petrini, M. *J. Org. Chem.* **1999**, *64*, 8970-8972.
3. (a) Kanazawa, A. M.; Denis, J.-N.; Greene, A. E. *J. Org. Chem.* **1994**, *59*, 1238-1240. (b) Wenzel, A. G.; Jacobsen, E. N. *J. Am. Chem. Soc.* **2002**, *124*, 12964-12965. (c) Song, J.; Wang, Y.; Deng. L.; *J. Am. Chem. Soc.* **2006**, *128*, 6048-6049.
4. For review, see: Mukherjee, S.; Yang, J. W.; Hoffmann, S.; List, B. *Chem. Rev.* **2007**, *107*, 5471-5569.
5. List, B. *J. Am. Chem. Soc.* **2000**, *122*, 9336-9337.
6. (a) Yang, J. W.; Stadler, M.; List, B. *Angew. Chem. Int. Ed.* **2007**, *46*, 609-611. (b) Yang, J. W.; Stadler, M.; List, B. *Nat. Protocols* **2007**, *2*, 1937-1942. (c) Yang, J. W.; Chandler, C.; Stadler, M.; Kampen, D.; List, B. *Nature* **2008**, *452*, 453-455.
7. Enders, D.; Grondal, C.; Vrettou, M. *Synthesis* **2006**, 3597-3604.

Appendix
Chemical Abstracts Nomenclature; (Registry Number)

tert-Butyl phenyl(phenylsulfonyl)methylcarbamate: Carbamic acid, *N*-[phenyl(phenyl-sulfonyl)methyl]-1,1-dimethylether ester; (155396-71-7)

(*E*)-*tert*-Butyl benzylidenecarbamate: Carbamic acid, *N*-(phenylmethylene)-1,1-dimethyl-ethyl ester, [*N*(*E*)]-; (177898-09-2)

Carbamic acid, *N*-[(1*S*,2*S*)-2-methyl-3-oxo-1-phenylpropyl]-1,1-dimethylethyl ester; (926308-17-0)

Benjamin List was born in 1968 in Frankfurt, Germany. He graduated from Freie University Berlin (1993) and received his Ph.D. (1997) from the University of Frankfurt (Mulzer). After postdoctoral studies (1997-1998) as a Feodor Lynen Fellow of the Alexander von Humboldt foundation at The Scripps Research Institute (Lerner), he became an Assistant Professor there in January 1999. Subsequently, he developed the first proline-catalyzed asymmetric intermolecular aldol-, Mannich-, Michael-, and α-amination reactions. He moved to the Max-Planck-Institut für Kohlenforschung in 2003, and currently is a director there and an honorary professor at the University of Cologne. His research interests are new catalysis concepts, bioorganic chemistry, and natural product synthesis.

Jung Woon Yang was born in Cheju, Korea, in 1973. He received his Ph.D. in 2003 under the supervision of Drs. Choong Eui Song and Hogyu Han, working on the heterogeneous asymmetric catalysts for dihydroxylation, aminohydroxylation, allylic substitution reactions, and synthesis of DNA triangles with vertexes of bis(terpyridine)iron(II) complexes at the Korea Institute of Science and Technology (KIST) and Korea University, respectively. He undertook postdoctoral studies (2003-2006) at Max-Planck-Institut für Kohlenforschung, with Prof. Benjamin List, where he worked on the development of asymmetric organocatalytic reactions, including transfer hydrogenation, the Mannich reaction, and cascade reactions. In 2006, he became a senior scientist in the List group.

Subhas Chandra Pan was born in 1980 in Hooghly (West-Bengal), India. He obtained his B.Sc. degree in Chemistry Honours in 2001 from Calcutta University and M.S. degree in 2004 from Indian Institute of Science, Bangalore. During his MS thesis he worked in Prof. Goverdhan Mehta's laboratory on the total synthesis of epoxyquinone natural products. He obtained his PhD degree (*summa cum laude*) with Prof. Benjamin List at the Max-Planck-Institut für Kohlenforschung, Mülheim an der Ruhr, Germany and his research interest focuses on the development of new organocatalytic reactions.

Hisashi Mihara was born in 1981 in Shiga, Japan. He received his M.S. degree in 2006 from the University of Tokyo. He is pursuing a Ph.D. degree at the Graduate School of Pharmaceutical Sciences, The University of Tokyo, under the guidance of Professor Masakatsu Shibasaki. His current interest is catalytic enantioselective total synthesis of biologically active compounds.

SYNTHESIS OF PYRAZOLO[1,5-a]PYRIDINES VIA AZIRINES: PREPARATION OF 2-(3-BROMOPHENYL)-6-(TRIFLUOROMETHYL)PYRAZOLO[1,5-a]PYRIDINE

Submitted by Stephen Greszler[1] and Kirk L. Stevens.[2]
Checked by Hiroshi Nakagawa and Jonathan A. Ellman.

Caution! *Hydroxylamine hydrochloride explodes with heating above 110 °C.*

1. Procedure

A.   *1-(3-Bromophenyl)-2-[5-(trifluoromethyl)-2-pyridinyl]ethanone*
(**1**).   A 2-L, oven-dried, three-necked, round-bottomed flask equipped with
two stoppers and nitrogen inlet is charged with sodium hydride (13.2 g, 330

18

mmol, 60% dispersion in oil) and a magnetic stir bar (Note 1). Hexanes (300 mL) (Note 2) are added, and the suspension is stirred, allowed to settle, and the supernatant (250 mL) is poured off by decantation. Dry THF (300 mL) (Note 3) is added to the flask, and a reflux condenser with nitrogen inlet, a thermometer, and a rubber septum are attached. A solution of 3'-bromo-acetophenone (30.0 g, 20 mL, 151 mmol) (Note 4) in dry THF (60 mL) is added over 5 min via syringe to the sodium hydride suspension. The resulting light brown suspension is stirred for 5 min, and a solution of 2-chloro-5-(trifluoromethyl)pyridine (24.9 g, 137 mmol) (Note 5) in dry THF (60 mL) is added over 10 min via syringe. Additional dry THF (120 mL) is used to rinse the sides of the flask. The resulting orange-brown solution is heated in an oil bath until the internal temperature is 60 °C. The solution darkens in color as the reaction proceeded. After stirring for 24 h, the reaction solution is cooled to room temperature. Only trace starting material remains as determined by TLC: product (**1**) $R_f = 0.5$ (4:1 hexanes:ethyl acetate). To allow more volume during quenching, the mixture is transferred equally into two 1-L Erlenmeyer flasks. To each portion is added brine (150 mL) (Note 6) and ethyl acetate (150 mL). After the reaction is safely quenched, all material is transferred to a 2-L separatory funnel. The aqueous layer is removed and the organic layer is washed with additional brine (100 mL). The combined aqueous layers are extracted with ethyl acetate (100 mL), and the combined organic layers are dried with $MgSO_4$ (150 g). The drying agent is filtered off and washed with additional ethyl acetate (200 mL). The combined filtrate is concentrated in vacuo (40 °C, 20 mmHg). Celite (24 g) and methylene chloride (50 mL) are added to the resulting brown solid and concentrated by rotary evaporation (35 °C, 20 mmHg). The adsorbed material is loaded onto a column and purified by silica gel chromatography (330 g silica gel, 25 cm x 6 cm, start 10:1 hexanes:ethyl acetate to 3:1 gradient, 100 mL/min, 30 min, 3 L total solvent volume) to give product **1** (37.5 g, 79%) as a bright yellow solid (Note 7).

B. *(1Z)-1-(3-bromophenyl)-2-[5-(trifluoromethyl)-2-pyridinyl]-ethanone oxime* (**2**). A 500-mL, three-necked, round-bottomed flask equipped with a thermometer and a mechanical stirrer is charged with ketone (**1**) (15.0 g, 43.7 mmol). Methanol (120 mL) (Note 8) is added, followed by slow addition of aqueous sodium hydroxide (2.5 M, 75 mL, 188 mmol) (Note 9) over 5 min, which completely dissolved the ketone to give an orange solution. Hydroxylamine hydrochloride (15.2 g, 219 mmol) (Note 10) is added in small portions over 15 min, during which the solution changed from orange to bright yellow (Note 11). An additional 10 mL of methanol is used to wash down the sides of the flask. Vigorous stirring is required due to the large quantity of solids present. A reflux condenser is attached and the reaction mixture is heated in an oil bath until the internal temperature is 70 °C. The solids dissolved and the bright yellow color faded as the reaction proceeded. Once an internal temperature of 70 °C is reached, the reaction mixture is stirred for 90 min at which point only trace starting material remained as determined by TLC: product (2) R_f = 0.33 (4:1 hexanes:ethyl acetate). The oil bath is removed, the reaction solution cooled to room temperature over 30 minutes, and a solid precipitated out of solution (Note 12). The reaction mixture is further cooled to 0 °C in an ice bath for 30 min, and all of the white solid material is collected by filtration using a fritted funnel (9 cm, 350 mL, M). The solid is washed with cold water (3 x 20 mL) and is dried under high vacuum (<1 mmHg) at 50 °C for 5 hours. The pale yellow solid is dissolved in methanol (75 mL) at 60 °C and water (25 mL) is added slowly at 60 °C until the solution became cloudy. The solution is cooled to room temperature over 1 h and the precipitate that formed is filtered using a fritted funnel (9 cm, 350 mL, M). The solid is washed with cold water (3 x 20 mL) and dried under high vacuum (<1 mmHg, 50 °C) overnight to give (2) as pale yellow solid (11.4 g, 72%) (Notes 13, 14).

C. *2-[3-(3-Bromophenyl)-2H-azirin-2-yl]-5-(trifluoromethyl)pyridine* (3). A 250-mL, oven-dried, three-necked, round-bottomed flask equipped with nitrogen inlet, a thermometer, a rubber septum, and a magnetic stir bar is charged with oxime (2) (12.0 g, 33.4 mmol) and dry dichloromethane (50 mL) (Note 15). The cloudy yellow solution is cooled in an ice bath and triethylamine (18.6 mL, 133 mmol) (Note 16) is added dropwise over 10 min via syringe. Following the triethylamine addition, all solids had dissolved, and stirring is continued at 0 °C for 15 min. Trifluoroacetic anhydride (5.7 mL, 41 mmol) (Note 17) is then added dropwise over 10 min via syringe (Note 18). The solution darkened in color as the reaction proceeded. After 30 min of stirring, the flask is allowed to warm to room temperature and stirring is continued for an additional 60 min. Only trace starting material remained as determined by TLC: product (3) R_f = 0.25 (9:1 hexanes:ethyl acetate). The reaction is then quenched by the addition of water (20 mL). The aqueous layer is extracted with methylene chloride (2 x 15 mL), and the combined organic layers are dried over $MgSO_4$ (20 g), filtered, and concentrated (35 °C, 20 mmHg). Celite (10 g) and methylene chloride (40 mL) are added to the resulting dark red oil and concentrated by rotary evaporation (35 °C, 20 mmHg). The adsorbed material is loaded onto a column and purified using silica gel chromatography (330 g silica gel, 25 cm x 6 cm, start 20:1 hexanes:ethyl acetate to 5:1 gradient, 100 mL/min, 30 min, 3 L total solvent volume) to give (**3**) as a dark orange oil (8.61 g 76%) (Note 19).

D. *2-(3-Bromophenyl)-6-(trifluoromethyl)pyrazolo[1,5-a]pyridine* (**4**). A 20-mL microwave vial is charged with azirine (**3**) (2.20 g, 6.45 mmol) dissolved in 1,2-dichloroethane (11 mL) (Note 20). A stir bar is added, the vial is sealed, and the resulting orange solution is heated for 1 h in a Biotage Initiator Eight Microwave Reactor held at a constant temperature of 175 °C (Notes 21, 22). Only trace starting material remained as determined by TLC: product (**4**) $R_f = 0.35$ (9:1 hexanes:ethyl acetate). A total of 6.60 g (3 x 2.20 g, 19.3 mmol) of (**3**) is treated in this manner and the resulting solutions are combined and concentrated in vacuo (40 °C, 20 mmHg). Celite (7 g) and methylene chloride (40 mL) are added and concentrated by rotary evaporation (35 °C, 20 mmHg). The adsorbed material is loaded onto a column and purified using silica gel chromatography (120 g silica gel, 25 cm x 3 cm, start 20:1 hexanes:ethyl acetate to 3:1 gradient, 40 mL/min, 30 min, 1.2 L total solvent volume) to give (**4**) (6.08 g, 92%) as a yellow powder (Notes 23, 24).

2. Notes

1.   Sodium hydride (dispersion, 60% in oil) was purchased from the Aldrich Chemical Company, Inc. and was used as received. When using 60% dispersions, the submitters have found the need to add 1.2 equiv. Consequently for the chemistry of Step A the submitters have found that the reaction proceeds to completion more reliably with a total of 2.4 equiv of NaH.

2.   Hexanes (HPLC Grade) was purchased from EMD and was used as received.

3.   Drisolv® THF (99.9%) was purchased from EMD and was used as received.

4.   3'-Bromo-acetophenone (99%) was purchased from the Aldrich Chemical Company, Inc. and was used as received.

5.   2-Chloro-5-(trifluoromethyl)pyridine was purchased from the Aldrich Chemical Company, Inc. and was used as received.

6. Aqueous saturated brine solution was added dropwise to minimize the exotherm.

7. Submitters made proton and carbon assignments using 2D NMR and found product **1** exists as an approximate 7:3 keto/enol tautomeric mixture: mp 85–86 °C; IR (film) cm^{-1} 1603, 1547, 1477, 1323, 1259, 1161, 1116, 1075, 1058; TLC: R_f = 0.52 (silica gel, 4:1 hexanes:ethyl acetate); ^{1}H NMR (400 MHz, DMSO-d_6) δ: 4.71 (s, 0.5 H), 6.58 (s, 0.75 H), 7.39 (m, 1.5 H), 7.49(t, 0.25 H, J = 8.0 Hz), 7.62 (m, 1 H), 7.82 (m, 1 H), 7.98 (m, 1 H), 8.07 (d, 0.75 H, J = 8.0 Hz), 8.14 (m, 0.5 H), 8.74 (br s, 0.75 H), 8.85 (br s, 0.25 H), 14.97 (br s, 0.75 H); ^{13}C NMR (100 MHz, DMSO-d_6) δ: 47.8, 94.9, 120.0 (q), 122.5, 122.6, 122.7, 122.9, 123.6, 124.0, 124.9, 125.4, 127.8, 128.1, 128.5, 131.1, 131.2, 131.4, 133.2, 134.2 (q), 134.9 (q), 136.5, 138.2, 138.7, 142.3 (q), 146.1 (q), 160.6, 164.9, 195.9; ^{19}F NMR (376 MHz, DMSO-d_6) δ: –60.03, -60.06; LRMS (ESI) m/z (%): 344 (100), 345 (17), 346 (100), 347 (17); HRMS (FAB) m/z M$^+$ calcd for C$_{14}$H$_9$BrF$_3$NO: 342.9820. Found: 342.9825. Anal. Calcd for C$_{14}$H$_9$BrF$_3$NO: C, 48.86; H, 2.64; N, 4.07. Found: C, 48.82; H, 2.69; N, 4.07.

8. Methanol was purchased from EMD and was used as received.

9. Sodium hydroxide (2.50 ± 0.02 M) was purchased from VWR and was used as received.

10. Hydroxylamine hydrochloride (98%) was purchased from the Aldrich Chemical Company, Inc. and was used as received.

11. The submitters found that extended reaction time is required when less than 5 equiv of hydroxylamine hydrochloride is used. For example, 3 equiv of hydroxylamine hydrochloride required 24 h at 70 °C to reach completion. In addition, test reactions demonstrated that the use of at least 2.5 equiv of hydroxylamine hydrochloride was necessary for the reaction to reach completion.

12. The submitters obtained a white flocculent solid (12.0 g, 76%) out of the reaction solution, which was used without purification in the next step.

13. The oxime (**2**) exhibits the following characteristics: mp 123–125 °C; IR (film) cm^{-1} 3002, 2835, 1610, 1324, 1176, 1164, 1127, 1081, 1059, 1028; TLC: R_f = 0.33 (silica gel, 4:1 hexanes:ethyl acetate); ^{1}H NMR (400 MHz, DMSO-d_6) δ: 4.35 (s, 2 H), 7.30 (t, 1 H, J = 8.0 Hz), 7.52 (m, 2 H), 7.68 (d, 1 H, J = 8.0 Hz), 7.87 (s, 1 H), 8.08 (d, 1 H, J = 8.0 Hz), 8.83 (br s, 1 H), 11.75 (br s, 1 H); ^{13}C NMR (100 MHz, DMSO-d_6) δ: 34.3, 122.2, 123.8, 125.5, 128.9, 131.0, 131.9, 134.5 (q), 138.7, 146.2 (q), 152.6, 162.5; ^{19}F NMR (376 MHz, DMSO-d_6) δ: –59.99; LRMS (ESI) m/z (%): 359 (100),

360 (17), 361 (100), 362 (17);   HRMS (FAB) m/z [M+H]$^+$ calcd for $C_{14}H_{11}BrF_3N_2O$:   359.0005   found: 359.0006.   Anal.   Calcd   for $C_{14}H_{10}BrF_3N_2O$: C, 46.82; H, 2.82; N, 7.80. Found: C, 46.55; H, 2.68; N, 7.67.

14. The submitters reported yields that ranged from 65-81% for this step, with decreasing returns as the reaction scale was increased. The submitters also reported that the oxime (**2**) could be efficiently converted back to ketone (**1**) by treatment with 1:1 water:acetone and a few drops of 4M HCl at 50°C for 24 h.

15. Dry dichloromethane (99.8%) was purchased from the Aldrich Chemical Company, Inc. and was used as received.

16. Triethylamine was purchased from the Aldrich Chemical Company, Inc. and was used as received.

17. Trifluoroacetic anhydride was purchased from the Aldrich Chemical Company, Inc. and was used as received.

18. Trifluoroacetic anhydride is volatile under these conditions; the addition should be made slowly and carefully.

19. The azirine (**3**) exhibits the following characteristics: IR (film) cm^{-1} 1736, 1604, 1323, 1162, 1123, 1077; TLC: R_f = 0.30 (silica gel, 9:1 hexanes:ethyl acetate); ^{1}H NMR (400 MHz, DMSO-d_6) δ: 3.63 (s, 1 H), 7.57 (t, 1 H, J = 8.0 Hz), 7.59 (d, 1 H, J = 8 Hz), 7.87 (d, 1 H, J = 8.0 Hz), 7.91 (d, 1 H, J = 8.0 Hz, 1 H), 8.07 (m, 1 H), 8.09 (dm 1 H, J = 8.0 Hz), 8.76 (s, 1 H); ^{13}C NMR (100 MHz, DMSO-d_6) δ: 35.3, 122.2, 122.9, 123.0, 123.9 (q), 125.5, 129.2, 132.1, 132.6, 134.2 (q), 136.8, 146.2 (q), 160.4, 164.4; ^{19}F NMR (376 MHz, DMSO-d_6) δ: –60.04; LRMS (ESI) m/z (%): 341 (100), 342 (17), 343 (100), 344 (17); HRMS (FAB) m/z [M+H]$^+$ calcd for $C_{14}H_9BrF_3N_2$: 340.9901 found: 340.9895. Anal. Calcd for $C_{14}H_8BrF_3N_2$: C, 49.29; H, 2.36; N, 8.21. Found: C, 49.37; H, 2.38; N, 7.96. The submitters observed that the product solidified over time: mp 123–125 °C;

20. 1,2-Dichloroethane (99.8%) was purchased from the Aldrich Chemical Company, Inc. and was used as received.

21. Safety precautions should always be taken with any "sealed flask" reaction. The microwave reactor has a built-in "blast shield" and an automatic shut-off when a pressure is over the instrument maximum setting. Solvent choice can be a very important factor in microwave chemistry as the vapor pressure buildup must be taken into consideration as the reaction is heated. 1,2-Dichloroethane was used in Step D because it can be heated

high enough to facilitate the desired transformation while maintaining a safe vapor pressure.

22. The submitters reported that 1,2,4-trichlorobenzene was a suitable solvent under conventional refluxing conditions and that diglyme and 1,2,4-trichlorobenzene proved to be suitable alternatives under microwave conditions. The detailed procedure described herein provided the highest conversion and yield of the desired product. Because the microwave reactor has a maximum volume of 20 mL, the checkers ran three identical reactions of 2.2 g, and the submitters ran two identical reactions of 3.3 g. The submitters also reported that the transformation of azirine (**3**) to pyrazolo[1,5-a]pyridine (**4**) is much cleaner when the microwave is used for 1 h instead of conventional heating for longer times.

23. The pyrazolo[1,5-a]pyridine (**4**) exhibits the following characteristics: mp 59–60 °C; IR (film) cm^{-1} 1647, 1460, 1334, 1321, 1157, 1112, 1076, 1052; TLC: R_f = 0.38 (silica gel, 4:1 hexanes:ethyl acetate); ^{1}H NMR (400 MHz, DMSO-d_6) δ: 7.24 (s, 1 H), 7.35 (d, 1 H, J = 9.2 Hz), 7.39 (t, 1 H, J = 8.0 Hz), 7.54 (d, 1 H, J = 8.0 Hz), 7.81 (d, 1 H, J = 9.2 Hz), 7.95 (d, 1 H, J = 8.0 Hz), 8.12 (m, 1 H), 9.26 (s, 1 H); ^{13}C NMR (100 MHz, DMSO-d_6) δ: 96.3, 115.3 (q), 119.7, 122.7, 122.7, 125.4, 125.6, 128.5 (q), 129.1, 131.4, 132.1, 134.7, 142.1, 153.6; ^{19}F NMR (376 MHz, DMSO-d_6) δ: –59.73; LRMS (ESI) m/z (%): 341 (100), 342 (25), 343 (83), 344 (17); HRMS (FAB) m/z M$^+$ calcd for $C_{14}H_8BrF_3N_2$: 339.9823 found: 339.9827. Anal. Calc for $C_{14}H_8BrF_3N_2$: C, 49.29; H, 2.36; N, 8.21. Found: C, 49.26; H, 2.33; N, 8.11.

24. The submitters made proton and carbon assignments based on correlations in the TOCSY, gCOSY, NOESY, gHSQC, and gHMBC experiments that fully support the structure (2D NMR data not shown).

Safety and Waste Disposal Information

All hazardous materials should be handled and disposed of in accordance with "Prudent Practices in the Laboratory"; National Academy Press; Washington, DC, 1995.

3. Discussion

Azirine intermediates in the synthesis of the complex pyrazolo[1,5-a]pyridine heterocycle are utilized because the method is facile and the intramolecular nature allows control of substituent regiochemistry on the resulting bicycle (the CF$_3$ group in **4**, for example). Thermal reactions of azirines involve regioselective ring opening to form highly reactive nitrene intermediates. With appropriately substituted pyridyl azirines, the electrophilic nitrene can be intercepted by the pyridyl nitrogen to give the bicyclic pyrazolo[1,5-a]pyridine with predictable substitution patterns based on the original pyridyl substitution. The most common side products derive from nitrene dimerization. The product resulting from formal C-H insertion is rarely seen in appreciable quantity.

A wide variety of intermolecular reactions have been successfully performed to produce unsubstituted pyrazolo[1,5-a]pyridines.[3] The intermolecular syntheses of an unsubstituted pyrazolo[1,5-a]pyridine dates back to at least 1962 when Huisgen synthesized a number of interesting heterocycles including pyrazolo[1,5-a]pyridines.[4] Huisgen was a pioneer of heterocyclic chemistry especially the use of unsubstituted 1-aminopyridinium iodide.[5] *Inter*molecular syntheses of pyrazolo[1,5-a]pyridines are perhaps more common than the *intra*molecular variety, but the intermolecular nature fails to control the regiochemical outcome.

Azirines themselves have attracted much research interest because they represent the smallest nitrogen-containing unsaturated heterocyclic system.[6] Most interest stems from the influence of ring strain of the azirine on the potential for elaboration to other heterocyclic systems. As with the synthesis of the pyrazolo[1,5-a]pyridines, azirine syntheses can also be categorized into intra- and intermolecular classes. Intramolecular variants as described herein (reaction of an *N*-functionalized imine) are usually termed the Neber reaction based on the first synthesis of an azirine in a cycloelimination reaction under basic conditions.[7] Other intramolecular syntheses of azirines begin with vinyl azides,[8] isoxazoles,[9] and oxazaphospholes.[10] The less common intermolecular approach to the synthesis of azirines usually relies on a cycloaddition between carbenes and nitriles.[11] Useful yields in this approach generally require stabilized phosphanylcarbenes which produce phosphorous substituted azirines.[12]

1. Department of Chemistry, University of North Carolina, Chapel Hill, North Carolina 27599-3290.

2. Department of Oncology Medicinal Chemistry, GlaxoSmithKline, 5 Moore Drive, 3.4174-4B, Research Triangle Park, North Carolina 27709; email: kls62784@gsk.com.

3. Stevens, K. L.; Jung, D. K.; Alberti, M. J.; Badiang, J. G.; Peckham, G. E.; Veal, J. M.; Cheung, M.; Harris, P. A.; Chamberlain, S. D.; Peel, M. R. *Org. Lett.* **2005**, *7(21)*, 4753-4756.

4. Huisgen, R.; Grashey, R.; Krischke, R. *Tetrahedron Lett.* **1962**, *3(9)*, 387-391.

5. Gösl, R.; A. Meuwsen, A. *Org. Synth. Coll. Vol. 5*, **1973**, 43-44; *Org. Synth. 43*, **1963**, 1-2.

6. Reviews: a) Palacios, F.; De Retana, A.M.O.; De Marigorta, E.M.; De Los Santos, J.M. *Targets in Heterocyclic Systems* **2003**, *7*, 206-245. b) Padwa, A.; Murphree, S. *Progress in Heterocyclic Chemistry* **2003**, *15*, 75-99. c) Palacios, F.; Ochoa de Retana, A.M.; Martinez de Marigorta, E.; De Los Santos, J. *Organic Preparations and Procedures International* **2002**, *34(3)*, 219-269. d) Zwanenburg, B.; Ten Holte, P. *Topics in Current Chemistry* **2001**, *216*, 93-124. e) Gilchrist, T.L. *Aldrichimica Acta*, **2001**, *34(2)*, 51-55.

7. Neber, P.W.; Huh, G. *Justis Liebigs Ann. Chem.* **1935**, *515*, 283-296.

8. Backes, J. *Methoden Org. Chem. (Houben-Weyl)* **1992**, *vol. E16c*, 317-369.

9. Lipshutz, B.H.; Reuter, D.C. *Tetrahedron Lett.* **1988**, *29*, 6067-6070.

10. Wentrup, C.; Fischer, S.; Berstermann, H.; Kuzaj, M.; Luerssen, H.; Burger, K. *Angew. Chem. Int. Ed. Engl.* **1986**, *25*, 85.

11. Anderson, D.J.; Gilchrist, T.L.; Gymer, G.E.; Rees, C.W. *J. Chem. Soc., Perkin Trans. 1* **1973**, 550-555.

12. Alcaraz, G.; Wecker, U.; Baceiredo, A.; Dahan, F.; Bertrand, G. *Angew. Chem. Int. Ed. Engl.* **1995**, *34*, 1246-1248.

Appendix
Chemical Abstracts Nomenclature (Registry Number)

3'-Bromoacetophenone: 1-acetyl-3-bromobenzene; (2142-63-4)
Sodium hydride; (7646-69-7)
2-Chloro-5-(trifluoromethyl)pyridine; (52334-81-3)
Hydroxylamine hydrochloride; (5470-11-1)
Trifluoroacetic anhydride; (407-25-0)

Kirk Stevens was born in Seattle, Washington. He received his undergraduate chemistry degree from Reed College (Portland, OR) where he conducted research under Prof. Pat McDougal. He then moved to UC-Santa Barbara for graduate studies with Prof. Bruce Lipshutz working on nickel-catalyzed reactions and their use in synthesis of prenylated natural products like CoQ_{10}. As an NIH postdoctoral fellow with Prof. Yoshito Kishi at Harvard, he was part of the team that completed the total synthesis of Spongistatin. In 1998, he joined GlaxoWellcome, now GlaxoSmithKline, and has worked primarily in oncology medicinal chemistry.

Stephen Greszler was born in Painesville, Ohio. He received his undergraduate degree in biochemistry from The Ohio State University in Columbus, where he was involved in research with Professor Sean Taylor. He then moved to The University of North Carolina at Chapel Hill, where he is currently pursuing graduate work in organic synthesis under the guidance of Professor Jeffrey S. Johnson. Stephen interned at GlaxoSmithKline in Research Triangle Park during the summer of 2007, where he worked on the pyrazolo[1,5-a]pyridine chemistry described in this account.

Hiroshi Nakagawa was born in Mie, Japan, in 1973. He received his B.S. in 1996 and M.S. in 1998 from University of Tokyo. He received his Ph.D. in 2001 under the direction of Professor Tetsuo Nagano and Professor Tsunehiko Higuchi from University of Tokyo. He has been working at Dainippon Sumitomo Pharma since 2001. In 2006, he began his visiting scholar studies at the University of California, Berkeley, in the laboratories of Professor Jonathan A. Ellman. His research interest is in the area of medicinal chemistry.

BENZYL ALCOHOL AS AN ALKYLATING AGENT USING THE RUTHENIUM-CATALYZED BORROWING HYDROGEN STRATEGY
(4,4-Dimethyl-3-oxo-2-benzylpentanenitrile)

Submitted by Tracy D. Nixon, Paul A. Slatford, Michael K. Whittlesey and Jonathan M. J. Williams.[1]

Checked by David A. Candito and Mark Lautens.[2]

1. Procedure

4,4-Dimethyl-3-oxo-2-benzylpentanenitrile (1). A 250-mL two-necked round-bottomed flask (Note 1) equipped with a magnetic stir bar, a rubber septum and a reflux condenser connected to an inert gas supply (Note 2) is flushed with inert gas for five minutes by venting through a needle in the rubber septum. The rubber septum is removed and the flask is charged with 4,4-dimethyl-3-oxopentanenitrile (5.82 g, 46.5 mmol, 1.0 equiv.), Ru(PPh$_3$)$_3$(CO)H$_2$ (0.21 g, 0.23 mmol, 0.005 equiv.) (Note 3), Xantphos (0.13 g, 0.23 mmol, 0.005 equiv.), piperidinium acetate (0.34 g, 2.3 mmol, 0.05 equiv.) (Note 4), anhydrous toluene (50 mL) (Note 5) and benzyl alcohol (4.82 ml, 5.03 g, 46.5 mmol, 1.0 equiv.) (Note 6). The rubber septum is replaced and the equipment flushed with inert gas for a further five minutes before replacing the rubber septum with a glass stopper. While still under an inert gas atmosphere, the solution is stirred and heated to reflux using an oil bath (Note 7). Heating is continued for 4 h after which the reaction is deemed to be complete by monitoring the consumption of starting material via TLC (Note 8). The resulting mixture is cooled to ambient temperature and transferred to a one-necked 250-mL round-bottomed flask (Note 9). Solvent evaporation using a rotary evaporator affords crude (1) as a yellow oil (Note 10). The crude product is purified by flash chromatography on silica gel (Note 11) to yield the product as a pale yellow oil (8.37 g, 84%) (Notes 12 and 13).

28

2. Notes

1.　All glassware was oven-dried at 120 °C overnight prior to use.

2.　Nitrogen or argon are both suitable. In this experiment, nitrogen was used.

3.　Ru(PPh₃)₃(CO)H₂ was purchased from Strem Chemicals Inc. and used as received (99% purity). Ru(PPh₃)₃(CO)H₂ can also be prepared via literature methods .[3]

4.　Piperidinium acetate was purchased from TCI and used as received. Piperidinium acetate can also be prepared via literature methods.[4]

5.　The submitters utilized dry toluene, which was dried using an anhydrous Engineering drying column. They also note that one equivalent of water is generated in the course of the reaction without any adverse effects. The Checkers utilized toluene which was distilled from sodium/benzophenone ketyl prior to use.

6.　4,4-Dimethyl-3-oxopentanenitrile (99% purity), Xantphos (97% purity) and benzyl alcohol (≥ 99% purity) were purchased from Aldrich. All chemicals were used as received.

7.　The oil bath temperature rose from ambient temperature to 120 °C over approximately 10 minutes. Furthermore, the solution darkens over the first 30 minutes of heating to give a dark yellow-colored solution, which remains throughout.

8.　Reaction monitoring was carried out by observing the consumption of starting material (R_f= 0.32, 1:1 ether:hexanes, visualized by KMnO₄ stain, Silia*Plate*™ silica gel plates, 250 μm, F254, available from Silicycle).

9.　Portions of dichloromethane (3 x 20 mL) were used to rinse the reaction flask and were added to the transferred solution.

10.　The bulk of the toluene was removed by rotary evaporation (80 °C, 100-200 mmHg). Furthermore, the crude material (12.84 g) showed minor impurities in the ¹H NMR spectrum.

11.　The submitters purified the product in the following manner: Column chromatography was carried out on an 8-cm diameter column packed with 220 g of Davisil LC 60 A silica gel using petroleum ether (bp. 40 – 60 °C) and diethyl ether (19:1) as the eluent. The column fractions were collected in 20 mL portions in boiling tubes. TLC analysis of the fractions (using 0.25 mm Macherey-Nagel silica gel G/UV₂₅₄ visualising at 254 nm; R_f 0.21) showed that fractions 22-90 contained product and solvent was removed from these fractions on a rotary evaporator to give product. The Checkers purified the product in the following manner: Column chromatography was carried out on a 6 x 60 cm column with a fused reservoir. The column was packed with 240 g of silica gel (Silia-P Flash silica gel, particle size 40-63 μm, pore diameter 60 Å) as a slurry using the eluent (1:9 ether:hexanes). The crude material was loaded onto the column and portions of the eluent were utilized to rinse the flask containing the crude (3 x 2 ml) and these washings were then transferred onto the column. Fractions were collected in 25 x 200 mm test tubes and analysis of the fractions via TLC (R_f= 0.31, 1:9 ether:hexanes, visualized by KMnO$_4$ stain, Silia*Plate*^{TM} silica gel plates, 250 μm, F254, available from Silicycle) revealed that fractions 7-17 contained pure product. These fractions were collected and the solvent removed by rotary evaporation (45 °C, 100-200 mmHg) followed by 12 hours under vacuum (22 °C, 6.5-7.0 mmHg) in order to yield the desired product.

12. The checkers conducted a trial in which significant modifications were made to the stated procedure. The run was carried out on half the scale of the stated procedure and no previous drying of glassware or solvent was employed. An attempt was made to purify the crude material via kugelrohr distillation. An early fraction was collected (161 °C, 5.10 mmHg) containing a clear oil which was later discarded. The product was distilled at a temperature of 190 °C and pressure of 7.20 mmHg, to leave behind an orange residue. It was observed that the distilled product was obtained in a purity of 86 %. Subsequent column chromatography provided the pure product in a yield of 78 %. The result obtained is not significantly different from the original yield reported by the submitters indicating that previous drying of glassware and solvent is unnecessary.

13. Spectroscopic and analytical data are as follows: ^{1}H NMR (400 MHz, CDCl$_3$) δ: 1.09 (s, 9 H), 3.11-3.23 (m, 2 H), 4.01 (t, J = 7.6 Hz, 1 H), 7.18-7.23 (m, 2 H), 7.24-7.35 (m, 3 H). ^{13}C NMR (400 MHz, CDCl$_3$) δ: 25.4, 35.8, 38.6, 45.3, 116.9, 127.4, 128.6, 128.9, 136.0, 204.8. IR (neat, cm^{-1}): 3036, 2979, 2946, 2879, 2244, 1723, 1605, 1501, 1482, 1458, 1401, 1373. Anal. Calcd. for C$_{14}$H$_{17}$NO: C, 78.03; H, 7.89; N, 6.50. Found: C, 77.83; H, 8.05; N, 6.55. MS (EI) m/z (rel. intensity): 215(18, M$^+$), 187(38), 131(10), 103(33), 91(65), 85(52), 57(100). HRMS [M$^+$] calcd for C$_{14}$H$_{17}$NO: 215.1310. Found: 215.1311. UV (hexane, 25 °C) λ$_{max}$, 307 nm (ε): 25.57 (L mol^{-1}cm^{-1}). A run on half of the reported scale (3.0 g of 4,4-

dimethyl-3-oxopentanenitrile) was conducted, which provided 4.41 g (85%) of the desired product.

Safety and Waste Disposal Information

All hazardous materials should be handled and disposed of in accordance with "Prudent Practices in the Laboratory"; National Academy Press; Washington, DC, 1995.

3. Discussion

The borrowing hydrogen strategy allows alcohols to be used as alkylating agents for the alkylation of suitable carbon nucleophiles. Transition metal catalysts, including ruthenium and iridium complexes,[5,6] have been developed that temporarily remove hydrogen from the alcohol substrate to generate an intermediate aldehyde. The aldehyde is converted *in situ* into an intermediate alkene via a Wittig reaction or a condensation reaction. The hydrogen is then returned by the catalyst to the alkene, affording a new C-C bond. A similar approach for the alkylation of amines with alcohols is also known. In these reactions, the intermediate aldehyde is converted into an imine prior to return of the hydrogen to generate the amine.[7] The use of alcohols as alkylating agents avoids the use of traditional alkylating agents such as alkyl halides which can be toxic or mutagenic.

We have previously reported that while $Ru(PPh_3)_3(CO)H_2$ is not a very active catalyst for the alkylation of keto nitriles by alcohols, the addition of the bidentate ligand Xantphos[8] provides a catalyst which can be used at 0.5 mol% loading.[9]

The $Ru(PPh_3)_3(CO)H_2$ / Xantphos combination has also been used to catalyze related reactions involving the alkylation of 4,4-dimethyl-3-oxopentanenitrile with other alcohols. As shown in Table 1, this chemistry is most effective with benzylic alcohols, although aliphatic alcohols can be successfully used at higher catalyst loadings. The exact role of Xantphos in these reactions has not been fully established, although the reaction of Xantphos with $Ru(PPh_3)_3(CO)H_2$ affords $Ru(Xantphos)(PPh_3)(CO)H_2$ which is slightly more catalytically active than the *in situ* combination, and a crystal structure of $Ru(Xantphos)(PPh_3)(CO)H_2$ has been obtained. Other bidentate ligands can also enhance the reactivity of the parent complex, and details of these studies have been published elsewhere.[9,10]

The Ru(PPh$_3$)$_3$(CO)H$_2$ / Xantphos combination has been found to be effective for other reactions including the dehydration of oxime ethers to give nitriles[11] and the isomerisation of 1,4-alkynediols with further reaction to give furans[12] or pyrroles.[13]

Table 1. Alkylation of 4,4-Dimethyl-3-oxopentanenitrile with alcohols.[a]

Alcohol	Product	Ru cat.[b] (mol%)	Yield[c] (%)
		0.5	89
		0.5	79
		0.5	82
		2.5	87
		0.5	85
		5.0	78[d]

[a]Reactions were run in toluene at reflux for 4 h, using piperidinium acetate as a co-catalyst (5 mol% for reactions using 0.5 mol% Ru, and 25 mol% for reactions using 2.5 or 5 mol% Ru). [b]Ru cat = Ru(PPh$_3$)$_3$(CO)H$_2$ / Xantphos (1:1). [c]Isolated yield. [d]Dibenzylmalonate was used in place of the ketonitrile in this experiment.

Org. Synth. **2009**, *86*, 28-35

1. Department of Chemistry, University of Bath, Claverton Down, Bath, BA2 7AY, UK

2. Department of Chemistry, University of Toronto, Toronto, Ontario, M5S 3H6, Canada

3. Ahmad, N.; Levison, J. J.; Robinson, S. D.; Uttley, M. F. *Inorg. Synth.* **1974**, *15*, 45-64.

4. Binovi, L. J.; Arlt, H. G. *J. Org. Chem.* **1961**, *26*, 1656-1657.

5. Examples of Ru catalysts; (a) Grigg, R.; Mitchell, T. R. B.; Sutthivaiyakit, S.; Tongpenyai, N. *Tetrahedron Lett.* **1981**, *22*, 4107-4110; (b) Cho, C. S.; Kim, B. T.; Kim, T.-J.; Shim, S. C. *J. Org. Chem.* **2001**, *66*, 9020-9022; (c) Motokura, K.; Nishimura, D.; Mori, K.; Mizugaki, T.; Ebitani, K.; Kaneda, K. *J. Am. Chem. Soc.*, **2004**, *126*, 5662-5663; (d) Martínez, R.; Brand, G. J.; Ramón, D. J.; Yus, M. *Tetrahedron Lett.* **2005**, *46*, 3683-3686; (e) Edwards, M. G.; Jazzar, R. F. R.; Paine, B. M.; Shermer, D. J.; Whittlesey, M. K.; Williams, J. M. J.; Edney, D. D. *Chem. Commun.* **2004**, 90-91; (f) Burling, S.; Paine, B. M.; Nama, D.; Brown, V. S.; Mahon, M. F.; Prior, T. J.; Pregosin, P. S.; Whittlesey, M. K.; Williams, J. M. J. *J. Am. Chem. Soc.* **2007**, *129*, 1987-1995.

6. Examples of Ir catalysts; (a) Fujita, K.; Yamaguchi, R. *Synlett* **2005**, 560-571; (b) Löfberg, C.; Grigg, R.; Keep, A.; Derrick, A.; Sridharan, V.; Kilner, C. *Chem. Commun.* **2006**, 5000-5002; (c) Morita, M.; Obora, Y.; Ishii, Y. *Chem. Commun.* **2007**, 2850-2852; (d) Edwards, M. G.; Williams, J. M. J. *Angew. Chem. Int. Ed.* **2002**, *41*, 4740-4743.

7. For reviews of the borrowing hydrogen approach, see; (a) Guillena, G.; Ramón, D. J.; Yus, M. *Angew. Chem. Int. Ed.* **2007**, *46*, 2358-2364; (b) Hamid, M. H. S. A.; Slatford, P. A.; Williams, J. M. J. *Adv. Synth. Catal.* **2007**, *349*, 1555-1575.

8. (a) Kamer, P. C. J.; van Leeuwen, P. W. N. M.; Reek, J. N. H. *Acc. Chem. Res.* **2001**, *34*, 895-904; b) Freixa, Z.; van Leeuwen, P. W. N. M. *Dalton Trans.* **2003**, 1890-1901.

9. Slatford, P. A.; Whittlesey, M. K.; Williams, J. M. J., *Tetrahedron Lett.*, **2006**, *47*, 6787-6789.

10. Ledger, A. E. W.; Slatford, P. A.; Lowe, J. P.; Mahon, M. F.; Whittlesey, M. K.; Williams, J. M. J. *Dalton Trans.,* submitted.

11. Anand, N.; Owston, N. A.; Parker, A. J.; Slatford, P. A.; Williams, J. M. J. *Tetrahedron Lett.* **2007**, *48*, 7761-7763.

12. Pridmore, S. J.; Slatford, P. A.; Williams, J. M. J. *Tetrahedron Lett.* **2007**, *48*, 5111-5114.

13. Pridmore, S. J.; Slatford, P. A.; Daniel, A.; Whittlesey, M. K.; Williams, J. M. J. *Tetrahedron Lett.* **2007**, *48*, 5115-5120.

Appendix
Chemical Abstracts Nomenclature; (Registry Number)

Benzyl alcohol; (100-51-6)

4,4-Dimethyl-3-oxopentanenitrile; (59997-51-2)

4,5-bis(Diphenylphosphino)-9,9-dimethylxanthene (Xantphos); (161265-03-8)

Piperidinium acetate; (4540-33-4)

Carbonyl(dihydrido)tris(triphenylphosphine)ruthenium (II); (25360-32-1)

Jonathan Williams was born in Worcestershire, England, in 1964. He received a B.Sc. from University of York, a D.Phil. from University of Oxford (with Prof. S G Davies), and was then a post-doctoral fellow at Harvard with Prof. D. A. Evans (1989-1991). He was appointed to a Lectureship in Organic Chemistry at Loughborough University in 1991, and was then appointed as a Professor of Organic Chemistry at the University of Bath in 1996, where his research has mainly involved the use of transition metals for the catalysis of organic reactions.

Tracy Nixon was born in Newcastle-upon-Tyne in 1979 and graduated from the University of Leeds in 2001, where she then remained for her Ph.D. under the supervision of Dr. Terry Kee. After a two-year postdoctoral position with Dr. Jason Lynam at the University of York, she started work at the University of Bath in October, 2007, as a postdoctoral research associate in the group of Professor Jonathan Williams.

Paul Slatford was born in Hornchurch, England, in 1977, and graduated from the University of Bristol in 2000. He completed his Ph.D. under the supervision of Professor Guy Lloyd-Jones, also at the University of Bristol, in 2004. Following a three-year post-doc with Professor Jonathan Williams at the University of Bath, he is now pursuing a career in industry.

Mike Whittlesey (born Nottingham, England, 1966) received a D.Phil. for work in organometallic photochemistry with Professor Robin Perutz and Dr. Roger Mawby at the University of York, before moving to post-doctoral work in organic photochemistry with Professor Tito Scaiano at the University of Ottawa in Canada. He returned to inorganic chemistry, working with Perutz at York once more on metal induced C-F bond activation. After a fixed-term Lectureship at the University of East Anglia, he moved to Bath in 1999, where he is now a Senior Lecturer. His research interests focus on the reactivity of transition metal-N-heterocyclic carbene complexes.

David A. Candito completed his undergraduate degree in pharmaceutical and biological chemistry at York University in 2007. During his undergraduate thesis, he had the opportunity to work in the group of Professor Michael G. Organ where his research focused on the application of N-heterocyclic carbenes to palladium mediated cross-coupling reactions. He is currently pursuing a Ph.D. at the University of Toronto in the research group of Professor Mark Lautens where his research is focused upon Domino reactions involving norbornene mediated C-H activation.

2-(2,2-DIBROMOETHENYL)-BENZENAMINE

Submitted by Christopher Bryan, Valentina Aurregi and Mark Lautens.[1]
Checked by Andreas Pfaltz and Ivana Fleischer.

1. Procedure

A. *1-(2,2-Dibromoethenyl)-2-nitrobenzene* (**2**). A 500-mL, three-necked round-bottomed flask equipped with a magnetic stirring bar and fitted with a 125-mL pressure-equalizing addition funnel (capped with a rubber septum), a thermometer and a two-tap Schlenk adaptor connected to a bubbler and an argon/vacuum manifold (Note 1) is sequentially charged at room temperature with 2-nitrobenzaldehyde (4.03 g, 26.4 mmol, 1.0 equiv) (Note 2) and tetrabromomethane (13.13 g, 39.5 mmol, 1.5 equiv) (Notes 3 and 4). Dichloromethane (100 mL) is added into the flask using the addition funnel and the system is purged with argon for 5 min (Note 5). The addition funnel is capped with a rubber septum, and the reaction is kept under argon atmosphere. The resulting yellow solution is cooled to 2–3 °C (internal temperature) with an ice-bath. The addition funnel is re-charged with dichloromethane (20 mL), and then triisopropyl phosphite (15.9 mL, 58 mmol, 2.2 equiv) is added *via* syringe through the septum and the resulting solution is added dropwise to the stirred mixture in the flask over a period of 60 min (Notes 6 and 7). During the addition of triisopropyl phosphite the internal temperature of the flask is maintained between 3 and 5 °C. The mixture is then stirred (internal temperature, 3 °C) until the starting material has been completely consumed as shown by thin layer chromatography (TLC) (Note 8), at which point the mixture is quenched with sodium

Org. Synth. **2009**, *86*, 36-46
Published on the Web 9/16/2008

bicarbonate (saturated aqueous solution, 85 mL) added through the addition funnel with the stopcock open (Note 9). The mixture is transferred into a 500-mL separatory funnel and the organic phase separated. The aqueous phase is extracted further with dichloromethane (2 x 40 mL) and the combined organic phase is washed once with water (80 mL), transferred into a 500-mL, one-necked, round-bottomed flask and concentrated by rotary evaporation (35 °C, 15 mmHg) to give 27.0–28.5 g of a brown oil (Note 10). The flask containing the crude material is equipped with a water-cooled condenser. Concentrated HCl (12 M, 30 mL) and glacial acetic acid (30 mL) are added (Note 11), the condenser is capped with a two-tap Schlenk adaptor connected to a bubbler and an argon/vacuum manifold, and the resulting mixture is heated to reflux in an oil bath (110 °C external temperature) and stirred for 14 h under an argon atmosphere. The reaction is then cooled to room temperature and distilled water (150 mL) is added. The mixture is neutralized by the portion-wise addition of sodium carbonate (45 g, 0.36 mol) (Note 12). Diethyl ether (150 mL) is added, the contents of the flask are stirred vigorously for 5 min and any solids are removed by vacuum filtration (water aspirator, 20 mbar) through a fritted glass funnel (8.0 cm diameter, porosity 3). The residue is washed with diethyl ether (2 x 20 mL), the filtrate is poured into a 500-mL separatory funnel and the phases are separated. The aqueous phase is extracted with diethyl ether (2 x 100 mL). The combined organic phases are washed with distilled water (2 x 100 mL) and brine (75 mL), dried ($MgSO_4$), filtered (gravity filtration through a fluted filter paper) and concentrated under reduced pressure with a rotary evaporator (30 °C, 15 mmHg) to give 8.01–8.06 g (99 %) of a brown oil (Note 13), which solidified under high vacuum (0.04 mmHg) (Notes 13 and 14). A stir bar is inserted into the flask, heptane (15 mL) is added, and the flask is heated with stirring to 50 °C in an oil bath (external temperature). Ethyl acetate (1 mL) is added, and the solution is decanted into a second 50-mL, round-bottomed flask which is pre-warmed to at least 50 °C to remove any insoluble material (Note 15). The flask is placed in the oil bath, which is allowed to cool slowly. When the temperature of the oil bath has reached 40 °C, the flask is scratched to induce crystallization (Note 16). The flask is then allowed to cool to room temperature, left at room temperature for 1 h and then cooled to 4 °C (refrigerator) for a further 14 h. The product is isolated by vacuum filtration through a fritted glass funnel (3 cm diameter, porosity 4). The crystals are washed with ice-cold heptane (2 x 5 mL) and then dried under vacuum (0.03 mmHg, 14 h) to afford pure 1-(2,2-

dibromoethenyl)-2-nitrobenzene (5.95–6.11 g, 73–75 %) as a brown crystalline solid (Note 17).

B. *2-(2,2-Dibromo-vinyl)-phenylamine* (**3**). A 250-mL Büchi autoclave glass vessel (Note 18) is charged with 1-(2,2-dibromovinyl)-2-nitrobenzene (5.80 g, 18.9 mmol) and methanol (100 mL) (Note 19). The solution is stirred until all of the solid is dissolved, at which point vanadium-doped platinum on activated carbon is added (580 mg, 0.015 mmol, 8.0 x 10^{-4} equiv) (Note 20). The vessel is attached to the autoclave and pressurized to 2.5 bar with hydrogen. The flask is thrice evacuated (via diaphragm pump) and back-filled with hydrogen to 2.5 bar, then stirred for 3 h at room temperature (Note 21). The hydrogen is removed by evacuation, and the suspension is filtered through a pad of celite (2.5 cm in height) (Note 22). The pad is washed with methanol (3 x 80 mL), and the filtrate is concentrated on a rotary evaporator (30 °C, 15 mmHg) and dried under vacuum (0.03 mmHg, 16 h) to afford 5.22 g (99.8 %) (Note 23) of the title compound as an orange oil (Notes 24 and 25).

2. Notes

1.   All glassware was oven-dried, quickly assembled and cooled under vacuum (0.03 mmHg) prior to use. The submitters performed the reaction under an atmosphere of nitrogen.

2.   2-Nitrobenzaldehyde was purchased from Aldrich (>99 %) and used as received.

3.   Tetrabromomethane (98%) was purchased from VWR. It was dissolved in dichloromethane (30 g in 150 mL), dried over $MgSO_4$ (10 g), filtered, concentrated by rotary evaporation (30 °C, 15 mmHg), and dried under vacuum (30 °C, 15 mmHg, 4 h) prior to use.

4.   Use of less than 1.5 equivalents of CBr_4 led to a reduced yield, and the formation of small amounts of unidentified by-products. For example, the use of 1.1 or 1.3 equivalents gave a 67–71 % yield after recrystallization.

5.   Submitters used dichloromethane purchased from Caledon (≥99.5 %) and checkers used dichloromethane from J. T. Baker (>99.5%). Checkers and submitters used the solvent as received.

6.   Triisopropyl phosphite was purchased from Alfa Aesar (90 %) and used as received. The reagent may be purified by vacuum distillation

over sodium (bp = 65 °C, 15 mmHg); no improvement was observed with purified material.

7.   After the addition of the solution of triisopropyl phosphite, the color of the reaction mixture changes from yellow to dark brown.

8.   Reaction time was 0.5 h. The progress of the reaction was followed by TLC analysis on silica gel plates cut from 20 x 20 aluminum sheets (Submitters used Silica gel 60 F_{254} from EMD Chemicals, checkers Polygram® SIL G/UV254 from Macherey-Nagel) eluting with pentane-diethyl ether 9:1. R_f values of the starting material and product are 0.25 and 0.52, respectively.

9.   The temperature rises to 40 °C over the course of the addition.

10.   A small amount of white crystalline material was sometimes observed in the crude product; the presence or absence of this material had no effect on the course or outcome of the purification.

11.   Hydrochloric acid and glacial acetic acid were purchased from Fisher (submitters) or VWR (checkers) and used as received.

12.   Sodium carbonate was added in 20-30 portions, with care taken to ensure that the evolution of CO_2 does not become too vigorous. The temperature rises to 38–40 °C during the addition.

13.   For acid-sensitive substrates, the material may also be purified by chromatography: the crude material is loaded via Pasteur pipette onto a pre-packed silica gel column (7 x 9 cm, ~200 g silica; silica gel 0.040-0.063 mm, pore diameter 6 nm purchased from Silicycle; column made by using a slurry of silica gel and pentane) and eluted with 9:1 pentane/diethyl ether. The product is collected in a single fraction containing the first 1300 mL of eluent, and isolated as a yellow solid after evaporation (30 °C, 20 torr).

14.   According to submitters, this material is sufficiently pure to be used in most reactions (>95% HPLC). However, for catalytic applications, recrystallization may be necessary.

15.   Warming the second flask is essential to preclude premature crystallization. If no insoluble material is observed, this step is omitted.

16.   Addition of a seed crystal is a preferred alternative to scratching; usually, a few crystals are formed during the decanting process.

17.   1-(2,2,-Dibromoethenyl)-2-nitrobenzene   has   the   following physical properties: mp = 60–61 °C, ^{1}H NMR (500 MHz, CDCl$_3$) δ: 7.54 (t, J = 7.6 Hz, 1 H), 7.59 (d, J = 7.8 Hz, 1 H), 7.68 (t, J = 7.6 Hz, 1 H), 7.78 (s, 1 H), 8.12 (d, J = 8.2 Hz, 1 H). ^{13}C NMR (126 MHz, CDCl$_3$) δ: 93.3, 125.0, 129.6, 131.5, 131.8, 133.7, 134.2, 146.9. IR (ATR) 3035, 1609, 1575, 1516,

1338, 1307, 1200, 965, 896, 863, 852, 831, 788, 732, 666 cm^{-1}. Anal. Calcd. for $C_8H_5Br_2NO_2$: C, 31.30; H, 1.64; N, 4.56. Found: C, 31.32; H, 1.59; N, 4.48.

18. The checkers used a pressure autoclave from Büchi Glas Uster (2.5 bar). The submitters used a Parr hydrogenation apparatus (36 psig = 2.48 bar).

19. The submitters purchased methanol from ACP chemicals (99%), and the checkers from J. T. Baker (99.8%); both used the solvent as received.

20. Checkers used 1 % Pt + 0.2 % V on activated carbon (50% wetted powder) available from Strem as catalyst. Submitters used 3% Pt + 0.6% V on activated carbon (66 % water, 290 mg, 0.015 mmol, 8.0×10^{-4} equiv), donated by Degussa. The catalysts were used as received.

21. The reaction can be followed by TLC. The R_f value of the product is 0.27 (9:1 pentane-diethyl ether).

22. The pad is prepared by slurrying Celite with methanol in a fritted funnel (8.0 cm diameter, porosity 3) and filtering the excess methanol.

23. The reaction was carried out also on the half scale in the same equipment yielding the product as a brown oil (2.59 g, 99%).

24. The submitters isolated the product as a dark brown solid with mp = 40–42 °C. The checkers isolated the product as an orange oil (at room temperature), which solidified in the refrigerator (4 °C).

25. 2-(2,2-Dibromoethenyl)-benzenamine has the following physical properties: ^{1}H NMR (500 MHz, CDCl$_3$) δ: 3.66 (bs, 2 H), 6.71 (d, $J = 7.7$ Hz, 1 H), 6.81 (t, $J = 7.6$ Hz, 1 H), 7.18 (t, $J = 7.6$ Hz, 1 H), 7.32 (d, $J = 7.6$ Hz, 1 H), 7.34 (s, 1 H). ^{13}C NMR (126 MHz, CDCl$_3$) δ: 92.9, 115.9, 118.5, 121.8, 129.3, 129.8, 134.1, 143.7. IR (ATR): 3465, 3377, 3059, 3024, 2993, 1615, 1574, 1486, 1453, 1308, 1157, 939, 878, 833, 745 cm^{-1}. Anal. Calcd. for $C_8H_7Br_2N$: C, 34.69; H, 2.55; N, 5.06. Found: C, 34.68; H, 2.59; N, 5.04.

Safety and Waste Disposal Information

All hazardous materials should be handled and disposed of in accordance with "Prudent Practices in the Laboratory"; National Academy Press; Washington, DC, 1995.

3. Discussion

This procedure can be used to prepare a wide variety of *gem*-dibromovinyl aniline substrates, which have been used in the synthesis of a number of indoles and indole derivatives by tandem cross-coupling reactions (Scheme 1).[2] This indole synthesis offers the advantages of low cost, mild conditions, and tolerance to a wide range of functionalities.

Scheme 1. Utility of *gem*-dibromovinylanilines

C-N/Suzuki

Pd/L

C-N/Heck

Pd/L

C-N/Sonagashira

Pd/L

C-N/C-N

Cu/L

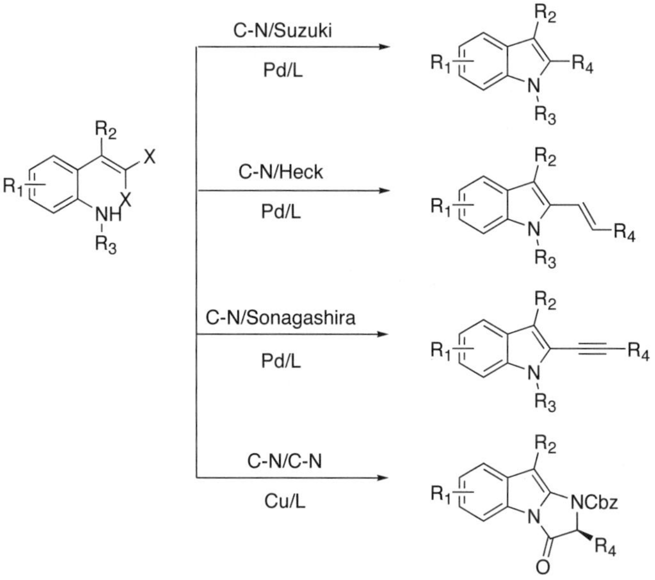

The Ramirez olefination used in the Corey-Fuchs procedure[3] has traditionally used triphenylphosphine, as well as an aldehyde and CBr_4. Modified methods have been developed to cope with the formation of PPh_3Br_2 as a by-product, with both zinc powder and Et_3N being used as scavenging reagents.[3b,4] The utility of the reaction towards less-reactive ketones, however, remains limited. Alternative non-ylide methods have been developed, including the use of organolithium or Grignard reagents[5] as well as copper-catalyzed redox chemistry between hydrazones and CBr_4.[6] These methods require multiple steps. The formation of triphenylphosphine oxide as a byproduct is also a limitation of this method, especially in large-scale preparations. Significant amounts of triphenylphosphine oxide leads to a tedious purification process, and the highly crystalline nature of triphenylphosphine oxide precludes product purification by recrystallization, requiring column chromatography, which is impractical on large scale.

We have recently reported the use of triisopropyl phosphite as an alternative reagent to triphenylphosphine in the Ramirez olefination.[7] It was found that the reaction displays comparable efficiency with both reagents when aldehydes are used as substrates, and the use of triisopropyl phosphite generally gave superior results with ketones (see Table 1).

Table 1. Scope of the P(O^{i}Pr)$_3$-mediated Ramirez olefination[7]

Entry	Substrate	Product	Yield [a] P(O^{i}Pr)$_3$	PPh$_3$
1			92	95
2			76	88
3			94	85
4			73	<40[b]
5			73	<40[b]
6			62[c]	<20[b]
7			98	69
8			44[c]	<5[b]
9			trace	trace
10			21	91

[a] Isolated yield. [b] Conversion measured by ^{1}H NMR. [c] Incomplete conversion.

The selective reduction of aromatic nitro groups in the presence of halides and olefins has been accomplished by using either SnCl$_2$·2H$_2$O or iron powder in the presence of a catalytic amount of $FeCl_3 \cdot 6H_2O$.[2a] Both of these methods, however, suffer from the twin drawbacks of large amounts of metal oxide waste (3–7 equiv) and difficulties with workup on large scale. Hydrogenation using traditional catalysts such as Pd/C (1–10%), Ru/C, Rh/C, Pearlman's catalyst (20%) and Raney nickel failed to furnish the desired product for *gem*-dibromovinylnitrobenzenes. The use of vanadium additives to promote the hydrogenation of aromatic nitro groups was first described by Baumeister.[8] The vanadium reduces the accumulation of hydroxylamines, which are intermediates in the reduction of nitro groups to amines. Little additional information has appeared in the literature regarding this strategy since the initial report, but our experience suggests that this is the method of choice for the selective reduction of aromatic nitro groups. Yields are comparable or superior to those obtained by redox chemistry with metals (see Table 2), and analytically pure products are obtained simply by removing the catalyst and solvent.

Table 2.   Reduction of *gem*-dihalovinylnitrobenzenes[2a]

Entry	R	R'	X	Yield[a]		
				H_2, Pt-C[V]	$SnCl_2 \cdot 2H_2O$	Fe/HOAc/FeCl$_3$
1	H	CO$_2$Me	Br	100	n.d.[b]	n.d.
2	CF$_3$	H	Br	82	89	n.d.
3	≡—Ph	H	Br	89	0	n.d.
4	⟨C$_6$H$_4$⟩—F	H	Br	93	40	83
5	H	H	Cl	94	n.d.	n.d.
6	Me	H	Cl	93	n.d.	n.d.

[a] Isolated yield. [b] Not determined.

1. Davenport Laboratories, Department of Chemistry, University of Toronto, Toronto, Ontario M5S 3H6; E-mail: mlautens@chem.utoronto.ca

2. a) Fang, Y.-Q.; Lautens, M. *J. Org. Chem.* **2008**, *73*, 538-549. b) Fang, Y.-Q.; Lautens, M. *Org. Lett.* **2005**, *7*, 3549-3552. c) Yuen, J.; Fang, Y.-Q.; Lautens, M. *Org Lett.* **2006**, *8*, 653-656. d) Fayol, A.; Fang, Y.-Q.; Lautens, M. *Org. Lett.* **2006**, *8*, 4203-4206. e) Nagamochi, M.; Fang, Y.-Q.; Lautens, M. *Org. Lett.* **2007**, *9*, 2955-2958.

3. a) Ramirez, F.; Desai, N. B.; McKelvie, N. *J. Am. Chem. Soc.* **1962**, *84*, 1745-1747. b) Corey, E. J.; Fuchs, P. L. *Tetrahedron Lett.* **1972**, *13*, 3769-3772.

4. Gradjean, D.; Pale, P.; Chuche, J. *Tetrahedron Lett.* **1994**, *35*, 3529-3530.

5. Rezaei, H.; Normant, J. F. *Synthesis* **2000**, 109-112.

6. Korotchenko, V.N.; Shastin, A.V.; Nenajdenko, V.G.; Balenkova, E. S. *J. Chem. Soc. Perkin. Trans. 1* **2002**, 883-887.

7. Fang, Y.-Q.; Lifchits, O.; Lautens, M. *Synlett* **2008**, 413-417.

8. Baumeister, P.; Blaser, H.-U.; Studer, M. *Catal. Lett.* **1997**, *49*, 219-222.

Appendix
Chemical Abstracts Nomenclature; (Registry Number)

2-Nitrobenzaldehyde; (552-89-6)

Tetrabromomethane; (558-13-4)

Triisopropyl phosphite; (116-17-6)

Sodium carbonate; (497-19-8)

Benzene, 1-(2,2-dibromoethenyl)-2-nitro-; (253684-24-1)

Benzenamine, 2-(2,2-dibromoethenyl)-; (167558-54-5)

Christopher Bryan was born in Winnipeg, Canada in 1982. He received his B.Sc. degree with distinction in 2005 from the University of Victoria, where he worked in the laboratory of Scott McIndoe. While an undergraduate, he worked as a Co-op student in the medicinal chemistry department at Boehringer-Ingelheim Pharmaceuticals in Laval, PQ. He is currently pursuing his Ph. D. at the University of Toronto under the supervision of Professor Mark Lautens. His research is focused on the synthesis of heterocycles via metal-catalyzed tandem processes.

Valentina Aureggi was born 1977 in Como, Italy. She obtained her Diploma in 1999 and MSci degree in 2003 at Insubria University. During this five-year course of study, her final year project was carried out at the University of Neuchâtel, investigating the synthesis of amido silyloxy dienes under the supervision of Professor Reinhard Neier. Following this, she pursued her PhD from 2003 – 2007 under the supervision of Professor Gottfried Sedelmeier in the Department of Process Research and Development of Novartis Pharma in Basel, obtaining her degree from the University of Neuchâtel. She is currently pursuing post-doctoral research in Professor Mark Lautens's group at the University of Toronto.

Ivana Fleischer was born in 1978 in Poprad, Slovakia. She received an M.Sc. degree from Commenius University, Bratislava in 2002, where she performed research under the direction of Professor Stefan Toma. In 2006, after a family break, she started her doctoral studies at the University of Basel, Switzerland, under the direction of Professor Andreas Pfaltz. Her graduate research focuses on mass-spectrometric screening of organocatalysts.

(S)-(−)-2-ALLYLCYCLOHEXANONE
(2-Allylcyclohexan-1-one)

Submitted by Manfred Braun, Panos Meletis, and Mesut Fidan.[1]
Checked by Ye Zhu and Viresh H. Rawal.

1. Procedure

A 1-L round-bottomed flask (Note 1) equipped with a 3-cm egg-shaped magnetic stir bar is charged with tris(dibenzylideneacetone)-dipalladium(0)-chloroform adduct (0.388 g, 0.375 mmol, 0.005 equiv), (S)-(−)-5,5'-dichloro-6,6'-dimethoxy-2,2'-bis(diphenylphosphino)-1,1'-biphenyl (0.977 g, 1.50 mmol, 0.02 equiv) and lithium chloride (7.63 g, 180 mmol, 2.4 equiv) (Note 2). The flask is closed with a rubber septum, connected to a combined nitrogen/vacuum line via a 16-gauge needle, evacuated (22 °C, 0.4 mmHg) for 5 h in order to remove traces of water from the lithium salt, and filled with nitrogen. To this flask is added a solution of allyl methyl carbonate (8.52 mL, 8.71 g, 75.0 mmol, 1.00 equiv) (Note 3) in dry tetrahydrofuran (THF) (195 mL) (Note 4) over 5 min through a 16-gauge cannula (Note 5). The deep purple homogeneous solution is stirred at 22 °C for 1 h. In the course of stirring, the color changes to yellow. The resulting mixture is cooled down to −78 °C in a dry ice/acetone bath with stirring. A 500-mL, three-necked, round-bottomed flask is equipped with a 3-cm egg-shaped magnetic stir bar, a glass stopper, a rubber septum fitted with the combined nitrogen/vacuum line via a 16-gauge needle, and a low-temperature thermometer via a thermometer adapter (Note 6). The flask is evacuated and refilled with nitrogen three times. To this flask, diisopropylamine (11.1 mL, 7.97 g, 78.7 mmol, 1.05 equiv) (Note 7) is injected by syringe, and dry THF (75 mL) is added via a 16-gauge cannula.

The flask is immersed in a dry ice/acetone bath and allowed to stir for 30 min. A 1.60 M solution of *n*-butyllithium in *n*-hexane (46.9 mL, 4.80 g, 75.0 mmol, 1.00 equiv) (Note 8) is added dropwise by syringe over 30 min. In the course of the addition, the internal temperature of the solution should not exceed –70 °C. The dry ice/acetone bath is replaced by an ice bath, and stirring is continued for 30 min. The mixture is then recooled in the dry ice/acetone bath, and a solution of freshly distilled cyclohexanone (7.78 mL, 7.36 g, 75.0 mmol, 1.00 equiv) (Note 9) in dry THF (75 mL) is added dropwise by a syringe over 1 h. In the course of the addition, the internal temperature of the solution should not exceed –70 °C. The mixture is stirred in an ice bath for 30 min and cooled again to –78 °C (internal temperature) in a dry ice/acetone bath. This colorless solution is then transferred over 10 min through a 16-gauge cannula into the 1 L flask (Note 5). After stirring for 40 h at –78 °C (Note 10), a yellow solution with white precipitate is obtained. This mixture is poured rapidly into a 4 L Erlenmeyer flask charged with 1.5 L of phosphate-buffered water (pH 7.00) (Note 11) stirred by a 7-cm egg-shaped magnetic stir bar. The resulting mixture is transferred into a 4 L separatory funnel, and extracted with three portions (170 mL, 170 mL, and 160 mL) of dichloromethane (CH_2Cl_2) (Note 11). The combined organic layers are dried over 20 g magnesium sulfate (Note 12), filtered, and concentrated by rotary evaporation (30 °C, 15 mmHg). The resulting brown-colored crude product is purified by Kugelrohr distillation (Note 13). A 50-mL pear-shaped flask containing the crude product is heated to 50 °C, and a 100-mL receiving flask is cooled in an ice-water bath. While maintaining a pressure of 0.05 mmHg, the pure product is collected as a colorless liquid; 8.08 g (78 %). The enantiomer ratio is determined by GC; *ee*: 94 % (Notes 14, 15).

2. Notes

1. The apparatus was oven-dried overnight, assembled hot and maintained under a positive pressure of nitrogen during the course of the reaction.

2. Tris(dibenzylideneacetone)dipalladium(0) chloroform adduct was obtained from Strem Chemicals, Inc. The submitters noted that distinctly lower enantioselectivity was obtained when the reagent of other suppliers was used. (*S*)-(–)-5,5'-Dichloro-6,6'-dimethoxy-2,2'-bis(diphenylphos-phino)-1,1'-biphenyl (min. 95 %) was also purchased by the checkers from

Strem Chemicals, Inc. The submitters obtained the ligand from LANXESS. Lithium chloride (SigmaUltra, min. 99.0 %) was obtained from Sigma-Aldrich, Inc. The submitters obtained lithium chloride (Normalpur) from VWR-Prolabo.

3. Allyl methyl carbonate is commercially available from Sigma-Aldrich, Inc. The submitters prepared it according to: H. O. L. Fischer, L. Feldmann, *Ber. Dtsch. Chem. Ges.* **1929**, *62*, 854. The checkers prepared it according to: I. Minami, J. Tsuji, *Tetrahedron* **1987**, *43*, 3903.

4. Tetrahydrofuran (Oprima®), was purchased from Fisher Scientific, Inc. It was purged with nitrogen and dried over activated alumina. The submitters purchased Tetrahydrofuran (technical grade) from Kraemer und Martin GmbH. It was refluxed over potassium hydroxide and distilled. Then the distillate was refluxed over sodium wire and distilled under nitrogen. Freshly distilled tetrahydrofuran was taken from the receiving flask via syringe or cannula.

5. The cannula was placed with one end immersed in the solution to be transferred and the other end in the receiver. A positive nitrogen pressure was maintained over the solution to be transferred, and the receiver was evacuated such that the solution flowed into the receiver at a steady rate. For a description of the technique, see: A. Salzer, in *Synthetic Methods of Organometallic and Inorganic Chemistry*, W. A. Herrmann, G. Brauer, Eds.; Vol. 1, W. A. Herrmann, A. Salzer, Volume Eds.; Thieme, Stuttgart **1996**, page 26.

6. The submitters used a thermocouple connected to a resistance thermometer introduced through a septum to indicate the internal temperature.

7. Diisopropylamine (purity >99.5 %) was purchased from Sigma-Aldrich, Inc. It was refluxed over calcium hydride and distilled under nitrogen. The submitters used 1.00 equiv (7.59 g, 10.5 mL, 75.0 mmol) of diisopropylamine.

8. *n*-Butyllithium (1.6 M solution in *n*-hexane) was purchased from Sigma-Aldrich, Inc., and titrated according to: *J. Org. Chem.* **1976**, *41*, 1879. The submitters purchased *n*-butyllithium (1.6 M solution in *n*-hexane) from Acros Organics.

9. Cyclohexanone (Selectphore®) was obtained from Sigma-Aldrich, Inc. The submitters obtained cyclohexanone (extra pure) from Riedel-de Haën. It was refluxed over calcium hydride and distilled under nitrogen.

10. In order to keep the bath temperature at -78 °C, the Dewar vessel was covered with a 3.5 cm thick insulating lid of Styropor®, cut in such a way that space is left only for the neck of the reaction flask. Solid dry ice was always maintained in the acetone bath. Additional dry ice was to be added every 6-8 hours. The submitters noted that shorter reaction time (15 h) led to a decrease in yield.

11. Buffer solution pH 7.00 with fungicide was supplied by Riedel-de Haën. Dichloromethane (Certified ACS) was purchased from Ficher Scientific, Inc. The submitters used dichloromethane (technical grade) from Ineos Chlor Ltd.

12. Magnesium sulfate (Anhydrous Certified) was purchased from Ficher Scientific, Inc. The submitters purchased magnesium sulfate (99% DAC) from Grüssing GmbH.

13. The submitters purified the product by trap-to-trap distillation. The flask containing the crude product was connected by glass tubes to two traps. The two traps were cooled to 0 °C and -196 °C, respectively, and the flask was heated to 50 °C. While maintaining a pressure of 0.05 mmHg, pure product was collected in the 0 °C-trap as a colorless liquid; 8.18 g (79 %); *ee*: 94 %. The trap-to-trap distillation is preferred to a conventional distillation in order to prevent partial racemization of the product; in addition, it provides a more efficient separation from starting materials. The checkers also tested the trap-to-trap distillation on a half-scale reaction. The pure product was isolated as a colorless liquid; 4.05 g (78 %); *ee*: 94 %.

14. GC column: Chiraldex™-beta-DP, 20 m x 0.25 mm x 0.25 μ m; column-temperature 80 °C; [(S)-enantiomer $t_R = 23.9$ min, (R)-enantiomer $t_R = 25.4$ min]. The submitters used different GC conditions; column: FS-Hydrodex-beta-TBDAc, 25m x 0.25 mm; column temperature 90 °C; [(S)-enantiomer $t_R = 17.4$ min, (R)-enantiomer $t_R = 20.0$ min].

15. Physical and spectroscopic data are as follows: $[\alpha]_D^{25}$ -17.4 (chloroform, *c* 1); the submitters reported: $[\alpha]_D^{20}$ -17.0 (chloroform, *c* 1); lit.[2g]: $[\alpha]_D^{20}$ -15.8 (methanol, *c* 3.0); IR (film): 3076, 2935, 2861, 1711, 1641, 1448, 1432, 1313, 1126, 996, 912 cm^{-1}; ^{1}H-NMR (500 MHz, CDCl$_3$) δ: 1.30-1.42 (m, 1 H), 1.61-1.71 (m, 2 H), 1.81-1.92 (m, 1 H), 1.94-2.09 (m, 2 H), 2.09-2.17 (m, 1 H), 2.25-2.44 (m, 3 H), 2.50-2.58 (m, 1 H), 4.96-5.05 (m, 2 H), 5.76 (dddd, $J = 6.5, 8.0, 10.2, 17.0$ Hz, 1 H); ^{13}C NMR (125 MHz, CDCl$_3$) δ: 24.9, 27.9, 33.4, 33.7, 42.0, 50.2, 116.2, 136.5, 212.5. Anal. Calcd. for C$_9$H$_{14}$O: C, 78.21; H, 10.21; O 11.58. Found: C, 78.01; H, 10.19;

O 11.89. HRMS: Calcd. for $C_9H_{14}ONa^+$ (M+Na): 161.093686. Found: 161.093303.

Waste Disposal Information

All hazardous materials should be handled and disposed of in accordance with "Prudent Practices in the Laboratory"; National Academy Press; Washington, DC, 1995.

3. Discussion

The enantiomeric (R)- and/or (S)-2-allylcyclohexanone has been chosen frequently in order to demonstrate the scope of different approaches in stereoselective synthesis. Among them, the allylic alkylation of azaenolates and enamines with chiral auxiliary groups has been used frequently.[2] In addition, access to the enantiomeric 2-allylcyclohexanones has been opened by deracemization through enzymic resolution[3] or formation of inclusion compounds.[4] Prochiral allyl-substituted enol acetates and enol carbonates have been converted into nonracemic 2-allylcyclohexanone by enantioface differentiating enzymatic hydrolysis.[5] Allyl-substituted enolates also lead to 2-allylcyclohexanone in an enantioselective way when treated with stoichiometric amounts of proton sources.[6] The allylation of the lithium enolate of cyclohexanone has been accomplished in the presence of chiral amines used in stoichiometric[7] or substoichiometric amounts.[8]

The palladium-catalyzed asymmetric allylic alkylation has developed into an exceptionally useful and versatile method for enantioselective carbon-carbon and carbon-heteroatom bond formation.[9] However, carbon nucleophiles have been limited to "soft", stabilized carbanions almost exclusively. Extending the asymmetric allylic substitution to include preformed, nonstabilized metal enolates[10] would significantly enhance its versatility, in particular because it would permit enantioselective alkylation in the homoallylic position. After early approaches of combining metal enolates with allylpalladium complexes,[11] the first enantioselective allylic alkylation of a nonstabilized ketone enolate was presented by Trost and Schroeder in 1999 using the tin enolate of 2-methyl-1-tetralone.[12a] Shortly thereafter, we reported the enantioselective and diastereoselective allylation of cyclohexanone through the lithium or magnesium enolates.[13] Additional examples using lithium enolates were also reported.[12c-e, 16] More recently, it has been shown that enantioselective palladium-catalyzed allylic alkylations are feasible not only with enol stannanes[12a, b] but also with enol silanes,[14] and enamines.[15] An alternative solution to the problem of the enantioselective allylic alkylation of nonstabilized enolates relies on their *in situ* generation, using allyl β-ketoesters or allyl enol carbonates.[17] Upon treatment with palladium(0) catalysts, carbon dioxide is liberated, and the enolate anion and the cationic palladium complex thus formed combine to give the allylated ketone. The method has been applied mostly for forming quaternary carbon centers, but was also used for the preparation of 2-allylcyclohexanone, starting from enol carbonate **1** as shown below.[17c]

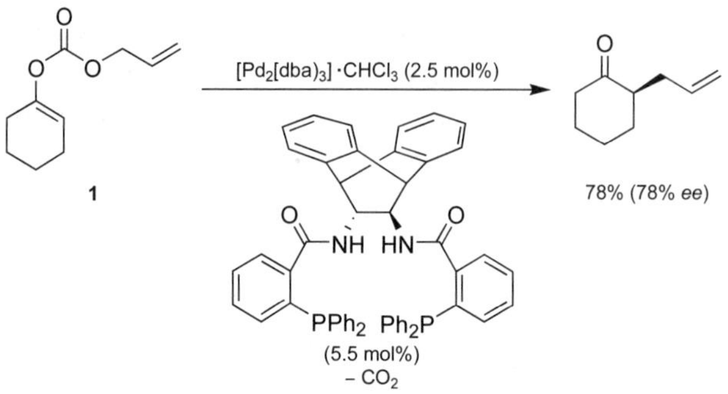

The allylic alkylation through the *in situ* generated enolates requires the previous preparation, isolation, and purification of either allyl enol carbonates or β-ketoesters. Compared to our direct allylic alkylation of preformed lithium or magnesium enolates, those protocols can be regarded as a detour in that an additional step is required. Additionally, in the case of 2-allylcyclohexanone, the route through the lithium enolate provides higher enantioselectivity than the enol carbonate route. The direct allylation of lithium and magnesium enolates has been extended to other ketones as well as to substituted allylic starting materials. As shown in Table 1,[18] both enantioselective and diastereoselective variants are possible. The protocol permits the generation of compounds with contiguous stereogenic centers in the allylic and the homoallylic position in an enantioselective and diastereoselective manner (entries 1-3) and can be applied to the asymmetric allylic alkylation of cyclopentanone and 1-tetralone (entries 4 and 5). When extended to diastereomerically and enantiomerically pure allylic substrates (entries 6-9), the achiral ligand bis(diphenylphosphino)ferrocene is suitable to bring about regioselective and diastereoselective allylic alkylations leading to enantiomerically pure products.

Table 1. Enantioselective and/or diastereoselective allylic alkylations of magnesium and lithium enolates.[18]

Entry	Enolate	Allyl Compound	Ligand	Product	Yield [%]
1	OMgCl (cyclohexenol)	Ph⌒CH(Ph)OAc	(R)-BINAP	d.r.:99 : 1 ee: 99%	67
2	OLi (cyclohexenol), LiCl	Me⌒CH(Me)OCO₂Me	(R)-BINAP	d.r.:97 : 3 ee: 96%	35[a]
3	OLi, Ar, Me	Ph⌒CH(Ph)OAc	(R)-BINAP	d.r.:90 : 10 ee: 88%	85[b]
4	OLi (cyclopentenol)	⌒OCO₂Me	(S)-BINAP	ee: 64%	76
5	OLi (dihydronaphthalenol)	⌒OCO₂Me	(R)-BINAP	ee: 66%	85
6	OLi (cyclohexenol), LiCl	Ph, AcO, MEMO, Me	dppf[c]	OMEM, Me / OMEM, Me d.r.:79 : 21	95[d]

Table 1. (continued) Enantioselective and/or diastereoselective allylic alkylations of magnesium and lithium enolates.[18]

Entry	Enolate	Allyl Compound	Ligand	Product	Yield [%]
7			dppf [c]	d.r.:57 :43	97[d]
8			dppf [c]	d.r.: >98%	95
9			dppf [c]	d.r.: >98%	97

Ar = 2,4,6-Me$_3$C$_6$H$_2$; MEM = CH$_2$OCH$_2$CH$_2$OCH$_3$
a) Starting material methyl (pent-3-en-2-yl) carbonate was recovered in 30%; b) yield of isolated major diastereomer; c) bis(diphenylphosphino)ferrocene; allyl compound was used as pure diastereomer and enantiomer; enantiomerically pure product(s); d) yield of diastereomeric mixture.

1. Institut für Organische und Makromolekulare Chemie, Heinrich-Heine-Universität, Universitätsstraße 1, D-40225 Düsseldorf, Germany. The procedure is based upon work supported by the Deutsche Forschungsgemeinschaft (Grant Br 604/16-1) and LANXESS Deutschland GmbH.

2. (a) M. Kitamoto, K. Hiroi, S. Terashima, S. Yamada, *Chem. Pharm. Bull.* **1974**, *22*, 459; (b) D. Enders, H. Eichenauer, *Angew. Chem. Int. Ed.* **1976**, *15*, 549; (c) A. I. Meyers, D. R. Williams, M. Druelinger, *J.*

Am. Chem. Soc. **1976**, *98*, 3032; (d) S. Hashimoto, K. Koga, *Tetrahedron Lett.* **1978**, *19*, 573; (e) D. Enders, H. Eichenauer, *Chem. Ber.* **1979**, *112*, 2933; (f) S. Hashimoto, K. Koga, *Chem. Pharm. Bull.* **1979**, *27*, 2760; (g) A. I. Meyers, D. R. Williams, G. W. Erickson, S. White, M. Druelinger, *J. Am. Chem. Soc.* **1981**, *103*, 3081; (h) K. Saigo, A. Kasahara, S. Ogawa, H. Nohira, *Tetrahedron Lett.* **1983**, *24*, 511; (i) K. Hiroi, K. Suya, S. Sato, *J. Chem. Soc., Chem. Commun.* **1986**, 469; (j) K. Hiroi, J. Abe, *Tetrahedron Lett.* **1990**, *31*, 3623; *Chem. Pharm. Bull.* **1991**, *39*, 616; (k) K. Hiroi, J. Abe, K. Suya, S. Sato, T. Koyama, *J. Org. Chem.* **1994**, *59*, 203; (l) D. Enders, T. Hundertmark, R. Lazny, *Synlett* **1998**, 721.

3.   (a) J. D. Stewart, K. E. Reed, J. Zhu, G. Chen, M. M. Kayser, *J. Org. Chem.* **1996**, *61*, 7652; (b) B. G. Kyte, P. Rouviere, Q. Cheng, J. D. Stewart, *J. Org. Chem.* **2004**, *69*, 12.

4.   (a) T. Tsunoda, H. Kaku, M. Nagaku, E. Okuyama, *Tetrahedron Lett.* **1997**, *38*, 7759; (b) H. Kaku, S. Ozako, S. Kawamura, S. Takatsu, M. Ishii, T. Tsunoda, *Heterocycles* **2001**, *55*, 847; (c) H. Kaku, N. Okamoto, A. Nakamaru, T. Tsunoda, *Chem. Lett.* **2004**, *33*, 516.

5.   (a) H. Ohta, K. Matsumoto, S. Tsutsumi, T. Ihori, *J. Chem. Soc., Chem. Commun.* **1989**, 485; (b) K. Matsumoto, S. Tsutsumi, T. Ihori, H. Ohta, *J. Am. Chem. Soc.* **1990**, *112*, 9614.

6.   (a) K. Matsumoto, H. Ohta, *Tetrahedron Lett.* **1991**, *32*, 4729; (b) A. Yanagisawa, K. Tetsuo, T. Kuribayashi, H. Yamamoto, *Tetrahedron* **1998**, *54*, 10253.

7.   M. Murakata, T. Yosukata, T. Aoki, M. Nakajima, K. Koga, *Tetrahedron* **1998**, *54*, 2449.

8.   M. Imai, A. Hagihara, H. Kawasaki, K. Manabe, K. Koga, *Tetrahedron* **2000**, *56*, 179.

9.   For reviews, see: (a) O. Reiser, *Angew. Chem. Int. Ed.* **1993**, *32*, 547; (b) J. M. J. Williams, *Synlett* **1996**, 705; (c) B. M. Trost, D. L. Van Vranken, *Chem. Rev.* **1996**, *96*, 395; (d) G. Helmchen, *J. Organomet. Chem.* **1999**, *576*, 203; (e) G. Helmchen, A. Pfaltz, *Acc. Chem. Res.* **2000**, *33*, 336; (f) B. M. Trost, M. L. Crawley, *Chem. Rev.* **2003**, *103*, 2921; (g) B. M. Trost, *J. Org. Chem.* **2004**, *69*, 5813.

10.   C. H. Heathcock, in *Modern Synthetic Methods 1992*, R. Scheffold, Ed.; VHCA, VCH: Basel, Weinheim, **1992**, 1, and references therein.

11.   (a) J.-C. Fiaud, J.-L. Malleron, *Chem. Soc., Chem. Commun.* **1981**, 1159; (b) B. Akermark, A. J. Jutand, *J. Organomet. Chem.* **1981**, *217*,

C41; (c) E. Negishi, H. Matsushita, S. Chatterjee, R. A. John, *J. Org. Chem.* **1982**, *47*, 3188; (d) B. M. Trost, E. Keinan, *Tetrahedron Lett.* **1980**, *21*, 2591; (e) B. M. Trost, C. R. Self, *J. Org. Chem.* **1984**, *49*, 468.

12. (a) B. M. Trost, G. M. Schroeder, *J. Am. Chem. Soc.* **1999**, *121*, 6759; (b) B. M. Trost, G. M. Schroeder, *Chem. Eur. J.* **2005**, *11*, 174; ; (c) S.-L. You, X.-L. Hou, L.-X. Dai, X.-Z. Zhu, *Org. Lett* **2001**, *3*, 149. (d) X.-X. Yan, C.-G. Liang, Y. Zhang, W. Hong, B.-X. Cao, L.-X. Dai, X.-L. Hou, *Angew. Chem. Int. Ed.* **2005**, *44*, 6544; (e) W.-H. Zheng, B. H. Zheng, Y. Zhang, X.-L. Hou, *J. Am. Chem. Soc.* **2007**, *129*, 7718.

13. M. Braun, F. Laicher, T. Meier, *Angew. Chem. Int. Ed.* **2000**, *39*, 3494

14. (a) T. Gaening, J. F. Hartwig, *J. Am. Chem. Soc.* **2005**, *127*, 17192; (b) E. Bélanger, K. Cantin, O. Messe, M. Tremblay, J.-F. Paquin, *J. Am. Chem. Soc.* **2007**, *129*, 1034.

15. D. J. Weix, J. F. Hartwig, *J. Am. Chem. Soc.* **2007**, *129*, 7720.

16. For reviews, see: (a) M. Braun, T. Meier, *Synlett* **2006**, 661; (b) M. Braun, T. Meier, *Angew. Chem. Int. Ed.* **2006**, *45*, 6952.

17. (a) E. C. Burger, J. A. Tunge, *Org. Lett.* **2004**, *6*, 4113; (b) D. C. Behenna, B. M. Stoltz, *J. Am. Chem. Soc.* **2004**, *126*, 15044; (c) B. M. Trost, J. Xu, *J. Am. Chem. Soc.* **2005**, *127*, 2846; (d) B. M. Trost, J. Xu, *J. Am. Chem. Soc.* **2005**, *127*, 17180; (e) B. M. Trost, R. N. Bream, J. Xu, *Angew. Chem. Int. Ed.* **2006**, *45*, 3109; (f) J. T. Mohr, D. C. Behenna, A. M. Harned, B. M. Stoltz, *Angew. Chem. Int. Ed.* **2005**, *44*, 6924; (g) J. A. Tunge, E. C. Burger, *Eur. J. Org. Chem.* **2005**, 1715; (h) S. Trudeau, J. P. Morken, *Tetrahedron* **2006**, *62*, 11470; (i) H. He, X.-J. Zheng, Y. Li, L.-X. Dai, S.-L. You, *Org. Lett.* **2007**, *9*, 4339; (j) for a recent review, see: S.-L. You, L.-X. Dai, *Angew. Chem. Int. Ed.* **2006**, *45*, 5246.

18. (a) M. Braun, T. Meier, *Synlett* **2005**, 2968; (b) M. Braun, T. Meier, F. Laicher, P. Meletis, M. Fidan, *Adv. Synth. Catal.* **2008**, *350*, 303.

Appendix
Chemical Abstracts Nomenclature; (Registry Number)

Tris(dibenzylideneacetone)-dipalladium(0)-chloroform adduct; (52522-40-4)
(*S*)-(–)-5,5'-Dichloro-6,6'-dimethoxy-2,2'-bis(diphenylphosphino)-1,1'-
biphenyl: Phosphine, [(1S)-5,5'-dichloro-6,6'-dimethoxy[1,1'-

56

biphenyl]-2,2'-diyl]bis[diphenyl-; (185913-98-8)

Lithium chloride; (7447-41-8)

Allyl methyl carbonate: Carbonic acid, methyl 2-propen-1-yl ester; (35466-83-2)

Diisopropylamine: 2-Propanamine, N-(1-methylethyl)-; (108-18-4)

n-Butyllithium; (109-72-8)

Cyclohexanone; (108-94-1)

(S)-(–)-2-Allylcyclohexanone: Cyclohexanone, 2-(2-propen-1-yl)-, (2S)-; (36302-35-9)

Manfred Braun, born in 1948 in Schwalbach near Saarlouis, studied chemistry at the University of Karlsruhe from 1966 until 1971, and completed his doctorate under Professor Dieter Seebach in Gießen in 1975. After a postdoc with Professor George H. Büchi at the Massachusetts Institute of Technology in 1975 and 1976, he joined Professor Hans Musso's research group at the University of Karlsruhe and completed his Habilitation there in 1981. Since 1985, he has been a professor of organic chemistry at the Heinrich-Heine-University Düsseldorf. His current research interests include the development of new synthetic methods (especially for asymmetric synthesis), organometallic chemistry, development of metal complexes as smart materials, and syntheses of biologically active compounds.

Panos Meletis was born 1979 in Neuss, Germany. He graduated in 2006 and received his diploma from the Heinrich-Heine-Universität in Düsseldorf. He prepared his diploma thesis with the topic "Palladium Catalyzed Allylation of Lactone- and Ester-Enolates" under the supervision of Professor M. Braun. In the same year he started his Ph. D. study. His current interest is still the palladium catalyzed allylic alkylation.

Mesut Fidan was born 1979 in Solingen, Germany. He graduated in 2006 and received his diploma at Heinrich-Heine-Universität in Düsseldorf under the supervision of Professor M. Braun. The topic of his diploma thesis was "Palladium catalyzed allylation with dianions". In the same year he started his Ph.D. study in Organic Chemistry. His current interest is palladium-catalyzed regio- and enantioselective allylation of lactams.

Ye Zhu was born in Jinan, China in 1983. He received his B.S. in chemistry from Peking University in 2006, and went to the University of Chicago for graduate school in the same year. He is currently a chemistry graduate student with Prof. Viresh Rawal, working on metal-catalyzed and organocatalyzed asymmetric reactions.

PREPARATION OF 4-SPIROCYCLOHEXYLOXAZOLIDINONE
BY C-H BOND NITRENE INSERTION
[3-Oxa-1-azaspiro[4.5]decan-2-one]

Submitted by Kim Huard and Hélène Lebel.[1]
Checked by Adam Rosenberg, Darla Seifried, and Kay Brummond.

1. Procedure

A. Rhodium II tetrakis(triphenylacetate) dimer (**2**). To a 250-mL, one-necked, round-bottomed flask equipped with a magnetic stirring bar, a short-path distillation apparatus and a 100-mL receiving bulb (Note 1) is added rhodium (II) acetate dimer (1.26 g, 2.85 mmol) (Note 2), triphenylacetic acid (**1**) (6.55 g, 22.7 mmol, 8.0 equiv) (Note 3) and chlorobenzene (120 mL) (Note 3). The mixture is heated with an oil bath that is set to a temperature at which the solvent distills at a rate of 10 mL/hour (approximately 155 °C) (Note 4). The green mixture becomes homogeneous upon heating. After 7 h, the reaction is cooled to 25 °C and the residual solvent is concentrated at 65 °C by rotary evaporation (20-25 mmHg) and then at 0.5 mmHg. The solid residue is dissolved in 200 mL of dichloromethane. The solution is washed with saturated aqueous sodium bicarbonate (3 x 200 mL), saturated aqueous sodium chloride (100 mL), dried over magnesium sulfate (15 g), filtered and

Org. Synth. **2009**, *86*, 59-69
Published on the Web 9/26/2008

concentrated at 40 °C by rotary evaporation (225 mmHg) and then at 0.5 mmHg. To the green solid, methanol (40 mL) is added and the residue is stirred in a 35 °C water bath (Note 5). After 5 minutes, dichloromethane is added portion wise with stirring and heating in a 35 °C water bath until the solid is completely dissolved (approximately 200 mL of DCM). The solution is left to rest at -19 °C for 72 h. The crystals are collected by suction filtration on a fritted glass funnel and dried at 0.5 mmHg to afford 3.01 g (78% yield) of the desired product (Notes 6 and 7).

  B. *Cyclohexylmethyl N-hydroxycarbamate* (**4**). A 500-mL, one-necked, round-bottomed flask equipped with a magnetic stirring bar and a 50-mL pressure-equalizing addition funnel fitted with a nitrogen inlet is flame-dried and maintained under nitrogen atmosphere. The flask is quickly opened to add 1,1'-carbonyldiimidazole (17.0 g, 105 mmol, 1.05 equiv) (Notes 8 and 9) followed by acetonitrile (200 mL) (Note 10) via cannula. The white suspension is cooled with an ice bath and cyclohexanemethanol (**3**) (12.3 mL, 100 mmol) (Note 11) is transferred to the addition funnel using a syringe. Cyclohexanemethanol (**3**) is added dropwise over a 30-minute time period. Addition of the alcohol results in the suspension becoming a homogeneous solution. The addition funnel is rinsed with acetonitrile (10 mL) and the solution is stirred in the ice bath for 1 h. The reaction of cyclohexanemethanol **3** (R_f = 0.63 in 30% EtOAc/DCM) with 1,1'-carbonyldiimidazole to give a carbonylated intermediate (R_f = 0.59 in 30% EtOAc/DCM) can be followed by TLC analysis using a potassium permanganate solution for visualization. Continued stirring does not improve the conversion as the intermediate can be hydrolyzed and the starting alcohol recovered. The solution is then warmed to ambient temperature. To the solution is then added hydroxylamine hydrochloride (20.9 g, 300 mmol, 3.0 equiv) (Note 12) followed by imidazole (13.6 g, 200 mmol, 2.0 equiv) (Note 12) and vigorous stirring is continued for 3.5 h (Note 13). The conversion of the intermediate (R_f = 0.59 in 30% EtOAc/DCM) to cyclohexylmethyl *N*-hydroxycarbamate **4** (R_f = 0.52 in 30% EtOAc/DCM) can be followed by TLC analysis using a potassium permanganate solution for visualization. The suspension is concentrated at 40 °C by rotary evaporation (130 mmHg) and at 0.5 mmHg. The white residue is dissolved in a 1:1 mixture of ethyl acetate and 10% hydrochloric acid (300 mL) and transferred to a 1-L separatory funnel. After vigorous shaking, the phases are separated and the aqueous phase is extracted with ethyl acetate (2 x 150 mL). The combined organic layers are washed with a saturated sodium chloride solution (100 mL), dried over magnesium sulfate (15 g), filtered and concentrated at 40 °C by rotary evaporation (75 mmHg) and then at 0.5 mmHg to afford 16.57 g of a light amber oil containing the title compound **4** and residual starting alcohol **3** in a ratio of roughly 7:1, as determined by NMR analysis (Note 14). This mixture is used without further purification for the next step. A 200 mg sample of the crude product is purified by silica gel chromatography (Note 15) to afford 167 mg (84% recovery) of the title compound as a white solid (Note 16).

C. *Cyclohexylmethyl N-tosyloxycarbamate* (**5**). A 500-mL, one-necked, round-bottomed flask containing the oily residue obtained from procedure B (15.9 g) is equipped with a magnetic stirring bar and a 50-mL pressure-equalizing addition funnel fitted with a nitrogen inlet. The system is flushed with nitrogen for 30 min and maintained under nitrogen atmosphere. Diethyl ether (200 mL) (Note 17) is added via cannula and the solution is cooled with an ice bath. The system is quickly opened to add *p*-toluenesulfonyl chloride (20.2 g, 106 mmol) (Note 18). Triethylamine (13.4 mL, 96.4 mmol) (Note 19) is introduced into the addition funnel using a syringe and is added dropwise over a 15-minute time period to the ethereal solution. The addition funnel is rinsed with diethyl ether (10 mL) and the solution is stirred in the ice bath for 1.5 h. Distilled water (75 mL) is added to the white suspension and the resulting clear mixture is transferred to a 500-mL separatory funnel. After vigorous shaking, the organic layer is separated and washed with a saturated sodium chloride solution (75 mL), dried over $MgSO_4$ (15 g), filtered and concentrated at 40 °C by rotary evaporation (225 mmHg and 20 mmHg) until the residue becomes solid and then at 0.5 mmHg for 12 h to afford a yellowish solid. The solid is dissolved in 50 mL of dichloromethane and is charged on a 6.5 x 11 cm silica gel column (200 g of dry silica) (Note 20). The system is eluted with a mixture of 5% ethyl acetate-dichloromethane. The first 150 mL are discarded and the next 800 mL are collected in a 1-L round-bottomed flask and concentrated at 40 °C by rotary evaporation (225 mmHg, then 20 mmHg) until the residue becomes a solid which is then dried at 0.5 mmHg. The inner surface of the flask is scratched with a spatula in order to loosen the product from the flask. To the 1-L flask is added a 6-cm egg-shaped magnetic stirring bar and 300 mL of hexane (Note 21). The suspension is vigorously stirred for 24 h and filtered on a fritted glass funnel. The white solid is dried at 0.5 mmHg for 8 h to afford 23.75 g (73% yield from **3**) of the desired product (Notes 22 and

23). The tosyloxycarbamate must be over 90% purity, as determined by GC/MS, to be used for the next step.

   D. 4-Spirocyclohexyloxazolidinone (**6**). Using a 250-mL one-necked, round-bottomed flask equipped with a magnetic stirring bar, a solution of cyclohexylmethyl *N*-tosyloxycarbamate (**5**) (16.4 g, 50.0 mmol) in dichloromethane (50 mL) (Note 24) is prepared. A 1-L three-necked, round-bottomed flask equipped with a 6-cm egg-shaped magnetic stirring bar is charged with dichloromethane (75 mL) and distilled water (10 mL). Potassium carbonate (7.60 g, 55 mmol, 1.1 equiv) (Note 25) is added to the flask, which is fitted with a rubber septum, an internal thermometer, glass stopper and a short needle (through the septum) opened to the atmosphere, and the mixture is stirred until the white solid is dissolved. The dichloromethane solution containing **5** (Note 26) is taken up in a 60-mL glass syringe, which is installed on a syringe pump system set for an addition time of 60 minutes. Tetrakis(triphenylacetate) rhodium dimer (**2**) (0.678 g, 0.5 mmol, 0.01 equiv) is added to the biphasic solution and the dropwise addition is begun. Upon the addition of the tosyloxycarbamate **5** solution, the green reaction mixture becomes thicker, a white solid forms, and the internal temperature slowly rises to reach 30 °C when approximately half of the starting material is added. Then the mixture slowly cools down to room temperature. The starting material is added dropwise to avoid any significant rise of internal temperature. When the complete volume is added, the 250-mL flask and the syringe are rinsed with 10 mL of dichloromethane and this volume is added to the mixture which is stirred for a further 3 h to ensure complete consumption of the starting material. The dichloromethane is evaporated at 40 °C by rotary evaporation (225 mmHg) until only the residue and the water are left in the flask. The aqueous residue is dissolved in a 1:1 mixture of ethyl acetate and water (400 mL) and transferred to a 1-L separatory funnel. The phases are separated and the aqueous layer is extracted with ethyl acetate (2 x 100 mL). The combined organic extracts are washed with a saturated sodium chloride solution (100 mL), dried over MgSO$_4$ (15 g), filtered and concentrated at 40 °C by rotary evaporation (75 mmHg) and then at 0.5 mmHg to afford a yellowish solid. The solid is dissolved in 20 mL of dichloromethane and charged on a 6.5 x 22 cm column of silica gel (350 g of dry silica) (Note 20). The system is eluted with a mixture of 30% ethyl acetate-dichloromethane. Because the product is difficult to visualize by TLC, the first 900 mL are discarded and the next 1400 mL are collected and concentrated at 40 °C by rotary evaporation (225 mmHg then 20 mmHg) until the residue solidifies. It is then dried at 0.5 mmHg to afford 6.1 g (79% yield) of the title compound as a yellowish solid of purity higher than 98% as determined by NMR and GC/MS analysis (Note 27).

2. Notes

1. All glassware was flame-dried under vacuum and allowed to cool under an atmosphere of nitrogen.

2. Rhodium (II) acetate dimer was purchased from Strem Chemicals. It was stored and weighed in a glove box and was used without further purification.

3. Triphenylacetic acid (99%) and chlorobenzene (99%) were purchased from Aldrich Chemical Company, Inc. and used as received.

4. Slow distillation of the solvent (bp 132 °C) removes the acetic acid (bp 117 °C) formed from the ligand exchange. If the distillation rate is higher than 10 mL/hour, chlorobenzene is added in order to keep the total volume of the reaction mixture over 40 mL.

5. The solid is insoluble in methanol and the color changes from green to purple-blue.

6. A half-scale run provided 1.75 g (89%) of the rhodium II tetrakis(triphenylacetate) dimer.

7. Analytical data for Rhodium II tetrakis(triphenylacetate) dimer (2): R_f = 0.19 (10% ethyl acetate/hexane); ^{1}H NMR (300 MHz, CDCl$_3$) δ: 3.51 (residual MeOH), 6.63 (d, J = 7.5 Hz, 24 H), 6.86 (t, J = 8.1 Hz, 24 H), 7.07 (t, J = 7.5 Hz, 12 H); ^{13}C NMR (75 MHz, CDCl$_3$) δ: 69.2 (residual MeOH), 126.8, 127.4, 130.7, 143.4, 192.9; IR (film) 3055, 1590, 1580, 1365 cm^{-1}; Analysis calc. for C$_{80}$H$_{60}$O$_8$Rh$_2$: C 70.90, H 4.46; found: C 67.96, H 4.73 %. The low value obtained for the carbon analysis is due to residual coordinated methanol from the recrystallization. Up to 2 equiv of MeOH per molecule of dimer (determined by ^{1}H NMR, δ = 3.51 and ^{13}C NMR, δ = 69.2) can be present and do not affect the reactivity of the catalyst.

8. 1,1'-Carbonyldiimidazole, reagent grade, was purchased from Aldrich Chemical Company, Inc. It was stored and weighed in a glove box to avoid hydrolysis and was used without further purification.

9. Short exposure to 1,1'-carbonyldiimidazole could cause serious temporary or residual injury. This reagent should be handled with careful attention to avoid contact with skin.

10. Acetonitrile was distilled over calcium hydride. If wet acetonitrile is used, hydrolysis of a reaction intermediate reduces the yield.

11. Cyclohexanemethanol (**3**), 99+%, was purchased from Aldrich Chemical Company, Inc. and was used as received.

12. Hydroxylamine hydrochloride (99%) and imidazole (99%) were purchased from Alfa Aesar and used as received.

13. An excess of hydroxylamine and imidazole is used in order to accelerate the reaction and minimize the hydrolysis of the intermediate.

14. The ^{1}H NMR methylene resonances of the *N*-hydroxycarbamate (doublet at 3.96 ppm) and the starting alcohol (doublet at 3.44 ppm) are used to determine the product ratio.

15. Flash chromatography is performed on a 3 x 10 cm silica gel column (33 g of dry silica). The product is charged on the column and the system is eluted with 50 mL of 20% ethyl acetate-dichloromethane. At that point, fraction collection (8-mL fractions) is begun and elution is continued with 450 mL of a 20% ethyl acetate-dichloromethane mixture. Collected fractions were analyzed by TLC, eluting with 30% ethyl acetate-dichloromethane (R_f = 0.52 for **4** and 0.63 for **3**). Visualization was accomplished by spraying with a potassium permanganate solution followed by heating. All the fractions (10-30) containing the desired product were combined and concentrated at 45 °C by rotary evaporation (40-50 mmHg) and then at 0.5 mmHg.

16. Analytical data for cyclohexylmethyl *N*-hydroxycarbamate (**4**): mp 33–34 °C; ^{1}H NMR (300 MHz, CDCl$_3$) δ: 0.85–1.04 (m, 2 H), 1.06–1.37 (m, 3 H), 1.40–1.90 (m, 6 H), 3.99 (dd, *J* = 6.3, 5.1 Hz, 2 H), 7.03 (br s, 1 H), 7.28 (br s, 1 H); ^{13}C NMR (75 MHz, CDCl$_3$) δ: 25.7, 26.4, 29.6, 37.4, 71.5, 159.9; IR (film) 3291, 2922, 2851, 1712, 1449, 1267, 1121 cm^{-1}; HRMS (TOF$^+$) calc. for C$_8$H$_{15}$NO$_3$Na [M+Na]$^+$: 196.0950; found: 196.0940; Analysis calc. for C$_8$H$_{15}$NO$_3$: C 55.47, H 8.73, N 8.09; found: C 55.49, H 8.81, N 7.93 %.

17. Anhydrous diethyl ether was obtained by filtration through a drying column on a Sol-Tec solvent purification system.

18. *p*-Toluenesulfonyl chloride, 99+%, was purchased from Aldrich Chemical Company, Inc. and was used as received.

19. Triethylamine was purchased from Fisher Scientific and was freshly distilled from CaH$_2$ (88–90 °C) under argon atmosphere prior to use.

20. UltraPure silica gel (32–63 µm) was purchased from EcoChrom.

21. Omnisolv grade hexane was purchased from EMD and was used as received.

22. A second run on approximately half-scale provided a 79% yield.

23. The physical properties of **5** are as follows: R_f 0.68 (5% ethyl acetate/dichloromethane); mp 88–90 °C; ^{1}H NMR (500 MHz, CDCl$_3$) δ: 0.82–0.85 (m, 2 H), 1.0-1.3 (m, 3 H), 1.46–1.57 (m, 3 H), 1.63–1.67 (m, 3 H), 2.46 (s, 3 H), 3.83 (d, J = 6.3 Hz, 2 H), 7.35 (d, J = 8.1 Hz, 2 H), 7.87 (d, J = 8.1 Hz, 2 H); ^{13}C NMR (125 MHz, CDCl$_3$) δ: 21.9, 25.7, 26.3, 29.3, 37.1, 72.2, 129.7, 129.9, 130.5, 146.2, 155.9; IR (film) 3230, 2927, 2853, 1740, 1597, 1380, 1191, 1179 cm^{-1}; HRMS (TOF$^+$) calc. for C$_{15}$H$_{21}$NO$_5$SNa [M+Na]$^+$: 350.1038; found: 350.1006; Analysis calc. for C$_{15}$H$_{21}$NO$_5$S: C 55.03, H 6.47, N 4.28, S 9.79; found: C 54.85, H 6.30, N 4.11, S 10.07 %.

24. Dichloromethane was purchased from Fisher Scientific and was used as received.

25. Granular potassium carbonate was purchased from J.T. Baker and was used as received.

26. The total volume added is approximately 60 mL.

27. Analytical data for 3-oxa-1-azaspiro[4.5]decan-2-one (**6**): R_f 0.53 (30% ethyl acetate/dichloromethane); mp 81–82 °C; ^{1}H NMR (300 MHz, CDCl$_3$) δ: 1.20–1.72 (m, 10 H), 4.08 (s, 2H), 6.82 (br s, 1H); ^{13}C NMR (75 MHz, CDCl$_3$) δ: 22.7, 24.9, 37.2, 57.9, 75.8, 159.7; IR (film) 3230, 2932, 2858, 1745, 1397, 1251, 1031 cm^{-1}; HRMS (TOF$^+$) calc. for C$_8$H$_{13}$NO$_2$Na [M+Na]$^+$: 178.0844; found: 178.0843; Analysis calc. for C$_8$H$_{13}$NO$_2$: C 61.91, H 8.44, N 9.03; found: C 61.99, H 8.61, N 8.94 %.

Safety and Waste Disposal Information

All hazardous materials should be handled and disposed of in accordance with "Prudent Practices in the Laboratory"; National Academy Press; Washington, DC, 1995.

3. Discussion

This procedure describes an attractive method for the preparation of oxazolidinones using a rhodium-catalyzed nitrene C-H bond insertion reaction. It is well known that nitrene species are accessible by photolysis or thermolysis of acyl azides. However, their employment in the formation of oxazolidinones is limited to some specific cases,[2] because of their lack of stability and selectivity.[3] Alternatively, metal nitrenes are prepared via the oxidation of carbamates using hypervalent iodine reagents which generate a stoichiometric amount of iodobenzene as byproduct.[4] In the described procedure, a tosyloxycarbamate is decomposed with a base to form a metal nitrene in the presence of a rhodium catalyst. The insertion reaction proceeds then smoothly, with total conversion. The only stoichiometric byproduct is potassium tosylate, which is simply removed by an aqueous work up. The starting material is a stable white solid easily prepared from the corresponding commercially available alcohol. The metal nitrene insertion is effective at ethereal, benzylic, tertiary, secondary, and even primary positions, providing an interesting route to various substituted oxazolidinones (Table 1).[5]

Table 1. Oxazolidinone formation[a,b]

Entry	Product	Isolated yield	Entry	Product	Isolated yield
1		92%	5		73%
2		84%	6		84%
3		87%	7		64%
4		71%	8		41%

[a] see reference 5; [b] 0.5 mmol scale

Chiral oxazolidinones can also be prepared by this method using an enantioenriched N-tosyloxycarbamate as starting material, as the C-H insertion occurs with total retention of configuration (Entry 5). The method has been initially developed on a 0.5 mmol scale, using 6 mol% of rhodium dimer and 3 equivalents of potassium carbonate in dichloromethane at room temperature. Under the optimized reaction conditions on a 50 mmol scale, the catalyst loading is lowered to 1 mol% and 1.1 equivalent of base is used for complete conversion. The intermolecular version of the reaction has also been reported using 2,2,2-trichloroethyl *N*-tosyloxycarbamate as the source of nitrene and Troc-protected amines derived from various unfunctionalized substrates were isolated in good yields.[6]

1. Département de Chimie, Université de Montréal, C.P. 6128, Succursale Centre-Ville, Montréal, Québec, Canada, H3C 3J7. helene.lebel@umontreal.ca
2. (a) Berndt, D. F.; Norris, P. *Tetrahedron Lett.* **2002**, *43*, 3961-3962; (b) Kan, C.; Long, C. M.; Paul, M.; Ring, C. M.; Tully, S. E. Rojas, C. M. *Org. Lett.* **2001**, *3*, 381-384; (c) Yuan, P.; Plourde, R.; Shoemaker, M. R.; Moore, C. L.; Hansen, D. E. *J. Org. Chem.* **1995**, *60*, 5360-5364; (d) Schneider, G.; Hackler, L.; Szanyl, J.; Sohar, P. *J. Chem. Soc. Perkin Trans. 1* **1991**, 37-42; (e) Banks, M. R.; Cadogan, J. I. G.; Gosney, I.; Gaur, S.; Hodgson, P. K. G. *Tetrahedron Asymm.* **1994**, *5*, 2447-2458.
3. Meth-Cohn, O. *Acc. Chem. Res.* **1987**, *20*, 18-27.
4. (a) Espino, C. G.; Du Bois, J. *Angew. Chem. Int. Ed.* **2001**, *40*, 598-600; (b) Cui, Y.; He, C. *Angew. Chem. Int. Ed.* **2004**, *43*, 4210-4212.
5. (a) Lebel, H.; Huard, K.; Lectard, S. *J. Am. Chem. Soc.* **2005**, *127*, 14198-14199; (b) Lebel, H.; Leogane, O.; Huard, K.; Lectard, S. *Pure Appl. Chem.* **2006**, *78*, 363-375.
6. Lebel, H.; Huard, K. *Org. Lett.* **2007**, *9*, 639-642.

Appendix
Chemical Abstracts Nomenclature; (Registry Number)

Rhodium II acetate: (15956-28-2)
Triphenylacetic acid: Benzeneacetic acid, α,α-diphenyl-; (595-91-5)
Rhodium (II) tetrakis(triphenylacetate);(142214-04-8)
Cyclohexylmethyl *N*-hydroxycarbamate: Carbamic acid, hydroxy-,
    cyclohexylmethyl ester; (869111-30-8)
1,1'-Carbonyldiimidazole: Methanone, di-1*H*-imidazol-1-yl-; (530-62-1)
Cyclohexanemethanol; (100-49-2)

Hydroxylamine hydrochloride; (5470-11-1)

Imidazole; (288-32-4)

p-Toluenesulfonyl chloride: Benzenesulfonyl chloride, 4-methyl-; (98-59-9)

Triethylamine; (121-44-8)

Cyclohexylmethyl N-tosyloxycarbamate: Benzenesulfonic acid, 4-methyl-, [(cyclohexylmethoxy)carbonyl]azanyl ester; (869111-41-1)

4-Spirocyclohexyloxazolidinone: 3-Oxa-1-azaspiro[4.5]decan-2-one: (81467-34-7)

Potassium carbonate; (584-08-7)

Hélène Lebel received her B.Sc. degree in biochemistry from the Université Laval in 1993. She conducted her Ph.D. studies in organic chemistry at the chemistry department of the Université de Montréal under the supervision of professor André B. Charette as a 1967 Science and Engineering NSERC Fellow. In 1998, she joined the research group of professor Eric Jacobsen at Harvard University as a NSERC Postdoctoral Fellow. She started her independent career in 1999 at the Université de Montréal, where her research program focuses on the development of novel synthetic methods.

Kim Huard was born in Granby, Québec, Canada in 1979. She received a B.Sc. degree in chemistry in 2004 from Université de Montréal. She then received her Ph.D. in Hélène Lebel's research group at Université de Montréal where she studied rhodium-catalyzed C-H bonds amination. She has been recipient of NSERC Postgraduate Scholarships. She is currently a postdoctoral fellow at University of California at Irvine under the guidance of Professor Larry Overman.

Adam Rosenberg obtained his B.S. in chemistry from the University of Rochester while performing undergraduate research in Dr. Boeckman's research group. He is currently pursuing graduate studies in the laboratory of Professor Kay Brummond at the University of Pittsburgh. His graduate work has included investigations into the total synthesis of 3α-hydroxy-15-rippertene as well as novel multi-component cycloaddition reactions.

Darla Seifried is a graduate of the University of Pittsburgh with a Bachelor of Science in biology. She is currently earning her B.S. in chemistry from the University of Pittsburgh. Darla plans to attend graduate school beginning in the fall of 2009.

PREPARATION OF (S)-4-ISOPROPYL-N-PROPANOYL-1,3-THIAZOLIDINE-2-THIONE

Submitted by Erik Gálvez, Pedro Romea, and Fèlix Urpí.[1]
Checked by John C. Jewett and Viresh H. Rawal.

1. Procedure

A. *(S)-4-Isopropyl-1,3-thiazolidine-2-thione* (**1**). A 500-mL single-necked round-bottomed flask equipped with a magnetic stir bar and a 250-mL pressure-equalizing addition funnel fitted with a Y-connector joint with one line fitted with a nitrogen inlet adaptor and the other with a rubber septum is charged with (S)-valinol (10.32 g, 0.10 mol, 1.0 equiv, Note 1), followed by ethanol (30 mL, Note 2) and carbon disulfide (18 mL, 0.26 mol, 2.6 equiv) (Notes 3, 4). The addition funnel is flushed with nitrogen and charged with a 2.25 M KOH (15.2 g, 0.27 mol, 2.7 equiv) in 1:1 EtOH/H$_2$O solution (120 mL), which is added dropwise at room temperature over 15–20 min. The addition funnel is replaced by a Graham condenser fitted with a nitrogen adaptor, and the reaction mixture is stirred and heated at reflux for 72 h under a nitrogen atmosphere (Notes 5, 6). After cooling, the volatiles are removed with a rotary evaporator (Notes 7, 8). The resulting liquid is acidified with 0.5 M aqueous HCl (550 mL), added slowly at room temperature, and the mixture is transferred to a 1-L separatory funnel. The slightly acidic mixture is extracted with CH$_2$Cl$_2$ (3 x 100 mL). The combined organic layers are dried over 5 g of MgSO$_4$, then vacuum filtered (water aspirator, ~20 mmHg) through a cotton plug into a side-arm flask. The solvent is removed with a rotary evaporator to give 14.92-15.05 g (92-93% yield) of (S)-4-isopropyl-1,3-thiazolidine-2-thione (**1**) as a yellowish solid (mp 68–69 °C), which is used in the next step without further purification (Notes 7, 9).

Org. Synth. **2009**, *86*, 70-80
Published on the Web 10/2/2008

B.   *(S)-4-Isopropyl-N-propanoyl-1,3-thiazolidine-2-thione* (**2**).   An oven-dried three-necked 250-mL round-bottomed flask equipped with a magnetic stir bar is charged with (*S*)-4-isopropyl-1,3-thiazolidine-2-thione (**1**) (8.08 g, 50.0 mmol, 1.0 equiv).   The necks are fitted with a low temperature thermometer (with an adaptor), nitrogen line, and a rubber septum.   The flask is flushed with nitrogen, and charged with anhydrous THF (33 mL, Note 10). The resulting solution is cooled in a liquid nitrogen-ethyl acetate bath (–83 °C bath temperature; –78 °C internal temperature) and treated with a 1.52 M solution of *n*-BuLi in hexanes (36.5 mL, 55.5 mmol, 1.1 equiv; Note 11), added dropwise by syringe over 25 min so as to keep the internal temperature below –70 °C.   After 15 min, the heterogeneous solution is treated with propanoyl chloride (5.50 mL, 63.5 mmol, 1.3 equiv; Note 12), added dropwise by syringe over 10 min, keeping the internal temperature below –70 °C. The resulting clear solution is stirred for 5 min, the cold bath is removed and the solution is allowed to warm to room temperature and then stirred for 1.5 h. During this time, the solution becomes dark yellow/orange, and solids precipitate. The reaction mixture is cooled with an ice-water bath and then quenched with a saturated solution of NH$_4$Cl (20 mL) and water (50 mL). The resulting mixture is transferred to a 500-mL separatory funnel and extracted with CH$_2$Cl$_2$ (3 x 100 mL).   The combined organic layers are dried over 5 g of MgSO$_4$ then vacuum filtered (water aspirator, ~20 mmHg) through a cotton plug into a side-arm flask. Concentration of the filtrate with a rotary evaporator gives a bright yellow oil (Note 7). This residue is loaded on a column (6.5-cm diameter glass column) of silica gel (*ca* 150 g; Note 13) and eluted with 1:1 CH$_2$Cl$_2$/hexanes (2 L, Note 14) to afford 8.38-8.63 g (77-79% yield) of (*S*)-4-isopropyl-*N*-propanoyl-1,3-thiazolidine-2-thione (**2**) (Notes 15, 16).

2. Notes

1.   (*S*)-Valinol (96%) was purchased from Aldrich Chemical Company, Inc., and used as supplied. Alternatively, it can be prepared according to the experimental procedures reported by D. A. Dickman, A. I. Meyers, G. A. Smith, R. E. Gawley. *Org. Synth., Coll. Vol. 7*, **1990**, 530.

2.   Unless otherwise indicated, the solvents were of reagent quality and were used without further purification.

3.   Carbon disulfide (99%) was purchased from Aldrich Chemical Company, Inc., and used as supplied.

4. ***Caution:*** Carbon disulfide must be manipulated in a well-ventilated fume hood because of its stench and flammability.

5. The initially deep red solution becomes orange and pale yellow after 24 h and 36 h, respectively.

6. The reaction of (*S*)-valinol and carbon disulfide also affords (*S*)-4-isopropyl-1,3-oxazolidin-2-thione, which is slowly converted into the corresponding 1,3-thiazolidine-2-thione. Thus, long reaction times are required to fully transform 1,3-oxazolidin-2-thione intermediate into the desired heterocycle. After 48 h, only tiny amounts of 1,3-oxazolidin-2-thione are detected and the reaction time is expanded to three days to get a quantitative conversion. For a mechanistic analysis of this process, see reference 3.

7. Rotary evaporation was performed at 15 mmHg (vacuum pump) and the water bath temperature was 25 °C.

8. In order to minimize the unpleasant stench of carbon disulfide, a bubble trap filled with 50 mL of 1:10 bleach:water mixture was placed between the rotary evaporator and the vacuum pump. The bleach was purchased from Fisher and used as supplied.

9. No impurities were observed in the ^{1}H NMR of this material. The physical properties and spectral data for **1** are as follows: R_f = 0.25 (CH$_2$Cl$_2$); mp 68-69 °C [lit.[2] mp 67–68 °C, lit.[3] mp 66–67 °C]; $[\alpha]_D^{25.6}$ – 34.76 (*c* 1.11, CHCl$_3$), [lit.[2] $[\alpha]_D$ –36.81 (*c* 1.16, CHCl$_3$), lit.[3] $[\alpha]_D$ –34.6 (*c* 0.94, CHCl$_3$)]; IR (film): 3185, 2962, 1490, 1389, 1372, 1305, 1271, 1228, 1165, 1144, 1112, 1056, 1032, 980 cm^{-1}; ^{1}H NMR (500 MHz, CDCl$_3$) δ: 0.99 (d, J = 7.2 Hz, 3 H), 1.02 (d, J = 7.0 Hz, 3 H), 1.92–2.02 (m, 1H), 3.31 (dd, J = 11.0, 8.2 Hz, 1 H), 3.50 (dd, J = 11.0, 8.2 Hz, 1 H), 4.05 (td, J = 8.2, 7.8 Hz, 1 H), 8.13 (br s, 1 H); ^{13}C NMR (125 MHz, CDCl$_3$) δ: 18.3, 18.9, 32.1, 36.0, 70.1, 201.2; Anal. Calcd. for C$_6$H$_{11}$NS$_2$: C, 44.68; H, 6.87; N, 8.68; S, 39.76. Found: C, 45.01; H, 7.02; N, 8.77; S, 39.68; HRMS [M + H$^+$] calcd for C$_6$H$_{11}$NS$_2$: 162.0406. Found: 162.0402; MS (*m/z*) 160.8 (100%), 117.8 (55%), 58.9 (18%), 49.0 (47%).

10. Checkers purchased optima grade THF from Aldrich Chemical Company, Inc., and purified it over activated alumina. Submitters purchased reagent grade THF from Scharlau S.A., and freshly distilled it from sodium-benzophenone ketyl under nitrogen before use.

11. The *n*-BuLi was purchased as a 1.6 M solution in hexanes from Aldrich Chemical Company, Inc., and titrated with 4-biphenylmethanol before use.[4]

12.   Propionyl chloride (98%) was purchased from Aldrich Chemical Company, Inc., and distilled at atmospheric pressure under nitrogen before use.

13.   Silica gel 60 Å C.C. for column chromatography (35–70 μm) was purchased from SDS and used as received.

14.   The purification can be followed easily by the bright yellow color of the product. Care should be taken as a yellow colored impurity comes out of the column immediately after the product band.

15.   The submitter reported a yield of 98% for this reaction. The product should be kept in the fridge under nitrogen atmosphere to avoid undesired decompositions.

16.   The physical properties and spectral data for **2** are as follows: R_f = 0.45 (1:1 CH_2Cl_2/hexanes); HPLC (97:3 hexanes/EtOAc) t_R = 12.0 min; $[\alpha]_D^{25.6}$ +420.84 (*c* 1.01, $CHCl_3$); IR (film): 2964, 2938, 2875, 1700, 1462, 1394, 1350, 1314, 1280, 1260, 1169, 1123, 1086, 1038, 996, 914, 850, 806 cm^{-1}; ^{1}H NMR (500 MHz, $CDCl_3$) δ: 0.96 (d, J = 6.6 Hz, 3 H), 1.04 (d, J = 6.6 Hz, 3 H), 1.14 (t, J = 7.2 Hz, 3 H), 2.30–2.40 (m, 1 H), 3.02 (dd, J = 11.4, 1.2 Hz, 1 H), 3.13 (dq, J = 18.0, 7.2 Hz, 1 H), 3.34 (dq, J = 18.0, 7.2 Hz, 1 H), 3.49 (dd, J = 11.4, 8.1 Hz, 1 H), 5.15 (ddd, J = 8.1, 6.1, 1.2 Hz, 1 H); ^{13}C NMR (125 MHz, $CDCl_3$) δ: 9.1, 17.8, 19.1, 30.5, 30.9, 32.2, 71.7, 175.0, 202.8; Anal. Calcd. for $C_9H_{15}NOS_2$: C, 49.73; H, 6.96; N, 6.44; S, 29.51. Found: C, 49.54; H, 6.80; N, 6.28; S, 29.25; HRMS [M + Na$^+$] calcd for $C_9H_{15}NOS_2$: 240.0487. Found: 240.0486; MS (*m/z*) 217.0 (100%), 162.0 (54%), 117.9 (9%)

Safety and Waste Disposal Information

All hazardous materials should be handled and disposed of in accordance with "Prudent Practices in the Laboratory"; National Academy Press; Washington, DC, 1995.

3. Discussion

Since Nagao and Fujita established the synthetic utility of 1,3-thiazolidine-2-thiones as chiral auxiliaries,[2] they have commonly been prepared by heating an alkaline solution of the corresponding β-amino alcohol and carbon disulfide.[5] However, this apparently simple transformation must be carried out in a strongly basic medium, with an excess of carbon disulfide for extended reaction times. According to Le

Corre,[3] these stringent experimental conditions are required to convert the initially formed 1,3-oxazolidin-2-thione back into the β-amino alcohol and the intermediate **I** (see Scheme 1), which can undergo an intramolecular nucleophilic substitution that eventually leads to the desired 1,3-thiazolidine-2-thione.

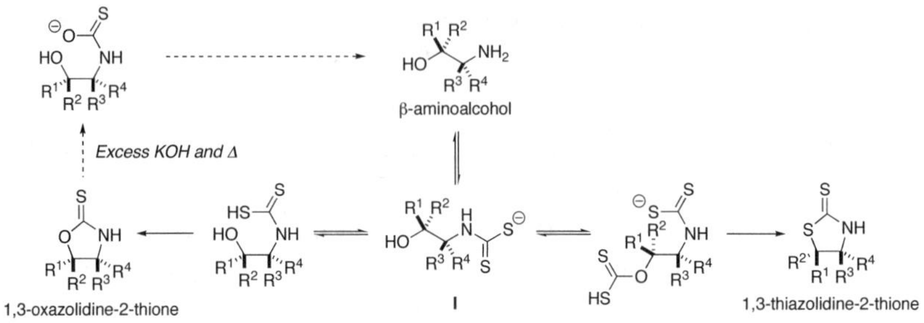

Scheme 1

The procedure disclosed herein has been adapted from the experimental conditions originally reported by Nagao and Fujita,[2] which involve refluxing an aqueous ethanol solution of a β-amino alcohol in the presence of a moderate excess of carbon disulfide and potassium hydroxide. The use of five equivalents of carbon disulfide in boiling aqueous potassium hydroxide has also been reported.[3] Both procedures allow the preparation of a wide range of enantiomerically pure 1,3-thiazolidine-2-thiones in good yields, except for those starting materials containing a tertiary alcohol (R^1, $R^2 \neq H$ in Scheme 1).[6] Interestingly, the synthesis of the 1,3-thiazolidine-2-thione derived from *tert*-leucinol (Scheme 2) requires harsher conditions due to the steric hindrance imposed by the bulky *tert*-butyl group.[7]

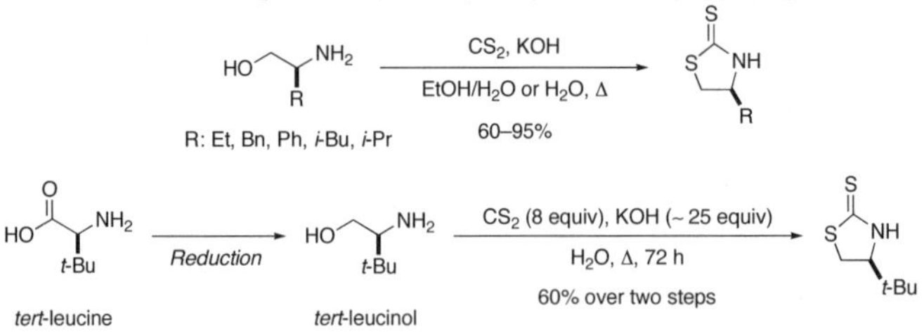

Scheme 2

Acylation of these chiral 1,3-thiazolidine-2-thiones can be accomplished using different methodologies. The most straightforward

Org. Synth. **2009**, *86*, 70–80

procedures take advantage of the acidity of the NH proton. Thus, treatment of these heterocycles with bases as *n*-BuLi,[5,7,8] NaH,[9] or Et$_3$N[10] and the appropriate acyl chloride furnish the corresponding *N*-acyl derivatives in high yields.[11]

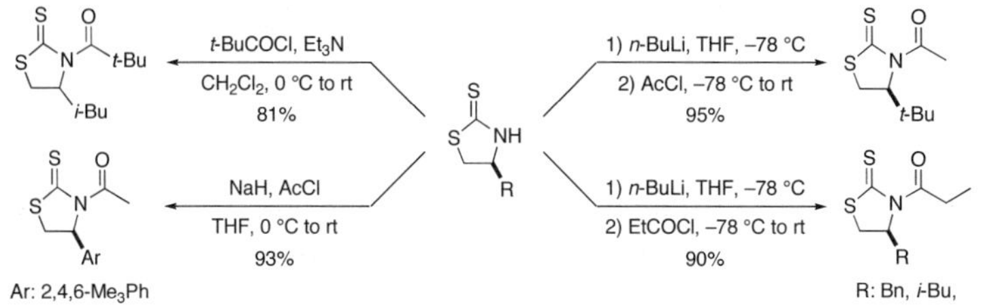

Scheme 3

Ar: 2,4,6-Me$_3$Ph

R: Bn, *i*-Bu,

Carbodiimides can also be used to prepare the desired acylated auxiliaries from the corresponding thiazolidinethiones and carboxylic acids. Nagao and Fujita first applied these coupling reactions to the acylation of thiazolidinethiones[10b,12] and, more recently, some communications have generalized such procedures.[13]

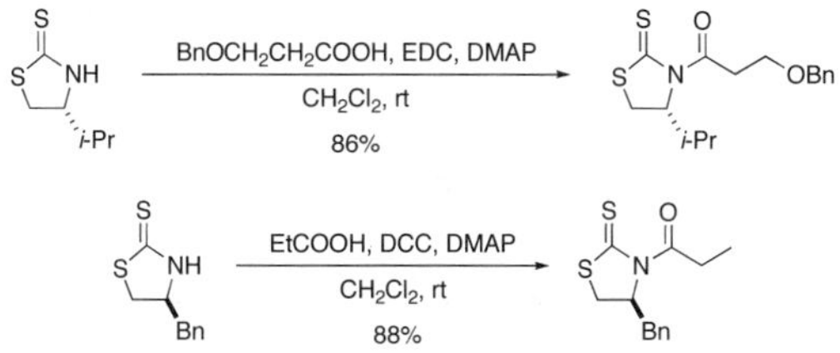

Scheme 4

The synthetic utility of the thiazolidinethiones as chiral auxiliaries is mainly in the stereoselective construction of carbon-carbon bonds, particularly in the aldol arena. Indeed, early reports on this area dealt with the tin-mediated aldol addition of *N*-acetyl-4-isopropyl-1,3-thiazolidine-2-thione to α,β-unsaturated aldehydes, finding that this auxiliary imparted a higher degree of stereocontrol than the corresponding 1,3-oxazolidin-2-one.[2] Later, Urpí and Vilarrasa expanded this methodology to titanium enolates.[14] More recently, outstanding diastereoselectivities have been achieved with a broad array of aldehydes using boron[5,7] or titanium[9] enolates from different thiazolidinethiones.

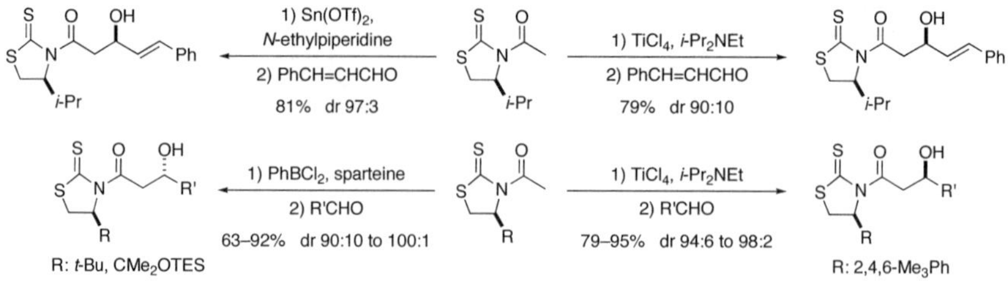

Scheme 5

Aldol reactions from other N-acyl-1,3-thiazolidine-2-thiones have also been highly successful. Interestingly, Crimmins proved that titanium enolates give access to both *syn*-aldol adducts depending on the enolization conditions,[8,15] whereas an *anti*-aldol adduct is available from aromatic and unsaturated aldehydes through a magnesium-catalyzed procedure developed by Evans.[16] Thus, up to three of the four possible aldol adducts can be prepared from a single precursor by the appropriate choice of enolization procedure.[17]

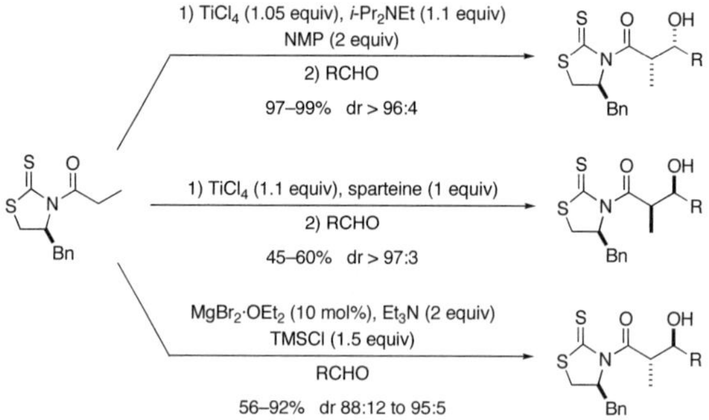

Scheme 6

Other applications of the chiral thiazolidinethiones involve the stereoselective construction of carbon-carbon bonds through the addition of their metal enolates to iminium intermediates,[12c,18] oximes,[19] acetals,[20] and glycals.[21] Moreover, simple N-acyl-1,3-thiazolidine-2-thiones lacking any substituent on the heterocycle can also participate in asymmetric transformations in the presence of catalytic amounts of chiral Lewis acids.[22]

Finally, it is worth mentioning that the value of the 1,3-thiazolidine-2-thiones as auxiliaries in asymmetric synthesis is not restricted to the stereochemical control imparted in the aforementioned transformations. Indeed, additional advantages of these auxiliaries rely in the ease of removal

Org. Synth. **2009**, *86*, 70-80

under very mild conditions and the bright yellow color that characterize such acylated heterocycles, which facilitates the analysis and purification of the reaction mixtures and the isolation of a broad array of enantiomerically pure synthons.[10c,20,23]

1. Departament de Química Orgànica, Universitat de Barcelona, Martí i Franqués 1-11, 08028 Barcelona, Catalonia, Spain. E-mail: pedro.romea@ub.edu; felix.urpi@ub.edu.

2. Nagao, Y.; Hagiwara, Y.; Kumagai, T.; Ochiai, M.; Inoue, T.; Hashimoto, K.; Fujita, E. *J. Org. Chem.* **1986**, *51*, 2391–2393.

3. Delaunay, D.; Toupet, L.; Le Corre, M. *J. Org. Chem.* **1995**, *60*, 6604–6607.

4. Juaristi, E.; Martínez-Richa, A.; García-Rivera, A.; Cruz-Sánchez, J. S. *J. Org. Chem.* **1983**, *48*, 2603–2606.

5. Occasionally, 1,3-thiazolidine-2-thiones are prepared from β-amino thiols, see: Zhang, Y.; Sammakia, T. *Org. Lett.* **2004**, *6*, 3139–3141.

6. For a review, see: Velázquez, F.; Olivo, H. F. *Curr. Org. Chem.* **2002**, *6*, 303–340.

7. a) Zhang, Y.; Phillips, A. J.; Sammakia, T. *Org. Lett.* **2004**, *6*, 23–25. b) Zhang, Y.; Sammakia, T. *J. Org. Chem.* **2006**, *71*, 6262–6265.

8. Crimmins, M. T.; King, B. W.; Tabet, E. A.; Chaudhary, K. *J. Org. Chem.* **2001**, *66*, 894–902.

9. Crimmins, M. T.; Shamszad, M. *Org. Lett.* **2007**, *9*, 149–152.

10. a) Yamada, S. *J. Org. Chem.* **1996**, *61*, 941–946. See also: b) Nagao, Y.; Kawabata, K.; Seno, K.; Fujita, E. *J. Chem. Soc., Perkin I* **1980**, 2470–2473. c) Crimmins, M. T.; Chaudhary, K. *Org. Lett.* **2000**, *2*, 775–777. d) Sun, Y.-P.; Wu, Y. *Synlett* **2005**, 1477–1479.

11. Mixed anhydrides have also been used as acylating reagents, see: Wu, Y.; Shen, X.; Yang, Y.-Q.; Hu, Q.; Huang, J.-H. *J. Org. Chem.* **2004**, *69*, 3857–3865.

12. a) Nagao, Y.; Ikeda, T.; Yagi, M.; Fujita, E.; Shiro, M. *J. Am. Chem. Soc.* **1982**, *104*, 2079–2081. b) Nagao, Y.; Miyasaka, T.; Seno, K.; Fujita, E.; Shibata, D.; Doi, E. *J. Chem. Soc., Perkin Trans. 1* **1984**, 2439–2446. c) Nagao, Y.; Nagase, Y.; Kumagai, T.; Matsunaga, H.; Abe, T.; Shimada, O.; Hayashi, T.; Inoue, Y. *J. Org. Chem.* **1992**, *57*, 4243–4249.

13. a) Andrade, C. K. Z.; Rocha, R. O.; Vercillo, O. E.; Silva, W. A.; Matos, R. A. F. *Synlett* **2003**, 2351–2352. b) Franck, X.; Langlois, E.; Outurquin, F. *Synthesis* **2007**, 719–724.

14. González, A.; Aiguadé, J.; Urpí, F.; Vilarrasa, J. *Tetrahedron Lett.* **1996**, *37*, 8949–8952.

15. Crimmins, M. T.; She, J. *Synlett* **2004**, 1371–1374.

16. a) Evans, D. A.; Tedrow, J. S.; Shaw, J. T.; Downey, C. W. *J. Am. Chem. Soc.* **2002**, *124*, 392–393. b) Evans, D. A.; Downey, C. W.; Shaw, J. T.; Tedrow, J. S. *Org. Lett.* **2002**, *4*, 1127–1130.

17. For boron-mediated aldol reactions from chiral 1,3-thiazolidine-2-thiones, see: Hsiao, C.-N.; Liu, L.; Miller, M. J. *J. Org. Chem.* **1987**, *52*, 2201–2206.

18. a) Nagao, Y.; Kumagai, T.; Tamai, S.; Abe, T.; Kuramoto, Y.; Taga, T.; Aoyagi, S.; Nagase, Y.; Ochiai, M.; Inoue, Y.; Fujita, E. *J. Am. Chem. Soc.* **1986**, *108*, 4673–4675. b) Barragán, E.; Olivo, H. F.; Romero-Ortega, M.; Sarduy, S. *J. Org. Chem.* **2005**, *70*, 4214–4217.

19. Ambhaikar, N. B.; Snyder, J. P.; Liotta, D. C. *J. Am. Chem. Soc.* **2003**, *125*, 3690–3691.

20. a) Cosp, A.; Romea, P.; Talavera, P.; Urpí, F.; Vilarrasa, J.; Font-Bardia, M.; Solans, X. *Org. Lett.* **2001**, *3*, 615–617. b) Cosp, A.; Larrosa, I.; Vilasís, I.; Romea, P.; Urpí, F.; Vilarrasa, J. *Synlett* **2003**, 1109–1112.

21. Larrosa, I.; Romea, P.; Urpí, F.; Balsells, D.; Vilarrasa, J.; Font-Bardia, M.; Solans, X. *Org. Lett.* **2002**, *4*, 4651–4654.

22. a) Evans, D. A.; Miller, S. J.; Lectka, T. *J. Am. Chem. Soc.* **1993**, *115*, 6460–6461. b) Evans, D. A.; Downey, C. W.; Hubbs, J. L. *J. Am. Chem. Soc.* **2003**, *125*, 8706–8707. c) Evans, D. A.; Thomson, R. J. *J. Am. Chem. Soc.* **2005**, *127*, 10506–10507. d) Evans, D. A.; Thomson, R. J.; Franco, F. *J. Am. Chem. Soc.* **2005**, *127*, 10816–10817.

23. Wu, Y.; Sun, Y.-P.; Yang, Y.-Q.; Hu, Q.; Zhang, Q. *J. Org. Chem.* **2004**, *69*, 6141–6144.

Appendix
Chemical Abstracts Nomenclature; (Registry Number)

(*S*)-Valinol: 1-Butanol, 2-amino-3-methyl-, (2*S*)-; (2026-48-4)
Carbon disulfide; (75-15-0)
(*S*)-4-Isopropyl-1,3-thiazolidine-2-thione: 2-Thiazolidinethione, 4-(1-

methylethyl)-, (4*S*)-; (76186-04-4)

n-Butyllithium; (109-72-8)

Propanoyl chloride; (79-03-8)

(*S*)-4-Isopropyl-*N*-propanoyl-1,3-thiazolidine-2-thione: 2-
   Thiazolidinethione, 4-(1-methylethyl)-3-(1-oxopropyl)-, (4*S*)-;
   (102831-92-5)

Pedro Romea completed his B.Sc. in Chemistry in 1984 at the University of Barcelona. That year he joined the group of Professor Jaume Vilarrasa, at the University of Barcelona, receiving his Masters Degree in 1985, and he followed Ph.D. studies in the same group from 1987 to 1991. Then, he joined the group of Professor Ian Paterson at the University of Cambridge (UK), where he participated in the total synthesis of oleandolide. Back to the University of Barcelona, he became Associate Professor in 1993. His research interests have focused on the development of new synthetic methodologies and their application to the stereoselective synthesis of naturally occurring molecular structures.

Fèlix Urpí received his B.Sc. in Chemistry in 1980 at the University of Barcelona. In 1981, he joined the group of Professor Jaume Vilarrasa, at the University of Barcelona, receiving his Masters Degree in 1981 and Ph.D. in 1988, where he was an Assistant Professor. He then worked as a NATO postdoctoral research associate in titanium enolate chemistry with Professor David A. Evans, at Harvard University in Boston. He moved back to the University of Barcelona and he became Associate Professor in 1991. His research interests have focused on the development of new synthetic methodologies and their application to the stereoselective synthesis of naturally occurring molecular structures.

 Erik Gálvez was born in Barcelona, Spain, in 1982. He received his B.Sc. in Chemistry in 2005 at the Universitat de Barcelona and joined the group of Pedro Romea and Fèlix Urpí. In 2006 he received his Masters Degree and is now pursuing the Ph.D. in the same group. His research concerns asymmetric methodologies involving cross-coupling reactions using 1,3-thiazolidine-2-thiones as source of chirality.

 John Jewett was born in Vermont in 1980. He received his A.B. in biophysical chemistry from Dartmouth College in 2003, and in the same year went to the University of Chicago for graduate school. He completed his Ph.D. in chemistry with Prof. Viresh Rawal in 2008, working on the total synthesis of various members of pederin family of natural products. In 2008 he began a postdoctoral position at the University of California, Berkeley with Prof. Carolyn Bertozzi.

STEREOSELECTIVE SYNTHESIS OF
anti α-METHYL-β-METHOXY CARBOXYLIC COMPOUNDS

A.

B.

C.

Submitted by Erik Gálvez, Pedro Romea, and Fèlix Urpí.[1]
Checked by Vijaya Bhasker Gondi and Viresh H. Rawal.

1. Procedure

A. *[(E)-3,3-dimethoxy-2-methyl-1-propenyl]benzene* (**1**).   An oven-dried 25-mL round-bottomed flask equipped with a magnetic stir bar is charged with (*E*)-2-methyl-3-phenylpropenal (6.0 mL, 43 mmol, 1.0 equiv) (Note 1) and Amberlyst 15 (50 mg) (Note 2). The flask is fitted with a rubber septum, flushed with nitrogen, cooled in an ice-water bath, and charged with trimethyl orthoformate (5.8 mL, 53 mmol, 1.2 equiv) (Note 3) and dry methanol (1.0 mL, 25 mmol, 0.56 equiv) (Note 4). The reaction mixture is stirred at room temperature for 36 h. The resin is removed by filtration through a cotton plug, and the volatiles are removed using a rotary evaporator (Note 5). The resultant oil is purified by short path vacuum distillation. The main fraction is collected at 120–130 °C (2.7 mmHg) and provides 6.48 g (78% yield) of the desired *[(E)-3,3-dimethoxy-2-methyl-1-propenyl]benzene* (**1**) (Notes 6, 7, 8).
B.   *(4S)-N-[(2R,3S,4E)-2,4-Dimethyl-3-methoxy-5-phenyl-4-pentenoyl]-4-isopropyl-1,3-thiazolidine-2-thione* (**2**). An oven-dried 250-mL round-bottomed flask equipped with a magnetic stir bar is charged with

(S)-4-isopropyl-N-propanoyl-1,3-thiazolidine-2-thione (2.20 g, 10.1 mmol, 1.0 equiv) (Note 9). The flask is fitted with a rubber septum, flushed with nitrogen, and charged with anhydrous CH_2Cl_2 (80 mL) (Note 10). The stirred solution is cooled in an ice-water bath, and neat $TiCl_4$ (1.2 mL, 11 mmol, 1.1 equiv) (Note 11) is added dropwise by syringe over 1 min, which causes the formation of a yellow solid. The resulting suspension is stirred for 5 min, cooled with a liquid nitrogen-ethyl acetate bath (–83 °C), and a solution of dry diisopropylethylamine (1.83 mL, 11.0 mmol, 1.08 equiv) (Note 12) in CH_2Cl_2 (5 mL) is added dropwise *via* canula over 5 min, which produces a deep red homogeneous solution. An additional 5 mL of CH_2Cl_2 are used to transfer the last traces of diisopropylethylamine to the reaction flask. The reaction mixture is stirred in the liquid nitrogen-ethyl acetate bath for 30 min, then transferred to a cryocool (or acetone-dry ice bath) (bath temperature –50 °C to –55 °C) for 2 h. The reaction mixture is cooled again in a liquid nitrogen-ethyl acetate bath, and $BF_3·OEt_2$ (1.4 mL, 11 mmol, 1.1 equiv) (Note 13) is added dropwise by syringe over 1 min. After 5 min, a cooled (liquid nitrogen-ethyl acetate bath) solution of *[(E)-3,3-dimethoxy-2-methyl-1-propenyl]benzene* (1) (2.11 g, 10.9 mmol, 1.08 equiv) in CH_2Cl_2 (5 mL) is added dropwise *via* cannula over 5 min. An additional 5 mL of CH_2Cl_2 are used to transfer the last traces of 1 to the reaction flask. After stirring for 2 h in the liquid nitrogen-ethyl acetate bath, the reaction mixture is quenched with a saturated solution of NH_4Cl (80 mL). The mixture is allowed to warm to room temperature, then transferred to a 250-mL separatory funnel. The aqueous layer is separated and extracted with CH_2Cl_2 (3 × 50 mL). The combined organic layers are dried ($MgSO_4$), filtered through a cotton plug under water aspirator pressure and concentrated using a rotary evaporator to afford a bright yellow oil (Note 5), a 95:5 ratio of diastereomers (Note 14). The crude product is dissolved in 5 mL of CH_2Cl_2 and loaded on a flash chromatography column of deactivated silica gel (6.5-cm diameter glass column and *ca* 450 g of silica) (Note 15). The column is eluted using a gradient solvent system of hexanes-CH_2Cl_2, 4:1 to 3:1 to 2:1. The fractions containing the desired product are combined and concentrated by rotary evaporator to afford 3.42 g (9.06 mmol, 89% yield) of 2 as a bright yellow oil (Notes 16, 17, 18).

C. *(2R,3S,4E)-N,3-Dimethoxy-N,2,4-trimethyl-5-phenyl-4-pentenamide* (3). An oven-dried 25-mL round-bottomed flask equipped with a magnetic stir bar is charged with (4S)-N-[(2R,3S,4E)-2,4-dimethyl-3-methoxy-5-phenyl-4-pentenoyl]-4-isopropyl-1,3-thiazolidine-2-thione   (2)

(1.85 g, 4.89 mmol, 1.0 equiv). The flask is fitted with a rubber septum, flushed with nitrogen, and charged with anhydrous CH_2Cl_2 (10 mL, Note 10). The septum is temporarily removed and N,O-dimethylhydroxylamine hydrochloride (733 mg, 7.50 mmol, 1.53 equiv) (Note 19) and 4-dimethylaminopyridine (610 mg, 5.0 mmol, 1.02 equiv) (Note 20) are quickly added. The septum is replaced, the flask is flushed again with nitrogen, and dry triethylamine (0.75 mL, 5.0 mmol, 1.0 equiv) (Note 21) is added by syringe. The resulting mixture is stirred at room temperature for 15 h, over which period the initial deep yellow solution gradually fades to become almost colorless. The mixture is diluted with CH_2Cl_2 (35 mL), transferred to a 100-mL separatory funnel, and washed successively with 10% citric acid (3 × 30 mL), 1 M NaOH (4 × 30 mL) (Note 22) and brine (30 mL). The organic layer is dried ($MgSO_4$), filtered through a cotton plug under water aspirator pressure, and concentrated with a rotary evaporator to afford a yellowish oil (Note 5). The oily residue is dissolved in 2.5 mL of CH_2Cl_2 and charged on a column (4.1-cm diameter) of silica gel (*ca* 50 g) and eluted with 3:2 hexanes/EtOAc. The desired product is obtained in fractions 10–25 (*ca* 20 mL/fraction), which are concentrated by rotary evaporator (Note 5) to deliver 1.15 g (4.14 mmol, 83% yield) of (2*R*,3*S*,4*E*)-*N*,3-dimethoxy-*N*,2,4-trimethyl-5-phenyl-4-pentenamide (**3**) as a viscous colorless oil that solidifies upon standing in the refrigerator (Note 23). To recover the chiral auxiliary, the basic aqueous layer is treated with 2 M aqueous HCl (60 mL), and the resultant mixture is transferred to a 250-mL separatory funnel and extracted with CH_2Cl_2 (3 × 50 mL). The combined organic layers are dried ($MgSO_4$), filtered through a cotton plug under water aspirator pressure, and concentrated with a rotary evaporator (Note 5) to afford the thiazolidine auxiliary (638 mg, 81%) as a yellowish solid (mp 67–68 °C) that can be reused.

2. Notes

1.   (*E*)-2-Methyl-3-phenylpropenal was purchased from Aldrich Chemical Company, Inc., and used as supplied.

2.   Amberlyst 15 was purchased from Aldrich Chemical Company, Inc., and used as supplied.

3.   Trimethyl orthoformate was purchased from Aldrich Chemical Company, Inc., and used as supplied.

4. Methanol was purchased from Aldrich Chemical Company, Inc., and distilled over magnesium.

5. Rotary evaporation was performed at 10 mmHg (vacuum pump) with the water bath temperature at 25 °C.

6. By HNMR analysis, the crude product is an 18:1 mixture of $E{:}Z$ diastereomers. After distillation, a 10:1 ratio of $E{:}Z$ diastereomers is present. Fractional distillation using a Vigreux column gives the product in higher E/Z ratio (>20:1) but in lower yield (38-43%). However, similar yield and dr were obtained when the second step was performed with acetals of 10:1 or 20:1 dr.

7. The submitters reported a lower bp: 75–80 °C (2.5 mmHg).

8. The physical properties and spectral data for the major diastereomer (1) are as follows: ^{1}H NMR (500 MHz, CDCl$_3$) δ: 1.89 (d, J = 1.2 Hz, 3 H), 3.38 (s, 6 H), 4.65 (d, J = 1.2 Hz, 1 H), 6.63 (br s, 1 H), 7.24–7.36 (m, 5 H); ^{13}C NMR (75.4 MHz, CDCl$_3$) δ: 13.0, 53.6, 107.7, 126.8, 128.1, 128.5, 129.1, 134.4, 137.0; IR (film) cm^{-1}: 2987, 2932, 2827, 1601, 1445, 1347, 1196, 1073; HRMS calcd for C$_{12}$H$_{16}$O$_2$Na (M$^+$+Na): 215.1043, found 215.1041. Anal. calcd for C$_{12}$H$_{16}$O$_2$: C, 74.97; H, 8.39; O, 16.64. Found: C, 75.22; H, 8.42; O, 16.58.

9. The thiazolidine thione was prepared by the method described in the accompanying procedure.

10. Dichloromethane was freshly distilled over calcium hydride. Checkers used the dichloromethane from a solvent purification system (activated alumina column).

11. TiCl$_4$ (reagent plus, 99.9%) was purchased from Aldrich Chemical Company, Inc., and used as supplied.

12. Diisopropylethylamine was purchased from Aldrich Chemical Company, Inc., and freshly distilled over calcium hydride or KOH (checkers).

13. BF$_3$·OEt$_2$ (purified, redistilled grade) was purchased from Aldrich Chemical Company, Inc., and used as supplied.

14. The checkers analyzed the crude product by ^{1}H NMR to determine the ratio of diastereomers. The submitters observed a 96:4 ratio of diastereomers by HPLC. The following HPLC conditions were used by the submitters. Detector: 254 nm; column: Tracer (250 × 4 mm) Spherisorb W Silica 5 μm; eluent: 97:3 hexanes/EtOAc; flow rate: 0.9 mL min^{-1}; retention times: t_R (major diastereomer) = 14.3 min; t_R (minor diastereomer) = 19.1 min.

15. The column is wet-loaded with flash grade silica gel (40-63 μm) using 4:1 hexanes-CH$_2$Cl$_2$ solvent mixture containing 3% Et$_3$N by volume. Further solvent mixtures used for running the column are also treated with 3% Et$_3$N. The submitters followed a different method for loading and running the column: Deactivated silica is prepared by addition of CH$_2$Cl$_2$ (300 mL) to SiO$_2$ (200 g) kept in a 1-L round-bottomed flask. After shaking carefully, dry triethylamine (5 mL) was added followed by additional CH$_2$Cl$_2$ (200 mL). The mixture is carefully shaken and the solvent is removed with a rotary evaporator (*Attention*: a cotton plug is placed at the neck of the flask to keep silica gel from blowing into the evaporator).

16. The product should be kept in the refrigerator under a nitrogen atmosphere to minimize decomposition.

17. The physical properties and spectral data for **2**: $[\alpha]\frac{25}{D}$ +189.7 (*c* 1.05, CHCl$_3$), $[\alpha]_D$ +178.7 (*c* 1.2, CHCl$_3$, submitters); TLC R$_f$ 0.83 (CH$_2$Cl$_2$); HPLC (97:3 hexanes/EtOAc) t_R = 14.3 min (submitters); ^{1}H NMR (400 MHz, CDCl$_3$) δ: 1.01 (d, *J* = 6.8 Hz, 3 H), 1.03 (d, *J* = 6.6 Hz, 3 H), 1.08 (d, *J* = 6.6 Hz, 3 H), 1.85 (br s, 3 H), 2.32–2.40 (m, 1 H), 2.99 (dd, *J* = 11.4, 2.1 Hz, 1 H), 3.16 (s, 3 H), 3.45 (dd, *J* = 11.4, 8.7 Hz, 1 H), 3.92 (d, *J* = 9.9 Hz, 1 H), 5.21 (dq, *J* = 9.9, 7.0 Hz, 1 H), 5.34 (ddd, *J* = 8.7, 5.4, 2.1 Hz, 1 H), 6.52 (br s, 1 H), 7.35–7.24 (m, 5 H); ^{13}C NMR (125 MHz, CDCl$_3$) δ: 12.1, 14.3, 16.8, 19.0, 28.7, 30.3, 41.3, 55.9, 71.9, 91.8, 126.7, 128.1, 129.0, 131.1, 135.0, 137.0, 177.8, 202.6; IR (film) cm^{-1}: 2940, 1690, 1440, 1365, 1240, 1150; HRMS calcd for C$_{20}$H$_{27}$NO$_2$S$_2$Na (M$^+$+Na): 400.1375, found: 400.1366. Anal. calcd for C$_{20}$H$_{27}$NO$_2$S$_2$: C, 63.62; H, 7.21; N, 3.71; S, 16.99. Found: C, 63.42; H, 6.97; N, 3.69; S, 17.00.

18. Checkers found that later fractions contain the starting aldehyde, (*E*)-2-methyl-3-phenylpropenal, and a minor diastereomer of **2**. The R$_f$ for the aldehyde is 0.77 (CH$_2$Cl$_2$). Properties of the minor diastereomer: R_f = 0.71 (CH$_2$Cl$_2$); ^{1}H NMR (500 MHz, CDCl$_3$) δ: 0.84 (d, *J* = 6.9 Hz, 3 H, CH$_3$CC**H**$_3$), 0.89 (d, *J* = 6.8 Hz, 3 H, COCHC**H**$_3$), 1.28 (d, *J* = 6.8 Hz, 3 H, C**H**$_3$CCH$_3$), 1.87 (br s, 3 H, (C**H**$_3$)C=CHPh), 2.12–2.16 (m, 1 H, C**H**(CH$_3$)$_2$), 2.90 (d, *J* = 11.5 Hz, 1 H, SCH$_a$C**H**$_b$), 3.28 (s, 3 H, OC**H**$_3$), 3.42 (m, 1 H, SC**H**$_a$H$_b$), 4.07 (d, *J* = 8.1 Hz, 1 H, C**H**OCH$_3$), 5.28 (m, 2 H, NC**H** and COC**H**CH$_3$), 6.51 (br s, 1 H, PhC**H**=C), 7.20–7.33 (m, 5H, Ar**H**). In one run, the checkers also isolated a very small amount (<1%) of another diastereomer, tentatively assigned to be the product of aldol reaction with the *Z* isomeric acetal.

19. *N,O*-Dimethylhydroxylamine hydrochloride (98%) was purchased from Aldrich Chemical Company, Inc., and kept overnight under vacuum before use.

20. 4-Dimethylaminopyridine (99%) was purchased from Aldrich Chemical Company, Inc., and used as supplied.

21. Triethylamine was purchased from Aldrich Chemical Company, Inc., and freshly distilled over calcium hydride.

22. The basic aqueous solution contains the deprotonated chiral auxiliary.

23. The physical properties and spectral data for **3** are as follows: $[\alpha]_{D}^{25}$ +58.0 (*c* 1.18, CHCl$_3$); TLC R$_f$ 0.40 (hexanes/EtOAc, 3:2); HPLC (85:15 hexanes/EtOAc) t_R = 23.4 min; chiral HPLC (97:3 hexanes/*i*-PrOH) t_R = 8.2 min; mp 40–42 °C; ^{1}H NMR (400 MHz, CDCl$_3$) δ: 0.99 (d, *J* = 7.0 Hz, 3 H), 1.83 (br s, 3 H), 3.20 (s, 3 H), 3.25 (s, 3 H), 3.23–3.25 (m, 1 H), 3.76 (s, 3 H), 3.86 (d, *J* = 10.2 Hz, 1 H), 6.54 (br s, 1 H), 7.23–7.37 (m, 5 H); ^{13}C NMR (125 MHz, CDCl$_3$) δ: 11.9, 14.3, 32.1, 37.5, 56.2, 61.4, 89.9, 126.7, 128.1, 129.0, 131.0, 135.0, 137.1, 175.0; IR (film) cm^{-1}: 2976, 2935, 2819, 1660, 1448, 1418, 1386, 1178; HRMS calcd. for C$_{16}$H$_{23}$NNaO$_3$ [M+Na]$^+$ 300.1576, found 300.1563. Anal. calcd for C$_{16}$H$_{23}$NO$_3$: C, 69.29; H, 8.36; N, 5.05; Found: C, 68.87; H, 8.13; N, 5.11.

Safety and Waste Disposal Information

All hazardous materials should be handled and disposed of in accordance with "Prudent Practices in the Laboratory"; National Academy Press; Washington, DC, 1995.

3. Discussion

The reported experimental procedure can be also applied to other dimethyl acetals from aromatic and α,β-unsaturated aldehydes (see entries 1–6 in Table 1), whereas less reactive substrates, such as acetals from deactivated aromatic or aliphatic aldehydes, require a stronger Lewis acid (SnCl$_4$) and higher temperatures to obtain similar yields and diastereomeric ratios (see entries 7–10 in Table 1). Thus, the addition of the titanium enolate from (*S*)-4-isopropyl-*N*-propanoyl-1,3-thiazolidine-2-thione to a wide range of dimethyl acetals in the presence of a Lewis acid (BF$_3$·OEt$_2$ or

$SnCl_4$) constitutes an efficient one-step entry to the stereoselective synthesis of *anti* β-methoxy-α-methyl carboxylic adducts.[2]

Table 1. Stereoselective synthesis of *anti* β-methoxy-α-methyl carboxylic adducts

Entry	R	Lewis acid	Reactions conditions (T, t)	dr[a] 2,3-*anti*/2,3-*syn*	Yield (%)[b]
1	(*E*)-PhCH=C(CH₃)	BF₃·OEt₂	−78 °C, 2.5 h	96:4	94
2[c]	H—≡—Co₂(CO)₆	BF₃·OEt₂	−78 °C, 2.5 h	99:1	84
3	Ph	BF₃·OEt₂	−78 °C, 2.5 h	86:14	75
4	4-CH₃OPh	BF₃·OEt₂	−78 °C, 2.5 h	81:19	77
5	3-CH₃OPh	BF₃·OEt₂	−78 °C, 2.5 h	92:8	79
6	4-ClPh	BF₃·OEt₂	−78 °C, 2.5 h	91:9	81
7	4-NO₂Ph	SnCl₄	−78 °C, 2 h	86:14	70
8	CH₃CH₂CH₂	SnCl₄	−50 °C, 2 h	93:7	64[d]
9	(CH₃)₂CHCH₂	SnCl₄	−20 °C, 2 h	92:8	76
10	(CH₃)₂CH	SnCl₄	−20 °C, 2 h	88:12	50

a Established by HPLC. *b* Isolated yield at 1 mmol scale. *c* Diethyl acetal was used. *d* Isolated yield of the corresponding ethyl ester.

As shown by the preparation of the Weinreb amide **3**, one of the most appealing advantages of the 1,3-thiazolidine-2-thione auxiliaries is that they are removed under very mild conditions.[3,4] Indeed, the above mentioned adducts are easily transformed into a large number of enantiopure 1,3-dioxygenated compounds with high interest in organic synthesis (Scheme 1).

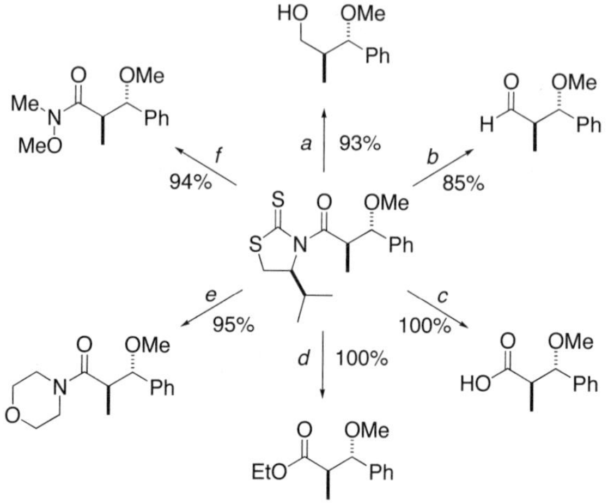

Scheme 1. *Experimental conditions: (a)* NaBH$_4$ (4.5 equiv), THF-H$_2$O, rt, 4 h. *(b)* DIBAL-H (1.1 equiv), CH$_2$Cl$_2$, -78 °C, 3 h. *(c)* LiOH·H$_2$O (6 equiv), CH$_3$CN-H$_2$O, rt, 12 h. *(d)* EtOH, DMAP cat., rt, 24 h. *(e)* Morpholine (4 equiv), THF, rt, 12 h. *(f)* MeONHMe·HCl (1.5 equiv), Et$_3$N (1 equiv), DMAP cat., CH$_2$Cl$_2$, rt, 24 h.

Further studies have established that this methodology can be generalized to other substrates. Dibenzyl acetals afford in high yields and diastereomeric ratios the corresponding *anti* adducts, which may be considered as protected *anti* aldol structures (Scheme 2). Importantly, *matched* pairs in double asymmetric reactions with chiral dibenzyl acetals deliver the Felkin adduct as the sole diastereomer.[5]

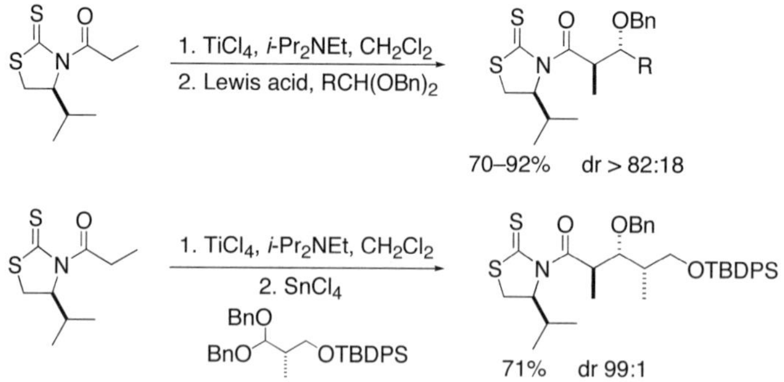

Scheme 2

88

Finally, this methodology has been applied to titanium enolates from (*S*)-*N*-acetyl-4-isopropyl-1,3-thiazolidine-2-thione (Scheme 3). Compared to the results summarized in Table 1, the diastereoselectivity of this process is slightly lower, but it achieves sufficiently good results to be used in asymmetric synthesis,[6] and has been successfully applied to the stereoselective synthesis of the C9–C21 fragment of debromoaplysiatoxin.[7]

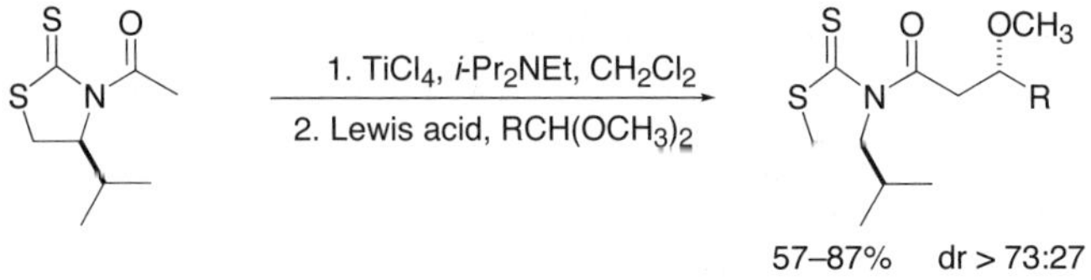

1. TiCl$_4$, *i*-Pr$_2$NEt, CH$_2$Cl$_2$
2. Lewis acid, RCH(OCH$_3$)$_2$

57–87%   dr > 73:27

Scheme 3

1. Departament de Química Orgànica, Universitat de Barcelona, Martí i Franqués 1-11, 08028 Barcelona, Catalonia, Spain. E-mail: pedro.romea@ub.edu; felix.urpi@ub.edu.
2. Cosp, A.; Romea, P.; Talavera, P.; Urpí, F.; Vilarrasa, J.; Font-Bardia, M.; Solans, X. *Org. Lett.* **2001**, *3*, 615–617.
3. Chiral 1,3-thiazolidine-2-thiones were introduced and identified as easy removable chiral auxiliaries in asymmetric synthesis by Nagao and Fujita. For instance, see: (a) Reference 2. (b) Nagao, Y.; Dai, W.-M.; Ochiai, M.; Tsukagoshi, S.; Fujita, E. *J. Org. Chem.* **1990**, *55*, 1148–1156.
4. For a review on the application of thiazolidinethiones in asymmetric synthesis, see: Velázquez, F.; Olivo, H. F. *Curr. Org. Chem.* **2002**, *6*, 303-340.
5. Cosp, A.; Larrosa, I.; Vilasís, I.; Romea, P.; Urpí, F.; Vilarrasa, J. *Synlett* **2003**, 1109–1112.
6. Cosp, A.; Romea, P.; Urpí, F.; Vilarrasa, J. *Tetrahedron Lett.* **2001**, *42*, 4629–4631.
7. Cosp, A.; Llàcer, E.; Romea, P.; Urpí, F. *Tetrahedron Lett.* **2006**, *47*, 5819–5823.

Appendix
Chemical Abstracts Nomenclature; (Registry Number)

(*E*)-2-Methyl-3-phenylpropenal; (15174-47-7)

Trimethyl orthoformate; (149-73-5)

(*E*)-3,3-Dimethoxy-2-methyl-1-propenyl]benzene: Benzene, [(1*E*)-3,3-dimethoxy-2-methyl-1-propen-1-yl]-: (137032-32-7)

(*S*)-4-(1-Methylethyl)-3-(1-oxopropyl)-2-thiazolidinethione; (102831-92-5)

Diisopropylethylamine: 2-Propanamine, *N*-ethyl-*N*-(1-methylethyl)-; (7087-68-5)

Titanium tetrachloride; (7550-45-0)

Boron trifluoride etherate: $BF_3 \cdot OEt_2$ (109-63-7)

(4*S*)-3-[(2*R*,3*S*,4*E*)-3-Methoxy-2,4-dimethyl-1-oxo-5-phenyl-4-pentenyl]-4-(1-methylethyl)-2-thiazolidinethione; (332902-42-8)

N,O-Dimethylhydroxylamine hydrochloride: Methanamine, *N*-methoxy-, hydrochloride; (6638-79-5)

4-Dimethylaminopyridine: 4-Pyridinamine, *N,N*-dimethyl-; (1122-58-3)

Triethylamine: thanamine, *N,N*-diethyl-; (121-44-8)

Pedro Romea completed his B.Sc. in Chemistry in 1984 at the University of Barcelona. That year he joined the group of Professor Jaume Vilarrasa, at the University of Barcelona, receiving his Masters Degree in 1985, and he followed Ph.D. studies in the same group from 1987 to 1991. Then, he joined the group of Professor Ian Paterson at the University of Cambridge (UK), where he participated in the total synthesis of oleandolide. Back to the University of Barcelona, he became Associate Professor in 1993. His research interests have focused on the development of new synthetic methodologies and their application to the stereoselective synthesis of naturally occurring molecular structures.

Erik Gálvez was born in Barcelona, Spain, in 1982. He received his B.Sc. in Chemistry in 2005 at the Universitat de Barcelona and joined the group of Pedro Romea and Fèlix Urpí. In 2006 he received his Masters Degree and is now pursuing the Ph.D. in the same group. His research concerns asymmetric methodologies involving cross-coupling reactions using 1,3-thiazolidine-2-thiones as source of chirality.

Fèlix Urpí received his B.Sc. in Chemistry in 1980 at the University of Barcelona. In 1981, he joined the group of Professor Jaume Vilarrasa, at the University of Barcelona, receiving his Masters Degree in 1981 and Ph.D. in 1988, where he was an Assistant Professor. He then worked as a NATO postdoctoral research associate in titanium enolate chemistry with Professor David A. Evans, at Harvard University in Cambridge, MA. He moved back to the University of Barcelona and he became Associate Professor in 1991. His research interests have focused on the development of new synthetic methodologies and their application to the stereoselective synthesis of naturally occurring molecular structures.

Vijaya Bhasker Gondi was born in Amrabad, India in 1979. He received his B.Sc. and M.Sc. in Industrial Chemistry from Indian Institute of Technology, Kharagpur, in 2002, and in the same year went to the University of Chicago for graduate studies. He completed his Ph.D. in chemistry with Prof. Viresh Rawal in 2008, working on the asymmetric catalysis of carbon-carbon bond forming reactions through hydrogen bonding. Soon thereafter, he began his postdoctoral studies at The Scripps Research Institute, La Jolla, with Prof. K. C. Nicolaou.

SYNTHESIS OF 2-ARYLINDOLE-4-CARBOXYLIC AMIDES: [2-(4-FLUOROPHENYL)-1*H*-INDOL-4-YL]-1-PYRROLIDINYLMETHANONE

Submitted by Jeffrey T. Kuethe[1] and Gregory L. Beutner.
Checked by Scott E. Denmark and Joseck M. Muhuhi.

1. Procedure

A.   2-[trans-2-(4-Fluorophenyl)vinyl]-3-nitrobenzoic Acid **2**.   A 250-mL, three-necked, round-bottomed flask, equipped with a magnetic stirring bar, a nitrogen inlet, a rubber septum, and a temperature probe is charged with methyl 2-methyl-3-nitrobenzoate **1** (7.50 g, 37.3 mmol) (Note 1) and anhydrous DMSO (52.5 mL) (Note 2). To the resulting mixture is charged sequentially 4-fluorobenzaldehyde (6.00 mL, 6.94 g, 1.5 equiv) (Note 3) and DBU (11.24 mL, 11.35 g, 2.00 equiv) (Note 4).   The homogeneous, brown solution is then stirred at room temperature for 20 h

92

Org. Synth. **2009**, *86*, 92-104
Published on the Web 10/15/2008

and then heated to an internal temperature of 55 °C for 4 h (Notes 5 and 6). The reaction mixture is then allowed to cool to room temperature. In a separate 500-mL, three-necked flask equipped with a magnetic stirring bar and a temperature probe is added water (100 mL) and MTBE (100 mL) (Note 7) and this biphasic mixture is cooled in an ice bath to 0 – 5 °C. The above reaction mixture is then transferred to a 125-mL pressure-equalizing addition funnel and the reaction flask is rinsed with MTBE (15–25 mL). The crude reaction mixture is then added dropwise to the water/MTBE solution at such a rate that the internal temperature of the biphasic quench solution is maintained < 5 °C (Note 8). The mixture is then transferred to a 500-mL separatory funnel and the layers are allowed to separate. The organic layer is discarded. The yellow aqueous layer is washed with MTBE (60 mL). The aqueous layer is then made acidic (pH 1) by the addition of 6 N H_2SO_4 (15 mL, 2.40 equiv) (Note 9) and is extracted with EtOAc (100 mL). The aqueous layer is extracted with an additional portion of EtOAc (60 mL) and the combined organic extracts (Note 10) are concentrated by rotary evaporation (30 °C, 40 mmHg) to give a yellow oil. The oil is re-dissolved in MeOH (55 mL) (Note 11), a magnetic stir bar is placed in the flask and water (175 mL) is added dropwise to the stirred solution over 20 min. The yellow, crystalline slurry of the product is allowed to stir at room temperature for 45 min. The slurry is then filtered through a 150-mL fritted-glass funnel of medium porosity. The wet solid is washed with water (60 mL) and then dried under a vacuum/N_2 sweep (Note 12) for 8 h to provide 2-[*trans*-2-(4-fluorophenyl)-vinyl]-3-nitrobenzoic acid **2** as a bright yellow powder (8.89–9.21 g, 83–86%) (Notes 13, 14).

   B.   *{2-[trans-2-(4-Fluorophenyl)vinyl]-3-nitrophenyl}-1-pyrrolidinylmethanone* **3**. A single-necked, 500-mL round-bottomed flask equipped with a magnetic stirring bar is charged 2-[*trans*-2-(4-fluorophenyl)vinyl]-3-nitrobenzoic acid (**2**) (8.50 g, 29.6 mmol) and CH_2Cl_2 (65 mL) (Note 15). The flask is closed with a rubber septum and placed under nitrogen by piercing the septum with an 18-gauge needle. Oxalyl chloride (3.37 mL, 4.88 g, 38.5 mmol, 1.30 equiv) (Note 16) is added via syringe through the septum followed by the addition of 2 drops of DMF (Note 17). The resulting mixture is stirred at room temperature for 1.5 h and then concentrated by rotary evaporation (23 °C, 40 mmHg). To the resulting crude, yellow acid chloride is added CH_2Cl_2 (25 mL) and the solution is concentrated by rotary evaporation (23 °C, 40 mmHg). In a separate 250-mL, three-necked, round-bottomed flask, equipped with a magnetic stirring bar, a nitrogen inlet, a rubber septum, 125-mL pressure-equalizing addition funnel, and a temperature probe is charged CH_2Cl_2 (50 mL), triethylamine (6.20 mL, 4.49 g, 44.4 mmol, 1.5 equiv) (Note 18), and pyrrolidine (3.18 mL, 2.74 g, 38.5 mmol, 1.30 equiv) (Note 19) and the mixture is cooled in an acetone/ice bath to an internal temperature of < 5 °C. The crude acid chloride is dissolved in CH_2Cl_2 (45 mL) and this solution is transferred to the addition funnel. The acid chloride is added dropwise to the stirred pyrrolidine solution at such a rate that the internal temperature is maintained < 23 °C. Upon completion of the addition of the acid chloride solution, the mixture is stirred at room temperature for 30 min (Note 5). The mixture is then diluted with 2 N HCl (60 mL) (Note 20) and is transferred to a 500-mL separatory funnel and the layers separated. The organic layer is washed with brine (60 mL) and is then concentrated under reduced pressure (23 °C, 40 mmHg) (Note 10), diluted with MeOH (60 mL) and re-concentrated under reduced pressure (23 °C, 40 mmHg) to give a crude, yellow solid. The solid is slurried in MeOH (150 mL) and water (300 mL) is added dropwise over 20 min (Note 21). The slurry is stirred at room temperature for 1 h and then is filtered through a 150-mL fritted-glass, medium porosity funnel. The wet solid is washed with water (60 mL) and dried under a vacuum/N_2 sweep (Note 11) for 8 h to provide {2-[*trans*-2-(4-fluorophenyl)vinyl]-3-nitrophenyl}-1-pyrrolidinyl-methanone (**3**) as a yellow, crystalline solid or a yellow powder (9.90-10.0 g, 98-99%) (Notes 22, 23).

C. *Caution! Due to the toxicity of carbon monoxide and the potential risk associated in handling pressurized glassware, it is recommended that this transformation be carried out in a well vented fume hood with a blast shield and the hood sash down or preferably conducted in an autoclave or metal reaction vessel. Users should exercise extreme caution at all times.*

[2-(4-Fluorophenyl-1H-indol-4-yl]-1-pyrrolidinylmethanone **4**. To a 1-liter glass-lined pressure tube was added sequentially {2-[*trans*-2-(4-fluorophenyl)-vinyl]-3-nitrophenyl}-1-pyrrolidinylmethanone **3** (8.00 g, 23.5 mmol), $Pd(OAc)_2$ (54 mg, 0.235 mmol, 0.01 equiv) (Note 24), 1,10-phenanthroline (300 mg, 1.65 mmol, 0.07 equiv) (Note 25), and DMF (80 mL) (Note 16). The resulting reaction vessel is inserted into a Rocker Assembly (Note 26) and purged with N_2 (60 psi) three times (Note 27). The mixture is then purged with CO (60 psi) three times, (Note 28) and the final pressure set to 50 psi at continuous flow. The mixture is then heated to a jacket temperature of 80 °C and stirred at this temperature for 16 h (Note 5).

The reaction mixture is allowed to cool to room temperature and the system is vented to atmospheric pressure. The vessel is pressurized with N_2 (60 psi) and vented three times to remove all residual amounts of CO. The reaction mixture is then filtered through 10 grams of Celite (Note 29) in a 150-mL fritted-glass funnel of medium porosity into a single-necked flask rinsing both the reaction flask and the filter cake with DMF (60 mL). To the yellowish-orange filtrate is added a stir bar and the flask is equipped with a 250-mL pressure-equalizing addition funnel. The addition funnel is charged with 1 M H_3PO_4 (225 mL, Note 30) and 75 mL of this solution is added to the DMF solution over 15 min (Note 31), at which point the solution becomes turbid and the product begins to crystallize. The resulting slurry is stirred for 30 min and the remaining 1 M H_3PO_4 (150 mL) is added over 15 min. The slurry of the product is stirred an additional 30 min and is filtered through a 150-mL, medium-porosity fritted-glass funnel. The wet solid is washed with water (2 x 100 mL) and then is dried under a vacuum/N_2 sweep (Note 12) for 8 h to provide [2-(4-fluorophenyl-1H-indol-4-yl]-1-pyrrolidinylmethanone **4** as an off-white crystalline solid (6.92 g, 96%) (Notes 32, 33).

2. Notes

1.   Methyl 2-methyl-3-nitrobenzoate **1** (97%) was purchased from Aldrich Chemical Company, Inc. and used as received.

2.   Anhydrous DMSO was purchased from Aldrich Chemical Company, Inc. and was used as received.

3.   4-Fluorobenzaldehyde (98%) was purchased from Aldrich Chemical Company, Inc. and was purified by distillation (175-177 °C, atmospheric pressure under nitrogen) prior to its use in order to eliminate any traces of 4-fluorobenzoic acid.

4.   DBU (1,8-diazabicyclo[5.4.0]undec-7-ene (98%)) was purchased from Aldrich Chemical Company, Inc. and used as received.

5.   All reactions were monitored by TLC using Merck silica gel 60 F_{254}, 250 μ, 1 × 10 cm, TLC plates using hexane/EtOAc, 1:1 as the eluent. The following R_f values were obtained: compound **1** (0.76), compound **2** (0.13), compound **3** (0.18), compound **4** (0.26). The submitters used reverse phase HPLC employing the following conditions: Zorbax Eclipse Plus C18 Rapid Resolution HT column (4.6 x 50 mm, 1.8-micron, Agilent part number: 959941-902) with a standard gradient of 10:90 MeCN/0.1% H_3PO_4

to 95:5 MeCN/0.1% H_3PO_4 over 5 min and hold at 95:5 MeCN/0.1% H_3PO_4 for 1 min, then back to 10:90 MeCN/0.1% H_3PO_4 at 8 min, detection at 210 nm. Retention times are as follows: compound **1** (3.78 min), compound **2** (3.92 min), compound **3** (4.03 min), compound **4** (3.73 min).

6. Typical conversion after 20 h was 80% by HPLC analysis. Heating for 4 h drives the conversion to > 90% by HPLC analysis.

7. Methyl *tert*-butyl ether was purchased from Aldrich Chemical Company, Inc. and was used as received.

8. The temperature should be maintained below 5 °C in order to avoid hydrolysis of any remaining methyl 2-methyl-3-nitrobenzoate **1** to 2-methyl-3-nitrobenzoic acid which will contaminate the product. Under these conditions, no hydrolysis occurs. The third neck is left open.

9. 6 N H_2SO_4 was purchased from Fisher Scientific.

10. The organic extracts will likely contain some water; however, there is no need to dry the extracts over any drying agent such as $MgSO_4$.

11. Methanol was purchased from Aldrich Chemical Company, Inc. and was used as received.

12. The product is dried under vacuum/N_2 sweep under house vacuum with house nitrogen at full flow using the apparatus below for the indicated time. Alternatively, the solid can be dried in a vacuum oven (10–25 mmHg, 35 °C) for the same amount of time.

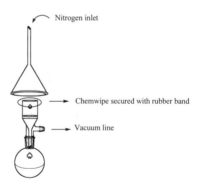

13. Product **2** displayed the following physicochemical properties: bright yellow solid; mp 166–168 °C (sealed tube); ^{1}H NMR (CDCl$_3$, 500 MHz) δ: 6.49 (d, J = 16.5 Hz, 1 H), 7.03 (tt, J = 8.9, 2.5 Hz, 2 H), 7.42 (dt, J = 6.1, 2.5 Hz, 2 H), 7.45 (d, J = 16.5 Hz, 1 H), 7.52 (t, J = 7.8 Hz, 1 H), 7.94 (dd, J = 8.0, 1.5 Hz, 1 H), 8.17 (dd, J = 8.3, 1.5 Hz, 1 H), 11.7 (br s, 1 H); ^{13}C NMR (CDCl$_3$, 125 MHz) δ: 115.7 (d, J = 21.2 Hz), 122.3 (d, J = 2.8

Hz), 127.7, 127.9, 128.5 (d, $J = 8.3$ Hz), 131.1, 132.5 (d, $J = 3.7$ Hz), 133.1, 134.1, 134.3, 150.9, 162.8 (d, $J = 246.6$ Hz), 171.5; ^{19}F NMR (CDCl$_3$, 500 MHz) δ: −113.9. Anal. Calcd. For C$_{15}$H$_{10}$FNO$_4$: C, 62.72; H, 3.51; N, 4.88. Found: C, 62.32; H, 3.45; N, 4.87.

14. The submitters reported the mp at 156–157 °C, 10 °C lower than that of the checkers. As both samples passed elemental analysis, the discrepancy most likely arises from two different crystal forms.

15. Methylene chloride was purchased from Aldrich Chemical Company, Inc. and was used as received.

16. Oxalyl chloride (98%) was purchased from Aldrich Chemical Company, Inc. and was used as received.

17. Anhydrous DMF was purchased from Aldrich Chemical Company, Inc. and was used as received.

18. Triethylamine (99.5%) was purchased from Aldrich Chemical Company, Inc. and was used as received.

19. Pyrrolidine (99%) was purchased from Aldrich Chemical Company, Inc. and was used as received.

20. 2 N HCl was prepared from 12 N HCl purchased from Fisher Scientific.

21. The submitters used MeOH (60 mL) and water (120 mL), which the checkers found gave a yellow powder. When checkers used MeOH (150 mL) and water (300 mL) the product obtained was a yellow, crystalline solid. The extra MeOH helps in forming more fine particles.

22. Product **3** displayed the following physicochemical properties: yellow powder or yellow crystalline solid; mp 123–124 °C (sealed tube); ^{1}H NMR (CDCl$_3$, 500 MHz) δ: 1.78 (m, 4 H), 3.05 (t, $J = 6.3$ Hz, 2 H), 3.52 (t, $J = 6.5$ Hz, 2 H), 6.91 (d, $J = 16.5$ Hz, 1 H), 7.03 (tt, $J = 8.9$, 2.5 Hz, 2 H), 7.24 (d, $J = 16.5$ Hz, 1 H), 7.42 (tt, $J = 7.3$, 2.5 Hz, 2 H), 7.46 (t, $J = 8.0$ Hz, 1 H), 7.59 (dd, $J = 7.5$, 1.0 Hz, 1 H), 7.94 (dd, $J = 8.5$, 1.0 Hz, 1 H); ^{13}C NMR (CDCl$_3$, 125 MHz) δ: 24.3, 25.8, 45.8, 47.7, 115.8 (d, $J = 22.1$ Hz), 120.1 (d, $J = 2.8$ Hz), 124.9, 128.2, 128.5 (d, $J = 8.3$ Hz), 12.2, 131.5, 132.5 (d, $J = 3.7$ Hz), 135.3, 139.2, 148.6, 162.9 (d, $J = 248.6$ Hz), 167.4; ^{19}F NMR (CDCl$_3$, 500 MHz) δ: −113.5. Anal. Calcd. For C$_{19}$H$_{17}$FN$_2$O$_3$: C, 67.05; H, 5.03; N, 8.23. Found: C, 66.99; H, 5.05; N, 8.19.

23. The submitters reported the mp at 94–95 °C, 30 °C lower than that of the checkers. As both samples passed elemental analysis, the discrepancy most likely arises from two different crystal forms.

24. Palladium (II) acetate (min. 98%) was purchased from Strem Chemicals, Inc. and was used as received.

25. 1,10-Phenanthroline (99%) was purchased from Aldrich Chemical Company, Inc. and was used as received.

26. Submitters used a 4300 mL Fluitron Rocker Assembly, while the checker's rocker assembly was purchased from American Instruments CC Inc. A similar type reaction apparatus may be utilized for this transformation, see Söderberg, B. C.; Shriver, J. A.; Wallace, J. M. *Org. Synth.* **2003**, *80*, 75-84; however, it is strongly recommended that the reaction be conducted in an autoclave or metal reaction vessel with glass liner as described in this procedure. http://www.fluitron.com/vessels.html

27. Submitters evacuated the reaction vessel under reduced pressure (10-30 mm Hg).

28. Carbon monoxide was purchased from Matheson Trigas Co. and was used as received. Submitters used CO (30 psi) while checkers used CO (30 psi) for small-scale reaction (4.7 mmol) and CO (50 psi) for the full scale runs.

29. Celite® 545 coarse was purchased from Aldrich Chemical Company, Inc. and was used as received.

30. 1 M H_3PO_4 was prepared by dilution with water of 85% o-H_3PO_4 (115.3 g, Fisher Scientific, HPLC grade) to a final volume of 250 mL.

31. The addition of 75 mL of 1 M H_3PO_4 to the DMF filtrate is slightly exothermic from room temperature to 35–40 °C, but is not detrimental to the crystallization of the product.

32. Compound **4** displayed the following physicochemical properties: off-white to slight yellow crystalline solid: mp 184–185 °C (sealed tube); ^{1}H NMR (CDCl$_3$, 500 MHz) δ: 1.84 (q, J = 6.6 Hz, 2 H), 1.98 (q, J = 7.0 Hz, 2 H), 3.39 (t, J = 6.8 Hz, 2 H), 3.76 (t, J = 7.3 Hz, 2 H), 6.75 (d, J = 2.0 Hz, 1 H), 7.00 (tt, J = 8.8, 2.3 Hz, 2 H), 7.08 (t, J = 7.5 Hz, 1 H), 7.16 (dd, J = 7.5, 0.5 Hz, 1 H), 7.33 (d, J = 9.0 Hz, 1 H), 7.56 (tt, J = 7.3, 2.3 Hz, 1 H), 9.43 (s, 1 H); ^{13}C NMR (CDCl$_3$, 125 MHz) δ: 24.6, 26.2, 45.9, 49.0, 98.7, 112.6, 115.7 (d, J = 22.1 Hz), 118.5, 121.3, 126.3, 127.1 (d, J = 7.4 Hz), 128.4 (d, J = 3.7 Hz), 128.6, 137.4, 138.3, 162.3 (d, J = 247.6 Hz), 170.1; ^{19}F NMR (CDCl$_3$, 500 MHz) δ: –115.0. Anal. Calcd. For $C_{19}H_{17}FN_2O$: C, 74.01; H, 5.56; N, 9.07. Found: C, 74.11; H, 5.61; N, 9.08.

33. The submitters reported the mp at 130–131 °C, over 50 °C lower than that of the checkers. As both samples passed elemental analysis, the discrepancy most likely arises from two different crystal forms.

Safety and Waste Disposal Information

All hazardous materials should be handled and disposed of in accordance with "Prudent Practices in the Laboratory"; National Academy Press; Washington, DC, 1995.

3. Discussion

The palladium-catalyzed reductive cyclization of aromatic nitrostyrene compounds employing carbon monoxide as the stoichiometric reductant has recently emerged as a highly versatile method for the construction of indoles due to superior yields, diminished amounts of reaction by-products, functional group compatibility, and favorable environmental impact.[2,3] The widespread use of *ortho*-nitrostyrenes as indole precursors has been limited by available methods for their preparation. Traditional approaches have relied on Wittig reactions of either 2-nitrobenzaldehydes or 2-nitrophosphonium and phosphonate salts.[4] Alternatively, cross-coupling approaches involving 2-halonitrobenzenes or 2-nitrophenylstannanes have received considerable attention as an attractive method for the preparation of a range of *ortho*-nitrostyrenes.[2] While each of these approaches offer certain advantages, they often require multiple steps for the construction of the appropriate starting materials and generally require purification by chromatography. These strategies also have a high environmental burden since they suffer from poor atom economy and generate a significant amount of phosphorous or tin byproducts. We have demonstrated that reactions of 2-nitrotoluenes or 2-trimethylsilylmethylnitrobenzenes with aromatic aldehydes via an addition/elimination protocol is an effective, high yielding method for the construction of *ortho*-nitrostyrenes and their subsequent conversion to indoles.[5]

The procedure described herein illustrates the concise preparation of [2-(4-fluorophenyl)-1*H*-indol-4-yl]-1-pyrrolidinylmethanone **4** in 75% overall yield from commercially available methyl 2-methyl-3-nitrobenzoate **1**.[6] Reaction of **1** with 4-fluorobenzaldehyde in the presence of DBU in DMSO is reversible leading to intermediate **5** (Scheme 1). Based on the juxtaposition of the methyl ester with the reacting center and its capacity to serve as an intramolecular trap, cyclization to lactone intermediate **6** occurs. In the presence of a second equivalent of DBU, deprotonation followed by elimination of the carboxylate anion furnishes nitrosytrene benzoic acid **2**.

After an extractive work up to remove excess 4-fluorobenzaldehyde, the product **2** is obtained in 84% isolated yield after crystallization from MeOH/water. The reaction sequence is general and allows for the preparation of an array of structurally diverse nitrostyrene benzoic acids in good to excellent yield (Table 1). Substrates containing electron donating (entries 1- 4) or electron withdrawing groups (entry 5) participate equally as well. In all cases, the *trans*-nitrostyrene benzoic acids were the exclusive products.

Scheme 1.

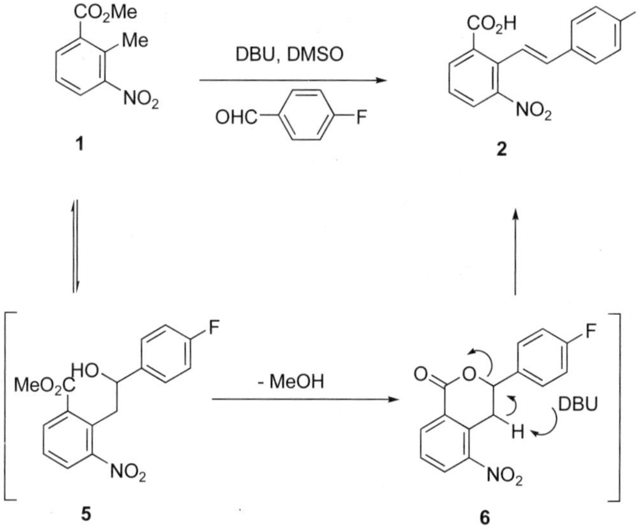

The preparation of 2-arylindole-4-carboxylic amide **4** involved conversion of **2** to the corresponding acid chloride with oxalyl chloride followed by reaction with pyrrolidine to give amide **3** followed by reductive cyclization. Reductive cyclization was carried out in the presence of 1 mol% Pd(OAc)$_2$, 7 mol% 1,10-phenanthroline, in DMF at 80 °C under an atmosphere of 30 psi CO for 16 h to give the desired product. Isolation of the product is accomplished by filtration through Celite and addition of the crude DMF filtrate to a solution of 1 M H$_3$PO$_4$, which precipitated the product in analytically pure form and in excellent overall yield. The sequence whereby the appropriately substituted nitrostyrene carboxylic acid was converted to the required amide followed by reductive cyclization provided an excellent, high yielding means of preparing the highly

100

Table 1. Synthesis of Nitrostyrene Benzoic Acids

entry	aldehyde	nitrostyrene benzoic acid	Yield
1	**7**	**8**	68%
2	**9**	**10**	79
3	**11**	**12**	90
4	**13**	**14**	82
5	**15**	**16**	88

functionalized derivatives shown in Table 2. The use of 1M H_3PO_4 for the isolation of the product aided in the removal of trace amounts of 1,10-phenanthroline from the product and was mild enough that sensitive functionalities such as a Boc-protecting group were preserved (see entry 3, Table 2).

Table 2. Synthesis of Nitrostyrene Amides and Reductive Cyclization to 2-Arylindole-4-carboxylic amides.

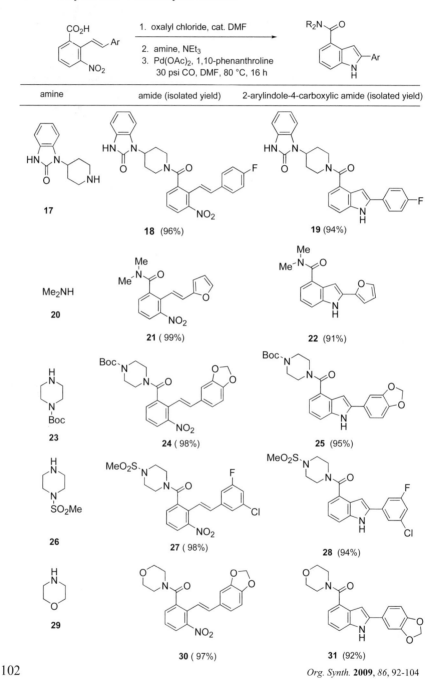

amine	amide (isolated yield)	2-arylindole-4-carboxylic amide (isolated yield)
17	**18** (96%)	**19** (94%)
Me₂NH **20**	**21** (99%)	**22** (91%)
23	**24** (98%)	**25** (95%)
26	**27** (98%)	**28** (94%)
29	**30** (97%)	**31** (92%)

Org. Synth. **2009**, *86*, 92-104

1. Department of Process Research, Merck & Co., Inc., Rahway, NJ, 07065, USA. E-Mail: Jeffrey_Kuethe@merck.com.
2. For leading references, see: (a) Dantale, S. W.; Söderberg, B. C. G. *Tetrahedron* **2003**, *59*, 5507-5514; (b) Scott, T. L.; Söderberg, D. C. G. *Tetrahedron Lett.* **2002**, *43*, 1621-1624; (c) Söderberg, B. C.; Shriver, J. A. *J. Org. Chem.* **1997**, *62*, 5838-5845; (d) Söderberg, B. C.; Rector, S. R.; O'Neil, S. N. *Tetrahedron Lett.* **1999**, *40*, 3657-3660.
3. Davies, I. W.; Smitrovich, J. H.; Sidler, R.; Qu, C.; Gresham, V.; Bazaral, C. *Tetrahedron* **2005**, *61*, 6425-6437.
4. (a) Sundberg, R. J. *J. Org. Chem.* **1965**, *30*, 3604-3610.; (b) Sundberg, R. J.; Yamazaki, T. *J. Org. Chem.* **1967**, *32*, 290-294. (c) Fresneda, P. M.; Molina, P.; Delgado, S. *Tetrahedron* **2001**, *57*, 6197-6202.
5. (a) Kuethe, J. T.; Wong, A.; Davies, I. W.; *Org. Lett.* **2003**, *5*, 3721-3723. (b) Kuethe, J. T.; Wong, A.; Davies, I. W. *Org. Lett.* **2003**, *5*, 3975-3978. (c) Wong, A.; Kuethe, J. T.; Davies, I. W.; Hughes, D. L. *J. Org. Chem.* **2004**, *69*, 7761-7764. (d) Kuethe, J. T.; Wong, A.; Qu, C.; Smitrovich, J. H.; Davies, I. W.; Hughes, D. L. *J. Org. Chem.* **2005**, *70*, 2555-2567.
6. For a complete account of our work in this area, see: Kuethe, J. T.; Davies, I. W. *Tetrahedron* **2006**, *62*, 11381-11390.

Appendix
Chemical Abstracts Nomenclature; (Registry Number)

Methyl 2-methyl-3-nitrobenzoate; (59382-59-1)

4-Fluorobenzaldehyde; (459-57-4)

2-[*trans*-2-(4-Fluorophenyl)vinyl]-3-nitrobenzoic Acid; (917614-64-3)

{2-[*trans*-2-(4-Fluorophenyl)vinyl]-3-nitrophenyl}-1-
 pyrrolidinylmethanone; (917614-83-6)

1,10-Phenanthroline; (66-71-7)

Pd(OAc)$_2$: Palladium acetate; (3375-31-3)

[2-(4-Fluorophenyl-1H-indol-4-yl]-1-pyrrolidinylmethanone; (917614-84-7)

Jeff Kuethe was born in Cincinnati, Ohio in 1965. He studied chemistry at Middle Tennessee State University where he received in Bachelor of Science in 1993. He then joined the group of Professor Albert Padwa at Emory University in Atlanta, Georgia where he received a Ph.D. in 1998. He continued as a postdoctoral fellow in the group of Professor Daniel Comins at North Carolina State University before joining the Department of Process Research at Merck & Co., Inc., Rahway, New Jersey in 2000. His research interests include Process Research, synthetic methodology, heterocyclic chemistry, alkaloid and natural product synthesis, and tandem transformations.

Gregory L Beutner was born in Malden, Massachusetts in 1976. He graduated in 1998 with a Bachelor of Science in chemistry from Tufts University where he worked with Professor Arthur Utz and Professor Marc d'Alarcao. He completed his Ph.D studies under the supervision of Professor Scott E. Denmark at the University of Illinois in May 2004. He continued as an NIH postdoctoral fellow in the laboratories of Professor Robert H. Grubbs at the California Institute of Technology. He is currently working in the Department of Process Research at Merck & Co., Inc., Rahway, New Jersey. His research interests include Process Research, synthetic methodology development, and asymmetric catalysis.

Joseck Muhuhi graduated with a B.Sc. in Chemistry from the University of Nairobi, Kenya, in 1997, and after working in industry for 3 years, joined Wayne State University in 2000, for graduate education under the guidance of Dr. Mark Spaller. His doctoral research study was on mechanistic study of the aza-Diels-Alder reaction in the synthesis of anticancer heterolignans, synthesis of reduced peptides, and E-alkene dipeptide isosteres that target HIV-1 protease enzyme and PDZ domain proteins respectively. In December 2006, he joined the Denmark group as a visiting postdoctoral research associate to work on the development of tandem ring-closing metathesis/silicon cross-coupling reactions and total synthesis of oximidine natural products. Dr. Muhuhi is a recipient of a NIH postdoctoral research fellowship.

PALLADIUM (II) ACETATE-BUTYLDI-1-ADAMANTYLPHOSPHINE CATALYZED ARYLATION OF ELECTRON-RICH HETEROCYCLES. PREPARATION OF 5-PHENYL-2-ISOBUTYLTHIAZOLE.

Submitted by Anna Lazareva, Hendrich A. Chiong, and Olafs Daugulis.[1]
Checked by Sirilata Yotphan and Jonathan A. Ellman.

1. Procedure

A 500-mL oven-dried Schlenk flask equipped with a magnetic stir bar and a rubber septum is charged with palladium(II) acetate (342 mg, 1.5 mmol, 0.05 equiv), sodium acetate (1.23 g, 15 mmol, 0.50 equiv), n-butyldi-1-adamantylphosphine (576 mg, 1.5 mmol, 0.05 equiv), and tribasic potassium phosphate (6.36 g, 30 mmol, 1.00 equiv). The Schlenk flask is evacuated and back-filled with dry argon three times. Anhydrous N,N-dimethylacetamide (120 mL), 2-isobutylthiazole (4.25 mL, 30 mmol, 1.00 equiv), and chlorobenzene (4.5 mL, 45 mmol, 1.50 equiv) are added via a gas tight syringe (Note 1). Under a continuous flow of argon the rubber septum is quickly replaced with a reflux condenser and bubbler to ensure an air-free system. The mixture is stirred in a preheated oil bath at 125 °C for 36 h (Note 2). A continuous argon flow is maintained throughout the reaction. The yellow color of the reaction mixture changes to dark brown after 5-10 min of heating. Potassium phosphate does not dissolve and a mild reflux is observed. The conversion is monitored by GC (Note 3). If complete conversion is not achieved after 36 h, the reaction mixture is heated for an additional 2-4 h.

After the reaction is complete as judged by GC analysis (Note 3), hydrazine hydrate (30 mL) is added in one portion and the flask is heated in an oil bath (125 °C) for 5 h (Note 4). At this point, the argon atmosphere is no longer required. The reaction mixture is cooled to room temperature followed by the addition of water (100 mL) and dichloromethane (50 mL). The mixture is transferred to a 2-L separatory funnel. Additional water (200 mL) and dichloromethane (100 mL) are added to the separatory funnel. After separation of the organic layer, the water layer is extracted with dichloromethane (3 x 150 mL) (Note 5). Magnesium sulfate (about 12 g) is directly placed in a 150-mL Büchner filter funnel with a medium porosity fritted disk. A round-bottomed flask (1-L) is connected to the filter funnel and the combined organic phase is filtered through the $MgSO_4$ layer (Note 6). The combined organic extracts are not treated with $MgSO_4$ beforehand. The reaction mixture is concentrated by rotary evaporation (40 °C, 15 mm Hg). Following concentration, 120–130 mL of the crude product solution in DMA is left in the flask. The reaction mixture is transferred to a 500-mL round-bottomed flask equipped with a magnetic stir bar. The solvent is removed by vacuum distillation (1.6 mm Hg) through a 4-cm column to a 200-mL round-bottomed receiver flask. The boiling point of DMA is 38–39 °C /1.6 mmHg; the temperature of the oil bath is kept at 65–70 °C (Note 7). After the distillation is complete the temperature inside the distillation apparatus drops to about 23 °C. The distillation residue is diluted with 10 mL of chromatography eluent mixture (hexanes/dichloromethane; 1/2). The crude product is purified by flash chromatography on silica gel (5 x 19 cm, 200 g of silica gel) (Note 8), using hexanes/dichloromethane (1/2) as the eluent (900 mL). The first fractions are blank, followed by fractions containing minor amounts of impure and then pure product. The elution is continued with 5% ethyl acetate/dichloromethane (about 1.1 L collected). After concentration of the fractions containing pure product, the residue is dried under reduced pressure (1.0 mmHg) at room temperature for 24 h to yield the arylated heterocycle. The pure product (6.15 g, 94%) is obtained as an air stable, light brown oil (Notes 9 and 10).

2. Notes

1. All reagents were used as received. Palladium acetate was obtained from Gelest, Inc. (purity > 95%). Potassium phosphate tribasic (reagent grade, ≥ 98%), anhydrous DMA (anhydrous, 99.8%), and hydrazine hydrate (reagent grade, N_2H_4 50–60%) were obtained from Aldrich. Butyldi-1-adamantylphosphine (min. 95%) was purchased from Strem. 2-Isobutylthiazole was obtained from Oakwood Products. Chlorobenzene was obtained from Aldrich (anhydrous, 99.8%), and sodium acetate (assay min. 99.0%) was purchased from Mallinckrodt Chemicals. Bulk of the air sensitive reagents (tribasic potassium phosphate and butyldi-1-

106

adamantylphosphine) was stored in an argon-filled glovebox. Small portions were taken out of the glovebox and stored on the bench in capped vials for up to one month. The submitters used chlorobenzene purchased from Matheson, Coleman & Bell Manufacturing Chemists and sodium acetate (assay min. 99.0%) purchased from Mallinckrodt Chemicals.

2.   If the reaction mixture is not vigorously stirred, complete conversion is not achieved. Stir bar size: 3.5 cm length and 1 cm diameter. Stirring rate: 1000-1100 rpm.

3.   GC analyses (by the submitters) were performed on a Shimadzu CG-2010 chromatograph equipped with a Restek column (Rtx®-5, 15 m, 0.25 mm ID). Initial temp: 50 °C (2 min), ramp at 50 °C/min to 170 °C, hold at 170 ° (3 min), ramp at 40 °C/min to 270 °C, hold at 270 (5 min). Retention times: 2-isobutylthiazole (3.17 min) and 5-phenyl-2-isobutylthiazole (6.30 min). GC analyses (by the checkers) were performed on an Agilent 6890N chromatograph equipped with an Agilent column (DB-1, polysiloxane, 15 m, 0.25 mm ID). Initial temp: 50 °C (2 min), ramp at 50 °C/min to 170 °C, hold at 170 ° (3 min), ramp at 40 °C/min to 270 °C, hold at 270 (5 min). Retention times: 2-isobutylthiazole (5.07 min) and 5-phenyl-2-isobutylthiazole (10.08 min). Aliquots from reaction are concentrated by rotary evaporation (50 °C, 15 mmHg) and diluted with 0.5 mL of CH_2Cl_2 for GC analyses. After 36 h, >99% of the starting material (2-isobutylthiazole) is consumed.

4.   Hydrazine workup removes palladium residue from the product.

5.   Organic and water layers must separate completely. Both phases must be clear, not cloudy, and there must be a sharp border between the two. Otherwise, the yield will be lower.

6.   The checkers used magnesium sulfate (anhydrous certified powder) purchased from Fisher Chemical.

7.   For efficient purification by flash chromatography **ALL** DMA must be removed by distillation.

8.   Flash chromatography was performed on 60Å silica gel (Sorbent Technologies). The checkers performed flash chromatography on 60Å silica gel (MP Silitech 32-63D). All solvents are HPLC grade purchased from Fisher Chemical.

9.   The characterization of the product is as follows: R_f=0.19 (hexanes/dichloromethane (1:2), visualization by UV; ^{1}H NMR (400 MHz, $CDCl_3$) δ: 1.00 (d, J = 6.8 Hz, 6 H), 2.07–2.19 (m, 1 H), 2.86 (d, J = 6.8 Hz, 2 H), 7.26–7.30 (m, 1 H), 7.34–7.38 (m, 2 H), 7.50–7.52 (m, 2 H), 7.81 (s, 1

H). ^{13}C NMR (75 MHz, CDCl$_3$) δ: 22.5, 30.0, 42.7, 126.8, 128.2, 129.2, 131.8, 137.8, 138.7, 169.9. FT-IR (neat, cm^{-1}) 2957, 1601, 1530, 1491, 1458. Anal calcd for C$_{13}$H$_{15}$NS: C, 71.84; H, 6.96; N, 6.44; Found: C, 71.49; H, 6.78; N, 6.13.

10. On the same scale, the submitters reported product yields of 5.99-6.26 g (92-96%). On a half-scale run, the checkers obtained 2.92 g of product (90%).

Safety and Waste Disposal Information

All hazardous materials should be handled and disposed of in accordance with "Prudent Practices in the Laboratory"; National Academy Press; Washington, DC, 1995.

3. Discussion

The combination of palladium(II) acetate with an electron-rich butyldi-1-adamantylphosphine ligand is a versatile catalyst for the direct arylation of electron-rich heterocycles with aryl chlorides.[2] The use of aryl bromides and aryl iodides in palladium-catalyzed direct heterocycle arylations is well-known.[3] However, a general method employing aryl chloride reagents has not been reported so far for such reactions.[4] Most of the published catalytic systems are efficient for only a few types of heterocycles thus limiting the generality of the reaction, and stoichiometric copper additives are often needed for successful arylation. This methodology overcomes most of the limitations described above. A number of structurally diverse electron-rich heterocycles are reactive (Table 1).[2] Thiophene, benzothiophene, 1,2- and 1,3-oxazole derivatives, benzofuran, thiazoles, benzothiazole, 1-alkylimidazoles, 1-alkyl-1,2,4-triazoles, and caffeine can be arylated. Both electron-rich and electron-poor aryl chlorides can be used; however, electron-poor chlorides are more reactive. Some steric hindrance is tolerated on the heterocycle and aryl chloride.

The arylation mechanism may be dependent on the heterocycle type. For example, triazole and imidazole arylations most likely proceed by an electrophilic aromatic substitution mechanism due to the observed regioselectivity.[5] Benzoxazole arylation may proceed via ring-opening pathways.[6]

108

While the method is very general, some additional optimization may be required to maximize reaction yields, since the heterocycles that can be arylated are very structurally different. The procedure for the phenylation of isobutylthiazole has been optimized and the reaction conditions have been modified from the original report[2] by decreasing the amount of phosphine ligand, adding sodium acetate reagent, and using DMA instead of NMP as the solvent. Under the conditions reported initially, about 80-85% conversion to phenylated derivative was observed as opposed to complete conversion by using these modified conditions. However, if these modified reaction conditions are used for the 2-methoxyphenylation of caffeine, lower conversion is obtained compared to the original procedure. The examples in Table 1 have been obtained by using the original conditions.[2] The limitations of the procedure are as follows: Arylation of NH-containing heterocycles such as indoles or pyrroles results in *N*-functionalization. Arylation of *N*-substituted indoles does not go to completion even if extended reaction times are used, and isomer mixtures are usually obtained. Benzofuran monoarylation results in the formation of a mixture of isomers.

Table 1. Heterocycle Arylation.[a]

Heterocycle	Aryl Chloride	Product	Yield
(thiophene)	Cl–C₆H₄–NHAc	(thiophen-2-yl) with NHAc aryl	54%[b]
(benzothiophene)	Cl–pyridine–OMe	(benzothiophene) with OMe pyridine	72%
(benzothiophene)	C₆H₄–Cl	(benzothiophene)–2–Ph	63%
(Me–isoxazole–Me)	(naphthyl chloride)	(Me, Me isoxazole)–Ar	76%
(benzofuran)	C₆H₄–Cl	(benzofuran) 3-Ph, 2-Ph	68%[c]
(benzoxazole)	Cl–C₆H₄–CO₂Et	(benzoxazole)–CO₂Et aryl	84%
(benzothiazole)	C₆H₄–Cl	(benzothiazole)–2–Ph	84%
(thiazole–NH–C(O)tBu)	Cl–C₆H₄–CF₃	F₃C aryl–thiazole–C(O)tBu	79%
(imidazole–Bu)	C₆H₄–Cl	Ph–imidazole–Bu	52%[d]
(triazole–Me)	MeO–C₆H₃–OMe–Cl	(MeO)₂C₆H₃–triazole–Me	76%
(caffeine, theophylline core)	Cl–C₆H₄–OMe	(xanthine)–C₆H₄–OMe	71%
(caffeine core)	Me–C₆H₃–Me–Cl	(xanthine)–C₆H₃(Me)₂	77%

[a] From Ref. 2. Substrate (1 equiv), ArCl (1.5 equiv), K_3PO_4 (2 equiv), $Pd(OAc)_2$ (5 mol %), 10 mol % $nBuAd_2P$, 24 h at 125 °C, NMP solvent. Isolated yields reported. [b] Thiophene (3 equiv), chloroarene (1 equiv). [c] Benzofuran (1 equiv), chloroarene (3 equiv). [d] 2,5-Diphenylated byproduct isolated in 13% yield.

1. University of Houston, Department of Chemistry, Houston, TX 77204-5003. Email: olafs@uh.edu
2. Chiong, H. A.; Daugulis, O. *Org. Lett.*, **2007**, *9*, 1449.
3. Reviews: a) Alberico, D.; Scott, M. E.; Lautens, M. *Chem. Rev.* **2007**, *107*, 174. (b) Seregin, I. V.; Gevorgyan, V. *Chem. Soc. Rev.* **2007**, *36*, 1173.
4. (a) Akita, Y.; Inoue, A.; Yamamoto, K.; Ohta, A. *Heterocycles* **1985**, *23*, 2327. (b) Gottumukkala, A. L.; Doucet, H. *Eur. J. Inorg. Chem.* **2007**, *23*, 3629. (c) Iwasaki, M.; Yorimitsu, H.; Oshima, K. *Chem. Asian J.* **2007**, *2*, 1430. (d) Rieth, R. D.; Mankad, N. P.; Calimano, E.; Sadighi, J. P. *Org. Lett.* **2004**, *6*, 3981.
5. (a) Gupta, R. R.; Kumar, M.; Gupta, V. *Heterocyclic Chemistry II;* Springer Publishing: New York, 1999, p. 388. (b) Park, C.-H.; Ryabova, V.; Seregin, I. V.; Sromek, A. W.; Gevorgyan, V. *Org. Lett.* **2004**, *6*, 1159.
6. Sanchez, R. S.; Zhuravlev, F. A. *J. Am. Chem. Soc.* **2007**, *129*, 5824.

Appendix
Chemical Abstracts Nomenclature; (Registry Number)

Palladium(II) acetate; (3375-31-3)
Sodium acetate; (127-09-3)
Butyl di-1-adamantylphosphine; (321921-71-5)
Tribasic potassium phosphate; (7778-53-2)
2-Isobutylthiazole; (18640-74-9)
Chlorobenzene; (108-90-7)
Hydrazine hydrate; (10217-52-4)
5-Phenyl-2-isobutylthiazole; (600732-10-3)

Olafs Daugulis was born in Riga, Latvia in 1968. He obtained his B.S. degree in chemical engineering from Riga Technical University in 1991. His Ph.D. research was performed at the University of Wisconsin-Madison in the group of Prof. E. Vedejs. After obtaining his Ph.D. in 1999 he joined the group of Prof. M. Brookhart at UNC-Chapel Hill as a postdoctoral associate. Since 2003, he has been an Assistant Professor of Chemistry at the University of Houston.

Anna Lazareva was born in Donetsk, Ukraine in 1979. She obtained her B.S. degree in chemistry from the University of Houston in 2006. After joining the Daugulis group in 2006, she received her MS degree in chemistry in 2008. Currently she is a Staff Chemist in the Medicinal Chemistry Department at Merck in Rahway, NJ. .

Hendrich A. Chiong was born in the Philippines in 1976 where he obtained his B.S. in chemistry in 1996. He joined the Daugulis group at the University of Houston in 2003 and received his Ph.D. in chemistry in 2007. Currently he is a Research Process Chemist at Sabic Innovative Plastics High Performance Polymers (HPP) in Mt. Vernon, IN.

Sirilata Yotphan grew up in Lampang, Thailand. She obtained her B.S. degree in chemistry from McGill University, Canada in 2006. She then began her doctoral studies at the University of California, Berkeley in the laboratories of Professor Jonathan A. Ellman and Professor Robert G. Bergman. Her graduate research has focused on metal-catalyzed C-H bond functionalization.

MILD AND EFFICIENT ONE-POT CURTIUS REARRANGEMENT: PREPARATION OF *N-tert*-BUTYL ADAMANTANYL-1-YL-CARBAMATE

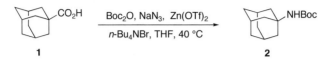

Submitted by Olivier Leogane and Hélène Lebel.[1]
Checked by Kay M. Brummond, Thomas O. Painter, and Matthew Klinge.

1. Procedure

Caution ! The reaction should be conducted in a well-ventilated hood.

N-tert-Butyl adamantan-1-yl-carbamate (**2**). A flame-dried 250-mL, three-necked, round-bottomed flask equipped with a mechanical stirrer and fitted with a thermometer and a rubber septum with an argon inlet is charged with adamantane-1-carboxylic acid (**1**, 5.40 g, 30.0 mmol) (Note 1), sodium azide (3.90 g, 60.0 mmol, 2.0 equiv)) (Note 2), tetra-*n*-butyl ammonium bromide (0.870 g, 2.70 mmol. 0.09 equiv) (Note 3), and finally zinc triflate (0.220 g, 0.60 mmol, 0.02 equiv) (Note 4). The flask is then purged with argon. After 10 min, 150 mL of THF (Note 5) is added via a syringe and the flask is heated in an oil bath at 40 °C. Once the internal temperature has reached 40 °C, the di-*tert*-butyl dicarbonate (7.58 mL, 33.0 mmol, 1.1 equiv) is added (Note 6). The reaction mixture is then stirred under argon at 40 °C until >95% conversion to product is observed by GC analysis (Note 7). The reaction mixture is cooled to room temperature and quenched with a 50 mL portion of a 10% aqueous solution of $NaNO_2$ (Note 8). The biphasic mixture is stirred 30 min at room temperature and transferred to a 500-mL separatory funnel. The reaction flask is washed with hexanes and water and the two layers are separated. The aqueous layer is extracted twice with hexanes (2 x 40 mL) and the combined organic layers are washed successively with saturated aqueous NH_4Cl (75 mL), brine (75 mL), and then dried over Na_2SO_4 (50 g). The organic solution is filtered into a round-bottomed 250-mL flask and concentrated at 40 °C by rotary evaporation (12–14 mmHg) to afford a white solid. The 250-mL flask is then equipped with a reflux condenser and the solid dissolved in a boiling mixture (T = 75 °C) of hexanes (90 mL) and ethyl acetate (2 mL) (Note 9). The suspension is heated in an oil bath at reflux for 20 min, complete dissolution of the solid occurs, along with the formation of a sluggish orange oil. The clear, colorless solution is decanted away from this orange residue into a 250 mL Erlenmeyer flask fitted with a 24/40 glass joint and then concentrated at 40 °C by rotary evaporation (12–14 mmHg) then under high vacuum (< 3 mmHg) for 30 min to afford a white solid, which is recrystallized from a mixture of hexanes (< 50-60 mL), and chloroform (< 2 mL) (Note 10). The resulting crystalline product is collected on a fritted funnel, the filtrate is washed with 10–15 mL of cold hexanes, and excess solvent is removed under vacuum to yield 6.03 g of product. The mother liquor is concentrated at 40 °C by rotary evaporation (12–14 mmHg) then under high vacuum (< 3 mmHg), and the resulting white solid recrystallized (Note 11) from hexanes (< 10 mL) to yield 0.55 g of additional product after filtration as above. The solids are combined to provide 6.58 g (87%) of the title compound as a crystalline white solid (Notes 12 and 13).

2. Notes

1.   Adamantane-1-carboxylic acid, 99% was purchased from Alfa Aesar, and was used without further purification.

2. NaN_3, 99% was purchased from Alfa Aesar, and was ground prior to use.

3.   Tetra-*n*-butyl ammonium bromide, 99% was purchased from Aldrich Chemical Company, Inc. and stored in a desiccator filled with drierite.

4.   The checkers purchased $Zn(OTf)_2$, 98% from Strem Chemicals and stored it in a glove-box under a nitrogen atmosphere upon receipt. $Zn(OTf)_2$ was weighed into a two-necked round-bottomed flask, sealed, then equipped with an argon inlet upon removal from the glove-box. $Zn(OTf)_2$ was quickly transferred to the reaction mixture by pouring it from the two-necked flask under an argon flow. The submitters report that for their best results, the $Zn(OTf)_2$ was purchased from Aldrich Chemical Company, Inc., handled under an argon atmosphere in a glove-box and was used without further purification affording an 85% yield. The checkers found that the $Zn(OTf)_2$ purchased from Aldrich resulted in incomplete reactions. The submitters report that $Zn(OTf)_2$ from Strem Chemicals is compatible with the reaction conditions (81% yield for **2**) but also comment that very low yields were observed if $Zn(OTf)_2$ from Alfa Aesar was used.

5. Anhydrous 99.9%, inhibitor free tetrahydrofuran was purchased from Aldrich and purified with alumina using the Sol-Tek ST-002 solvent purification system directly before use. The submitters report that THF was purified with a Glass Contour Seca Solvent Purification System.

6. Boc$_2$O, 99% was purchased from Aldrich Chemical Company, Inc and stored in the refrigerator. For convenience, the bottle of Boc$_2$O was warmed in a water bath to reach the melting point, and the colorless oil was then transferred quickly to the reaction flask via a syringe.

7. The internal reaction temperature must be maintained between 39 and 41 °C during the entire process. This was accomplished using a temperature controller and thermocouple connected to a variac and oil bath. After 120 hours, 96% conversion was observed by GC analysis in the first full scale reaction. The second full scale check did not reach >95% conversion until 240 hours. The submitters report that after 48 hours, 85% conversion for the desired carbamate was observed by GC - MS analysis. The product appeared to crystallize in the reaction mixture.

8. NaNO$_2$, 99% was purchased from Aldrich Chemical Company, Inc. and was used without further purification. Sodium nitrite is used to quench any residual azide derivatives from the reaction mixture.

9. Hexanes and ethyl acetate were purchased from EMD Chemicals and distilled prior to use. The submitters report that hexanes and ethyl acetate (ACS grade) were purchased from Fisher Scientific and were used as received.

10. Chloroform (ACS grade) was purchased from Fisher Scientific and was used as received. For the recrystallization, the white solid is dissolved in a boiling solution (75 °C) of hexanes (50–60 mL) and chloroform (< 2 mL) in a 250-mL Erlenmeyer flask. The mixture is heated on a hot plate until complete dissolution of the solid occurs. The solution is then concentrated with heating on the hot plate to a volume of less than 25 mL. The mixture was cooled to room temperature over 10 min, then placed into a cold room at 4 °C overnight.

11. The solution was concentrated to less than 5 mL by heating on a hot plate.

12. The physical properties of the purified material **2** are as follows: mp 114-115 °C; R$_f$ 0.49 (10% EtOAc/hexanes); ^{1}H NMR (500 MHz, C$_6$D$_6$, 70 °C) δ: 1.46 (s, 9 H), 1.50 (s, 6 H), 1.86 (s, 9 H), 4.10 (s (br), 1 H). ^{13}C NMR (100 MHz, CDCl$_3$) δ: 28.4, 29.4, 36.3, 41.9, 50.4, 78.5 (br), 154.1 (br); IR (neat) 3321, 2977, 2906, 2851, 1685, 1526, 1362, 1172, 1054, 874

cm^{-1}; HMRS (EI) calcd for C$_{15}$H$_{25}$NO$_2$ [M]$^+$: 251.1885. Found: 251.1883. Anal. Calcd. for C$_{15}$H$_{25}$NO$_2$: C, 71.67; H, 10.02; N, 5.57; O, 12.73; found C, 71.69; H, 10.20; N, 5.57.

13. A second run on the same scale provided 6.06 g (80%) of product with the same physical characteristics and melting point. The submitters reported a yield of 6.41 g (85%).

Safety and Waste Disposal Information

All hazardous materials should be handled and disposed of in accordance with "Prudent Practices in the Laboratory"; National Academy Press; Washington, DC, 1995.

3. Discussion

The Curtius rearrangement involves the concerted degradation of an acyl azide into an isocyanate, which can be trapped by a variety of nucleophiles (such as alcohols), providing a convenient method to synthesize amine derivatives (such as carbamates). Although a number of methods have been reported for the preparation of acyl azides, such a starting material is unstable and difficult to handle. Diphenylphosphorazidate (DPPA) was the first reagent reported for the direct conversion of carboxylic acids into carbamates, in a one-pot process without isolation of the acyl azide intermediate.[2] However, many drawbacks are associated with the use of this reagent including the requirement for high temperatures and the difficult separation of the desired product from phosphorus residues. The procedure described herein provides a practical and efficient method for the synthesis of aliphatic Boc-protected amines by one-pot zinc-catalyzed Curtius rearrangement starting from a variety of carboxylic acids (Table 1).[3] The method uses readily and commercially available reagents. A mixture of sodium azide, di-*tert*-butyl dicarbonate and an aliphatic carboxylic acid produces sodium *t*-butoxide and the corresponding acyl azide, which spontaneously rearranges at 40 °C to the corresponding isocyanate. In the presence of catalytic amounts of Zn(OTf)$_2$ and tetrabutyl ammonium bromide, the *tert*-butoxide species then reacts with the isocyanate intermediate to form the carbamate derivative.

116

Table 1. Curtius Rearrangement of Carboxylic Acids

$$R \overset{O}{\underset{}{\overset{\|}{C}}} OH \xrightarrow[\substack{Bu_4NBr\ (15\ mol\%),\ Zn(OTf)_2\ (3.3\ mol\%) \\ THF,\ 40\ °C}]{Boc_2O\ (1.1\ equiv.),\ NaN_3\ (3\ equiv.)} R \overset{H}{\underset{O}{\overset{|}{N}}} Ot\text{-}Bu$$

entry	carbamate	isolated yield (%)
1	(adamantyl)–NHBoc	90 (85)[a]
2	NHBoc	94
3	NHBoc	77
4	NHBoc	68
5	NHBoc	80 (86)[a]
6	NHBoc	57
7	NHBoc	72
8	NHBoc, Ph	58 (45)[a]

[a] In parenthese, yields for 10 mmol scale.

The by-products are easy to eliminate, thus the purification proceeds efficiently. Compared to the original procedure,[3] the catalyst loading of Zn(OTf)$_2$ was decreased to 2 mol % and only 2 equivalents of sodium azide were necessary, when the reaction was run on 30 mmol scale. Although azide species are known to be potentially explosive and hazardous compounds, the experimental procedure described here is safe at the temperatures described. Indeed, TGA (thermogravimetric analysis) showed decomposition of the reaction mixture at temperatures >100 °C. This new methodology was also used with malonates derivatives, giving access to unnatural 2,2-disubstituted protected amino acids (Table 2). Finally, an alternative procedure for the synthesis of anilines from aromatic carboxylic acids has been recently disclosed.[4]

Table 2. Curtius Rearrangement of Malonate Derivatives

entry	carbamate	isolated yield (%)
1	EtO$_2$C, NHBoc, Me, Ph	75 (70)[a]
2	EtO$_2$C, NHBoc	65
3	EtO$_2$C, NHBoc, Ph	60

[a] In parentheses, yields for 30 mmol scale.

1. Département de Chimie, Université de Montréal, P.O. Box 6128, Station Downtown, Montréal (Québec), Canada, H3C 3J7. Helene.lebel@umontreal.ca
2. (a) Shioiri, T.; Ninomiya, K.; Yamada, S. *J. Am. Chem. Soc.* **1972**, *94*, 6203-6205. (b) Ninomiya, K.; Shioiri, T.; Yamada, S. *Tetrahedron* **1974**, *30*, 2151-2157. (c) Murato, K.; Shioiri, T.; Yamada, S. *Chem. Pharm. Bull.* **1975**, *23*, 1738-1740. (d) Shioiri, T.; Yamada, S-i. *Org. Synth.* **1984**, *62*, 187-190.
3. Lebel, H.; Leogane, O. *Org. Lett.* **2005**, *7*, 4107-4110.
4. Lebel, H.; Leogane, O. *Org. Lett.* **2006**, *8*, 5717-5720.

Appendix
Chemical Abstracts Nomenclature (Registry Number)

N-tert-Butyl adamantan-1-yl-carbamate: Carbamic acid,
 tricyclo[3.3.1.1³,⁷]dec-1-yl-, 1,1-dimethylethyl ester; (151476-40-3)
Sodium azide; (26628-22-8)
Adamantine-1-carboxylic acid: Tricyclo[3.3.1.1³,⁷]decane-1-carboxylic
 acid; (828-51-3)
Tetra-*n*-butyl ammonium bromide; (1643-19-2)
Zinc triflate; (54010-75-2)
Di-*tert*-butyl dicarbonate: Dicarbonic acid, C,C'-bis(1,1-dimethylethyl)
 ester; (24424-99-5)

Hélène Lebel received her B.Sc. degree in biochemistry from the Université Laval in 1993. She conducted her Ph.D. studies in organic chemistry at the chemistry department of the Université de Montréal under the supervision of professor André B. Charette as a 1967 Science and Engineering NSERC Fellow. In 1998, she joined the research group of professor Eric Jacobsen at Harvard University as a NSERC Postdoctoral Fellow. She started her independent career in 1999 at the Université de Montréal, where her research program focuses on the development of novel synthetic methods.

Olivier Leogane is from Guadeloupe in the West Indies. He received a B.Sc. degree in chemistry in 2002 from Université de Nantes. He then pursued his Ms.C in organic chemistry at Université de Paris VI in 2003. His Ph.D. studies were conducted under supervision of Hélène Lebel at Université de Montréal where he developed new catalyzed-methodologies for the Curtius Rearrangement. He is currently an industrial NSERC Postdoctoral Fellow at Tranzyme Pharma.

Thomas Painter was born in 1980 in Pittsburgh, Pennsylvania. During his undergraduate studies he interned at Valspar Corporation in 2002, and obtained his B.S. degree in chemistry from the University of Pittsburgh in 2003. He is currently finishing graduate studies as a Bayer Fellow in the laboratory of Professor Kay Brummond at the University of Pittsburgh. His graduate research has included work on the synthesis of electron- deficient trienones and ε-lactams, and progress toward bicyclic analogs of irofulven using rhodium(I)-catalyzed cycloisomerization reactions. He will be pursuing future research endeavors as a post-doctoral associate under the guidance of Professor Jeffrey Aubé at the University of Kansas.

Matthew Klinge was born in 1986 in Erie, Pennsylvania. He is currently pursuing undergraduate studies at the University of Pittsburgh with a declared major in biology, and a minor in chemistry. He conducted undergraduate research in the laboratory of Professor Kay Brummond at the University of Pittsburgh during the 2007-2008 academic year. He looks forward to graduating from the University of Pittsburgh in April of 2009, and wishes to attend medical school following graduation.

ENANTIOSELECTIVE OXIDATION OF AN ALKYL ARYL SULFIDE: SYNTHESIS OF (*S*)-(–)-METHYL *P*-BROMOPHENYL SULFOXIDE

Submitted by Carmelo Drago, Emma-Jane Walker, Lorenzo Caggiano and Richard F. W. Jackson.[1]

Checked by Katherine Rawls and Jonathan A. Ellman.

1. Procedure

A.   (S)-(–)-2-(N-3,5-Diiodosalicyliden)amino-3,3-dimethyl-1-butanol [(S)-1]: To a solution of (L)-(+)-*tert*-leucinol (350 mg, 3.0 mmol, 1.13 equiv) (Note 1) dissolved in 15 mL of MeOH (Note 2) in a flame-dried 50-mL round-bottomed flask kept under positive nitrogen pressure during the reaction is added 3,5-diiodosalicylaldehyde (1.00 g, 2.66 mmol, 1.0 equiv) (Note 3). The reaction mixture is stirred at room temperature (25 °C) for 48 h (Note 4). The solvent is evaporated under reduced pressure (30 °C, 40 mmHg) and the crude product is transferred to a 25-mL Erlenmeyer flask and purified by dissolving in 5 mL of hot EtOH (Note 5). The resulting yellow solution is allowed to reach room temperature slowly over a two-hour period (25 °C), allowing partial solvent evaporation (~1 mL), and needle crystal formation is observed. At this point, the Erlenmeyer flask is plugged and left overnight (17 h) in a refrigerator at 3 °C. The crystals are filtered in a Büchner funnel and washed with ice-cold EtOH (3 x 2 mL), and then dried under vacuum (0.02 mmHg) to afford the ligand **(S)-1** as a yellow solid (1.04 g, 82% yield) (Note 6).

B. *(S)-(–)-p-Bromophenyl methyl sulfoxide (3).* A flame-dried, three-necked, 250-mL round-bottomed flask is equipped with a rubber septum, a 25-mL pressure-equalizing addition funnel, a thermometer adaptor fitted with a suitable thermometer, and a magnetic stir bar (30 mm length, oval) and is placed under positive nitrogen pressure. The apparatus is charged with *(S)*-(-)-2-(*N*-3,5-diiodosalicyliden)amino-3,3-dimethyl-1-butanol **(S)-1** (425.8 mg, 0.90 mmol, 1.5 mol%) and 15 mL of $CHCl_3$ (Note 7). A green solution of vanadyl acetylacetonate ($VO(acac)_2$, 159.0 mg, 0.60 mmol, 1.0 mol%) (Note 8) in 15 mL of chloroform is added dropwise via the addition funnel to the stirred bright yellow solution at room temperature over 3 minutes. The stirred translucent solution immediately turns from yellow to green and then finally dark brown as it is left stirring for 30 min at room temperature open to the atmosphere. *p*-Bromophenyl methyl sulfide (2) (12.19 g, 0.060 mol, 1.0 equiv) (Note 9) is added in one portion to the reaction mixture, followed by $CHCl_3$ (30 mL) and the solution is cooled to 0 °C by means of a cryocool with an isopropanol bath. Once the internal temperature reaches 0 °C (20–30 minutes), an aqueous solution of hydrogen peroxide (7.75 g of a 31.6% solution in H_2O, 0.072 mol, 1.2 equiv) (Note 10) is added dropwise over a period of 25 min via the dropping funnel, keeping the internal temperature between 0-5 °C. The reaction mixture is then kept at 0 °C (± 1 °C) and vigorously stirred for 16 h (Note 11).

The reaction is quenched at 0 °C by the addition of 10 % w/v aqueous solution of sodium thiosulfate (100 mL) added dropwise via the addition funnel and then transferred to a 500-mL separatory funnel, rinsing with CH_2Cl_2 (2 x 50 mL) (Note 12). The layers are separated and the aqueous fraction is washed with CH_2Cl_2 (2 x 50 mL). The combined organic fractions are washed with saturated aqueous sodium chloride (50 mL) and dried over magnesium sulfate (10 g). The solution is filtered, and the magnesium sulfate is washed with 100 mL CH_2Cl_2. The organic solvent is evaporated under reduced pressure (30°C, 40 mmHg) to give the crude product as a dark brown solid (13.83 g, Sample 1, Note 13). The solid is dissolved in CH_2Cl_2 (17 mL) and layered on a silica pad (50 g SiO_2, Note 14) in a 7-cm diameter glass-sintered filter funnel (Note 15) and a mixture of sulfoxide and sulfone is eluted using EtOAc and petroleum ether (50:50, 200 mL) (Notes 16 and 17). The mixture is collected and evaporated under reduced pressure (3.37 g brown solid, Sample 2.1, Note 18). Elution with EtOAc (300 mL) and

122

evaporation under reduced pressure gives pure sulfoxide as a pale yellow solid (9.3 g, Sample 2.2, 71% yield) (Note 19).

Column chromatography is performed on the mixed fraction Sample 2.1 (Note 20) and elution with a mixture of EtOAc and petroleum ether (50:50, 550 mL, until complete elution of ligand **(S)-1** and sulfone) followed by EtOAc (1.4 L) and evaporation of the EtOAc fraction under reduced pressure gives pure sulfoxide as a yellow-brown solid (0.93 g, 7% yield, Sample 3) (Note 21).

The pure sulfoxide fractions, Samples 2.2 and 3, are combined in a 250-mL Erlenmeyer flask and dissolved in a mixture of hot EtOAc and heptane (20:80, 75 mL, Notes 22 and 23). Recrystallization occurs as the solution sits at room temperature (12 h, 25 °C), allowing for partial solvent evaporation (5 mL). Filtration on a Büchner funnel gives the sulfoxide, which is dried under vacuum (0.02 mmHg) to afford off-white needles (8.07 g, 61%, Sample 4) (Note 24). Further purification is performed by sublimation of the solid at a pressure of 0.02 mmHg over a period of 4 h in a standard sublimation apparatus, with a water-cooled cold finger and with the external oil bath at 90 °C, which gives (S)-(-)-p-bromophenyl methyl sulfoxide as a white powder (8.07 g, 99 % ee, 61% yield) (Note 25).

2. Notes

1. The submitters received (L)-*tert* leucinol from Degussa AG Industrie, who prepared the reagent by $I_2/NaBH_4$ reduction of (L)-*tert* leucine.[3] {$[\alpha]_D^{20}$ +40.3, (EtOH, c 1.0), lit. +35.3 (EtOH, c 3.05)}[2] The checkers purchased (S)-*tert*-leucinol from Aldrich Chemical Company, and it was used as received.

2. HPLC Grade methanol from Fisher Scientific (submitters) or Spectro Grade from Fisher Scientific (checkers) was used as received.

3. 3,5-Diiodosalicylaldehyde (97%), from Aldrich Chemical Company, was used as received.

4. Reaction progress was monitored by silica gel TLC, using 2:1 hexanes:EtOAc as eluent, and UV and ninhydrin stain to visualize (checkers). The aldehyde starting material has an R_f value of 0.65 (UV only), the desired imine product has an R_f value of 0.38 (yellow), and (S)-*tert*-leucinol (pink) remains at the baseline.

5. HPLC Grade ethanol from Fisher Scientific was used as received. The checkers heated the product in ethanol in a 25 mL Erlenmeyer flask in an oil bath at 90 °C (bath temperature) until all product dissolved.

6. The submitters reported a yield of 1.08 g, 86%. Characterization of (S)-(–)-2-(N-3,5-diiodosalicyliden)amino-3,3-dimethyl-1-butanol [(S)-1] is as follows: mp 162–163 °C, lit. mp 163–164 °C;[3] $[\alpha]_D^{20}$ –16.0 (acetone, c 1.0), lit.[4] $[\alpha]_D^{20}$ –16.6 (acetone, c 1.0); ν_{max} (cm^{-1}) 3303 (O-H), 2965 (CH=N, C-H), 1637 (CH=N, C-N); ^{1}H NMR (400 MHz, CDCl$_3$) δ: 0.99 (s, 9 H), 2.73 (br s, 1 H), 3.09 (dd, 1 H, J = 9.2, 2.5 Hz), 3.69 (dd, 1 H, J = 11.2, 10.4 Hz), 4.00 (dd, 1 H, J = 11.5, 2.5 Hz), 7.50 (d, 1 H, J = 2.1 Hz), 7.99 (d, 1 H, J = 2.0 Hz), 8.09 (s, 1 H), 14.88 (br s, 1 H,); ^{13}C NMR (125 MHz, CDCl$_3$) δ: 26.8, 32.9, 61.8, 75.7, 78.0, 93.0, 116.8, 141.1, 150.0, 164.7, 167.0; m/z (EI) 474 (MH$^+$, 22%), 473 (M$^+$, 100), 442 (22), 416 (73) and 359 (30); Anal. Calcd. for C$_{13}$H$_{17}$I$_2$NO$_2$: C, 33.00; H, 3.62; N, 2.96. Found: C, 33.02; H, 3.73; N, 2.83.

The submitters synthesized (R)-(-)-2-(N-3,5-diiodosalicyliden)amino-3,3-dimethyl-1-butanol [(R)-1] from (D)-(–)-tert-Leucinol (kindly provided as a gift from Degussa AG Industrie) according to the procedures outlined above, in similar yields. Use of the (R)-1 ligand in the sulfoxidation reaction afforded the corresponding (R)-sulfoxide product in similar yields and enantioselectivity.

7. HPLC Grade Chloroform from Fisher Scientific, was used as received.

8. Vanadyl acetylacetonate [VO(acac)$_2$], from Alfa Aesar, was used as received.

9. p-Bromophenyl methyl sulfide (4-bromothioanisole, 98 %), from Alfa Aesar, was used as received.

10. Hydrogen peroxide solution (~30% w/w in water) from Aldrich Chemical Company was used. The concentration of H$_2$O$_2$ was determined to be 31.6 % by titration with KMnO$_4$/H$_2$SO$_4$ using a procedure based on the Degussa analytical method WM 09 / WM 11.[5]

11. Reaction progress was monitored by silica gel TLC, using diethyl ether as eluent, and UV or p-anisaldehyde to visualize (checkers). The desired sulfoxide product has an R$_f$ value of 0.29 (white-yellow), the overoxidation product sulfone has an R$_f$ of 0.67 (UV only), and the starting sulfide has an R$_f$ of 0.90 (UV only).

12. HPLC Grade dichloromethane from Fisher Scientific, was used as received.

13. The checkers obtained a crude ee of 93.4% using a Chiralpak AS chiral column from Chiral Technologies (250 x 4.6 mm) on an Agilent 1100 Series LC equipped with a multiwavelength detector (85:15

heptane:isopropanol, 1 mL/min, λ = 222, 254, and 280 nm, t_R (minor) = 35.96 min, t_R (major) = 43.20 min). The checkers obtained a ratio of sulfoxide:sulfone of 86:14 by ^{1}H NMR.

14. The submitters purchased silica gel from BDH. The checkers purchased silica gel from Sorbent Technologies (60 Å, 230–400 mesh).

15. The submitters used a fritted funnel with pore size 2. The checkers used a 150-mL fritted funnel, ASTM, 40–60 C.

16. HPLC Grade Ethyl Acetate from Fisher Scientific was used as received.

17. Analytical Reagent Grade petroleum ether (40–60 °C) from Fisher Scientific (submitters) or Certified ACS Grade petroleum ether (30–60 °C) from Fisher Scientific (checkers) was used as received.

18. The submitters report a ratio of sulfoxide:sulfone of 67:33 by ^{1}H NMR for 6.09 g of mixed fractions. The checkers obtained a ratio of sulfoxide:sulfone of 37:63 for 3.37 g of mixed fractions.

19. The submitters report collection of 7.21 g (55%) of pure product.

20. The checkers used a column with diameter = 4.5 cm, length = 38.1 cm, loaded with 100 g of silica gel (see Note 13). The compound was loaded onto the column with 13 mL of CH_2Cl_2.

21. The submitters report collection of 3.3 g of pure product, 25% yield.

22. HPLC Grade heptane from Fisher Scientific was used as received.

23. The submitters used 100 mL of 80:20 heptane:EtOAc for recrystallization. The checkers found that the enantiomeric purity and yield of the product was highly dependent on the amount of solvent used for recrystallization. The checkers heated the product in 80:20 heptane:EtOAc in a 250-mL Erlenmeyer flask in an oil bath at 90 °C (bath temperature) until all solid dissolved.

24. The submitters reported the isolation of 9.3 g (70.7 %).

25. The submitters reported the isolation of 9.2 g (70%) at >99% ee. The checkers collected the sublimation product in two batches due to the small size of the sublimation apparatus. **(S)-(–)-4-Bromophenyl methyl sulfoxide (3)**: mp 77–78 °C; $[\alpha]_D^{20}$ –106.9 (acetone, c 1.8), 99 % ee; lit. [α] $_D^{20}$ –97.5 (acetone, c 1.8) for 94 % ee;[4] ^{1}H NMR (500 MHz, $CDCl_3$) δ: 2.69 (s, 3 H), 7.49 (d, 2 H, J = 8.4 Hz), 7.64 (d, 2 H, J = 8.3 Hz); ^{13}C NMR (125 MHz, $CDCl_3$) δ: 43.9, 125.0, 125.3, 132.5, 144.8; Anal. Calcd. for C_7H_7BrOS: C, 38.37; H, 3.22; S, 14.63. Found C, 38.36; H, 3.17; S 14.85. Chiral HPLC using a Chiralpak AS chiral column from Chiral Technologies

(250 x 4.6 mm) on an Agilent 1100 Series LC equipped with a multiwavelength detector (85:15 heptane:isopropanol, 1 mL/min, $\lambda = 222$, 254, and 280 nm) retention times are as follows: p-bromophenyl methyl sulfide t_R 5 min.; p-bromophenyl methyl sulfone t_R 30 min.; (R)-p-bromophenyl methyl sulfoxide t_R 37 min. and (S)-p-bromophenyl methyl sulfoxide t_R 45 min.

Waste Disposal Information

All toxic materials were disposed of in accordance with "Prudent Practices in the Laboratory"; National Academic Press: Washington, DC, 1995.

3. Discussion

Asymmetric oxidation of alkyl aryl sulfides has received a substantial amount of attention, due to the synthetic utility of the product sulfoxides. An existing *Organic Syntheses* procedure describes the preparation of (S)-$(-)$-methyl p-tolyl sulfoxide by oxidation of methyl p-tolyl sulfide using a stoichiometric reagent derived from (S,S)-$(-)$-diethyl tartrate and titanium(IV) isopropoxide, employing cumene hydroperoxide as the oxidant.[6] Purification of the product is achieved by flash chromatography. Herein is presented an alternative catalytic procedure for the asymmetric oxidation of p-bromophenyl methyl sulfide, based on our optimization[7,8] of the asymmetric sulfur oxidation procedure discovered by Bolm,[9] which employs a catalyst derived from VO(acac)$_2$ and an amino alcohol Schiff base ligand. A significant feature of this procedure is that the enantiomeric excess of the initially formed product is enhanced by a subsequent kinetic resolution process,[10] in which the minor enantiomer of the product is selectively oxidized to the corresponding sulfone. The key parameter is the use of chloroform as solvent, which promotes initial asymmetric oxidation and subsequent efficient kinetic resolution at the same temperature (namely 0 °C). In addition to the catalytic nature of the procedure, the main advantage is the use of hydrogen peroxide as oxidant.

From a practical point of view, the challenge is not achieving high levels of enantiomeric excess, but rather finding a convenient purification procedure that allows the separation of the by-product sulfone that is always produced (typically 10%). While flash column chromatography is an

126

effective method, we have made a substantial effort to purify the product sulfoxide without recourse to chromatography. However, in the case of both methyl p-tolyl sulfoxide and *p*-bromophenyl methyl sulfoxide, it has proved impossible to remove the corresponding sulfone by-product using distillation, trituration, crystallization or sublimation. The compromise purification method presented in this procedure, namely the use of a pad of silica gel in a sintered filter funnel, allows separation of pure sulfoxide product in reasonable yield, due to the large R_f difference between the sulfone and sulfoxide. The yield can be enhanced by conventional flash chromatographic purification of the mixed fraction that is eluted first.

Another application of Bolm's system to the asymmetric synthesis of (*R*)-(+)-2-methyl-2-propanesulfinamide has been recently reported by Ellman in *Organic Syntheses.*[11]

1. Department of Chemistry, University of Sheffield, Sheffield S3 7HF, UK, r.f.w.jackson@sheffield.ac.uk
2. Romo, D.; Romine, J. L.; Midura, W.; Meyers, A. I. *Tetrahedron* **1990**, *46*, 4951-4994.
3. McKennon, M. J.; Meyers, A. I.; Drauz, K.; Schwarm, M. *J. Org. Chem.* **1993**, *58*, 3568-3571.
4. Legros, J.; Bolm, C. *Chem. Eur. J.* **2005**, *11*, 1086-1092.
5. http://www.degussa.com.au/australia/MCMSbase/Pages/ ProvideResource.aspx?; respath=/NR/rdonlyres/EC5E6FED-34E5-4AD1-9D59E2ADB8840764/0/; Application_Determination_of_ Hydrogen_Peroxide_Concentration.pdf
6. Zhao, S. H.; Samuel, O.; Kagan, H. B. *Org. Synth.* **1990**, *68*, 49-55.
7. Pelotier, B.; Anson, M. S.; Campbell, I. B.; Macdonald, S. J. F.; Priem, G.; Jackson, R. F. W. *Synlett* **2002**, 1055-1060.
8. Drago, C.; Caggiano, L.; Jackson, R. F. W. *Angew. Chem., Int. Ed. Engl.* **2005**, *44*, 7221-7223.
9. Bolm, C.; Bienewald, F. *Angew. Chem., Int. Ed. Engl.* **1996**, *34*, 2640-2642.
10. Mohammadpoor-Baltork, I.; Hill, M.; Caggiano, L.; Jackson, R. F. W. *Synlett* **2006**, 3540-3544.
11. Weix, D. J.; Ellman, J. A.; Wang, X.; Curran, D. P. *Org. Synth.* **2005**, *82*, 157-165.

(*S*)-(–)-2-(*N*-3,5-Diiodosalicyliden)amino-3,3-dimethyl-1-butanol; (477339-39-2)

(L)-(+)-tert-leucinol: 1-Butanol, 2-amino-3,3-dimethyl-, (2*S*)-; (112245-13-3)

3,5-Diiodosalicylaldehyde: Benzaldehyde, 2-hydroxy-3,5-diiodo-; (2631-77-8)

(*S*)-(–)-*p*-Bromophenyl methyl sulfoxide: Benzene, 1-bromo-4-[(*S*)-methylsulfinyl]-; (145266-25-0)

Vanadyl acetylacetonate; (3153-26-2)

p-Bromophenyl methyl sulfide: Benzene, 1-bromo-4-(methylthio)-; (104-95-0)

Richard Jackson obtained his Ph.D. in 1984 under the supervision of the late Professor Ralph Raphael, developing a new route to dihydrofuranones, and completing a synthetic approach to the pseudomonic acids. He spent a year at the ETH, Zürich, working with Professor Dieter Seebach on total synthesis, before starting his independent career at the University of Newcastle. His research interests range from applications of organometallic chemistry in non-proteinogenic amino acid synthesis to catalytic asymmetric oxidation reactions. In 2001 he moved to a Chair in Synthesis at the University of Sheffield, serving as Head of Department from 2003 to 2007.

Carmelo Drago obtained his first degree in Chemistry from the University of Catania in 2001, undertaking a final year project on asymmetric oxidation reactions mediated by enzymes under the supervision of Dr. Giovanni Nicolosi at the CNR (National Research Council). He moved to Sheffield for his Ph.D., working on asymmetric sulfur oxidation under the supervision of Professor Richard F. W. Jackson. In 2005, he joined the group of Professor Carlo Scolastico, at the CISI (Centre for Bio-molecular Interdisciplinary Studies and Industrial applications – University of Milan) where he is working on design, synthesis and characterization of proapoptotic agents for cancer therapy.

Emma Walker obtained her M.Chem. with Industrial Experience from the University of Edinburgh in 2003 before moving to the University of Sheffield to undertake her Ph.D. on the development of new methods for screening catalysts for kinetic resolution under the supervision of Professor Richard F.W. Jackson. In 2008 Emma joined the Whisky Technical Specialist Team within Diageo Scotland as a Project Chemist.

After obtaining his degree from the University of Liverpool in 1998, Lorenzo Caggiano performed his Ph.D. with Dr. Stuart Warren at the University of Cambridge. In 2002 as a European Network Fellow in the group of Professor Cesare Gennari at the University of Milan, he was involved in the formal synthesis of eleutherobin. In February 2004 he became a Research Officer in the group of Professor Richard F. W. Jackson at the University of Sheffield. In January 2007, he started his independent career as a RCUK Research Fellow at the Department of Pharmacy and Pharmacology, University of Bath, UK.

Katherine Rawls was born in 1983 in Wheat Ridge, Colorado. She graduated from Santa Clara University in 2005 with a B.S. in Chemistry and Mathematics. There she worked on chemoselective "post-translational" modifications of *N*-alkylaminooxy containing peptides under the mentorship of Professor Michael Carrasco. Currently, she is a fourth year graduate student at the University of California, Berkeley, working under the direction of Prof. Jonathan A. Ellman. Her research includes the development and application of the Substrate Activity Screening (SAS) method for the identification of nonpeptidic tyrosine phosphatase inhibitors.

PROTECTION OF DIOLS WITH
4-(*tert*-BUTYLDIMETHYLSILYLOXY)BENZYLIDENE ACETAL
AND ITS DEPROTECTION:
(4-((4*R*,5*R*)-4,5-DIPHENYL-1,3-DIOXOLAN-2-YL)PHENOXY)(*tert*-
BUTYL)DIMETHYLSILANE

Submitted by Hiroyuki Osajima,[1] Hideto Fujiwara,[2] Kentaro Okano,[2] Hidetoshi Tokuyama,[2] and Tohru Fukuyama.[1]
Checked by Scott E. Denmark and Wen-Tau T. Chang.

1. Procedure

A.   (4-(Dimethoxymethyl)phenoxy)(tert-butyl)dimethylsilane.   A 100-mL oven-dried, three-necked round-bottomed flask equipped with a Teflon-coated magnetic stir bar, a rubber septum, a thermocouple and an argon gas inlet is charged with *tert*-butylchlorodimethylsilane (6.08 g, 40.3 mmol) (Note 1) and DMF (20 mL) (Note 2). To the mixture are added

130

4-hydroxybenzaldehyde (4.92 g, 40.3 mmol) (Note 3), imidazole (4.12 g, 60.5 mmol) (Note 4), and 4-dimethylaminopyridine (0.248 g, 2.02 mmol) (Note 5) in separate portions with stirring. After being stirred for 2 h (Note 6) at ambient temperature (19 °C), the reaction mixture is poured into water (30 mL) in a 500-mL separatory funnel and the suspension is extracted with EtOAc (2 x 25 mL). The combined organic extracts are diluted with EtOAc (50 mL), and then washed with 5% aq. NaCl solution (100 mL), water (4 x 100 mL), and sat. aq. NaCl solution (100 mL), then are dried over anhydrous $MgSO_4$ (1.6 g), filtered, and concentrated on a rotary evaporator (20 °C, 30 mmHg). The residue is dissolved with toluene (50 mL) and the resulting solution is concentrated on a rotary evaporator (40 °C, 30 mmHg) (Note 7) to give a pale yellow oil (9.43 g) (Note 8). The crude product is used in the next step without further purification (Note 9).

A 500-mL, oven-dried, three-necked round-bottomed flask equipped with a Teflon-coated magnetic stir bar, a rubber septum, a thermocouple and an argon gas inlet is charged with the crude product prepared above (9.43 g), MeOH (60 mL) (Note 10), and trimethyl orthoformate (40 mL) (Note 11). To the mixture is added pyridinium p-toluenesulfonate (1.01 g, 4.03 mmol) (Note 12). After being heated in a 50 °C oil bath (internal temperature: 43 °C) for 3.5 h (Note 13), the solution is cooled to ambient temperature and is poured into sat. aq. $NaHCO_3$ solution (100 mL) in a 1-L separatory funnel. The two-phase mixture is extracted with EtOAc (3 x 100 mL) and the combined organic extracts are washed with sat. aq. NaCl solution (2 x 100 mL), then are dried over Na_2SO_4 (2.2 g), filtered, and concentrated on a rotary evaporator (20 °C, 30 mmHg). The residue is dissolved with toluene (50 mL) and the resulting solution is concentrated on a rotary evaporator (40 °C, 30 mmHg) (Note 7) to give a yellow oil (10.07 g). The oil is purified by silica gel column chromatography (elution with hexanes/EtOAc, 20:1) (Notes 14, 15) to yield (4-(dimethoxymethyl)phenoxy)(*tert*-butyl)dimethyl silane (9.66 g, 85%) (Note 16).

B.   (4- ((4R, 5R)-4,5- Diphenyl-1,3- dioxolan-2-yl) phenoxy) (tert-butyl)dimethylsilane. A 500-mL, oven-dried, three-necked round-bottomed flask equipped with a reflux condenser (fitted with an argon gas inlet), a Teflon-coated magnetic stir bar, a rubber septum, and a thermocouple is charged with (1R,2R)-1,2-diphenylethane-1,2-diol (6.43 g, 30.0 mmol) (Note 17), toluene (96.0 mL) (Note 18), MeCN (32.0 mL) (Note 19), and (4-(dimethoxymethyl)phenoxy)(*tert*-butyl)dimethylsilane (9.27 g, 32.8 mmol). After stirring the solution for 3 min, pyridinium p-toluenesulfonate (2.26 g,

9.0 mmol) (Note 11) is added. The reaction mixture is heated in a 50 °C oil bath (internal temperature: 45 °C) and is stirred for 1 h (Note 20). The mixture is cooled to ambient temperature and then is poured into sat. aq. NaHCO₃ solution (100 mL) in a 1-L separatory funnel, and the suspension is extracted with EtOAc (3 x 100 mL). The combined organic extracts are washed with sat. aq. NaCl solution (2 x 100 mL), then are dried over anhydrous Na₂SO₄ (20 g), filtered, and concentrated on a rotary evaporator under reduced pressure to give a yellow, viscous oil. The residue is purified by silica gel column chromatography (elution with hexanes/CH₂Cl₂, 7:3) (Notes 21, 22) to give a colorless, viscous oil (12.33 g) (Note 23) which is triturated with MeOH to yield (4-((4*R*,5*R*)-4,5-diphenyl-1,3- dioxolan-2-yl)phenoxy) (*tert*-butyl) dimethylsilane as a white solid (11.51 g, 89%) (Note 24).

C. *(1R,2R)-1,2-Diphenylethane-1,2-diol.* A 500-mL, three-necked round-bottomed flask equipped with a reflux condenser (fitted with an argon inlet), a Teflon-coated magnetic stir bar, a rubber septum, and a thermocouple is charged with (4-((4*R*,5*R*)-4,5-diphenyl-1,3- dioxolan-2-yl) phenoxy) (*tert*-butyl)dimethylsilane (10.8 g, 25.0 mmol), K₂CO₃ (17.3 g, 125 mmol) (Note 25), MeOH (200 mL) (Note 9), and water (50 mL). After addition of KF (1.46 g, 25.0 mmol) (Note 26) and stirring for 5 min followed by addition of NH₂OH·HCl (8.69 g, 125 mmol) (Note 27), the reaction mixture is heated in an 85 °C oil bath (internal temperature: 69 °C) and is stirred for 37 h (Note 28). The reaction mixture is cooled to ambient temperature and is decanted (Note 29) into sat. brine (150 mL) in a 1-L separatory funnel and is extracted with EtOAc (3 x 200 mL). The combined organic extracts are washed with 3 M NaOH (3 x 100 mL), H₂O (100 mL), and brine (100 mL), and then are dried over anhydrous Na₂SO₄ (20 g), filtered, and concentrated on a rotary evaporator (21 °C 30 mmHg) to afford 5.40 g of the crude diol as a white solid. Purification by recrystallization yields (1*R*,2*R*)-1,2-diphenylethane-1,2-diol as white crystals (4.65 g, 87%) (Notes 30, 31).

2. Notes

1. *tert*-Butylchlorodimethylsilane (97%) was purchased from Aldrich Chemical Company, Inc. and used as received without further purification.

2. DMF (99.9%, water content: < 0.02%) was purchased from Fisher Scientific and used as received without further purification.

3. 4-Hydroxybenzaldehyde (98%) was purchased from Aldrich Chemical Company, Inc. and used as received without further purification.

4. Imidazole (99%) was purchased from Aldrich Chemical Company, Inc. and used as received without further purification.

5. 4-Dimethylaminopyridine (99%) was purchased from Aldrich Chemical Company, Inc. and used as received without further purification.

6. The reaction typically requires 2 h to go to completion and is monitored by ^{1}H NMR spectroscopy. A small aliquot of the reaction mixture is removed and is treated with brine. The organic layer is separated and concentrated under reduced pressure (30 °C, 30 mmHg) to afford the NMR sample. Stirring is stopped when the ratio of the signal for the formyl proton of the alcohol (δ 9.85, 1 H) to that of the TBS ether (δ 9.89, 1 H) is less than 0.10.

7. Azeotropic removal of water was carried out with toluene (≥99.5%, water content: < 0.030%) purchased from Aldrich Chemical Company, Inc. and used as received without further purification.

8. The ratio of the alcohol (δ 9.85, 1 H) to the TBS ether (δ 9.89, 1 H) is 0.06.

9. The compound exhibits the following physicochemical properties: R_f = 0.61 (hexanes/EtOAc, 4:1; Merck silica gel 60F-254 plate (visualized with 254 nm UV lamp and stained with an ethanol solution of $Ce_2(SO_4)_3$ and phosphomolybdic acid (Ce-PMA)); IR (neat) cm^{-1}: 2956, 2930, 2859, 1701, 1599, 1576, 1508, 1274, 1211, 1156, 909, 841, 800, 783, 705; ^{1}H NMR (500 MHz, CDCl$_3$) δ: 0.25 (s, 6 H), 0.99 (s, 9 H), 6.94 (d, J = 8.5 Hz, 2 H), 7.79 (d, J = 8.5 Hz, 2 H), 9.89 (s, 1 H); ^{13}C NMR (126 MHz, CDCl$_3$) δ: –4.4, 18.2, 25.5, 120.4, 130.4, 131.9, 161.5, 190.9.

10. Methanol (≥99.8%, water content: < 0.100%) was purchased from Aldrich Chemical Company, Inc. and used as received without further purification.

11. Trimethyl orthoformate (99%) was purchased from Aldrich Chemical Company, Inc. and used as received without further purification.

12. Pyridinium p-toluenesulfonate (98%) was purchased from Aldrich Chemical Company, Inc. and used as received without further purification.

13. The reaction typically requires 3.5 h to go to completion and is monitored by ^{1}H NMR spectroscopy. A small aliquot of the reaction mixture is removed and treated with brine. The organic layer is concentrated under reduced pressure (30 °C, 30 mmHg) to afford the NMR sample. Heating and stirring is stopped when the ratio of the signal of the aldehyde (δ 9.89, 1 H) to that of the dimethyl acetal (δ 5.33, 1 H) is less than 0.05 for maximizing the yield and reproducibility.

14. Silica gel (grade 9385, 60 Å) was purchased from Merck (230–400 mesh).

15. The crude material is dissolved in eluent (5 mL) and then is charged onto a column (diameter = 5 cm, height = 5 cm) of 42 g of silica gel (slurry packed). The column was eluted with hexanes/EtOAc, 20:1 and 50-mL fractions were collected. Fraction 2-7 were combined and concentrated on a rotary evaporator under reduced pressure (30 °C, 30 mmHg).

16. The compound exhibits the following physicochemical properties: R_f = 0.71 (hexanes/EtOAc, 4:1; Merck silica gel 60F-254 plate (visualized with 254 nm UV lamp and stained with Ce-PMA)); IR (neat) cm^{-1}: 2955, 2931, 2896, 2858, 1610, 1511, 1472, 1362, 1264, 1165, 1099, 1056, 915, 840, 805, 781; ^{1}H NMR (500 MHz, CDCl$_3$) δ: 0.19 (s, 6 H), 0.98 (s, 9 H), 3.31 (s, 6 H), 5.34 (s, 1 H), 6.83 (d, J = 8.5 Hz, 2 H), 7.30 (d, J = 8.5 Hz, 2 H); ^{13}C NMR (126 MHz, CDCl$_3$) δ: –4.5, 18.2, 25.6, 52.6, 103.1, 119.8, 127.8, 130.9, 155.8; MS (EI) m/z (relative intensity): 282.1 (9), 251.1 (100), 225.0 (24), 195 (6), 179 (6), 151 (4), 137 (3), 97 (4), 75 (39); Anal. Calcd for C$_{15}$H$_{26}$O$_3$Si: C, 63.78; H, 9.28. Found: C, 63.50; H, 9.64.

17. (1R,2R)-1,2-Diphenylethane-1,2-diol (99%, 98% ee) was purchased from Aldrich Chemical Company, Inc. and used as received without further purification.

18. Toluene (≥99.5%, water content: < 0.030%) was purchased from Aldrich Chemical Company, Inc. and used as received without further purification.

19. MeCN (≥99.9%, water content: < 0.01%) was purchased from Fisher Scientific and used as received without further purification.

20. The reaction typically requires 1 h to go to completion and is monitored by ^{1}H NMR spectroscopy. A small aliquot of the reaction mixture is removed and treated with brine. The organic layer is separated and concentrated under reduced pressure (30 °C, 30 mmHg) to afford the NMR sample. Heating and stirring is stopped when the signal for the dimethyl acetal (δ 5.33, 1 H) is fully converted to the signal for the diphenylethane acetal (δ 6.36, 1 H).

21. The crude material is dissolved in eluent (5 mL) and then is charged onto a column (diameter = 6 cm, height = 22 cm) of 345 g of silica gel (slurry packed). The column was eluted with hexanes/CH$_2$Cl$_2$, 7:3 and 50-mL fractions were collected. Fractions 17-53 were combined and concentrated on a rotary evaporator under reduced pressure (30 °C, 30 mmHg).

134

22. Alternatively, the conversion to (4-((4R,5S)-4,5-diphenyl-1,3-dioxolan-2-yl)phenoxy)(*tert*-butyl)dimethylsilane could be carried out following the method of Noyori[3] using (4-(dimethoxymethyl)phenoxy)-(*tert*-butyl)dimethylsilane and the bis(trimethylsilyl) ether.

23. The checkers found that the product can be crystallized by addition of methanol (10 mL) followed by sonication for 15 min. The murky solution is cooled to –27 °C for 1 h to induce precipitation and the white solid is collected by filtration (10.79 g). A second crop (0.72 g) is obtained by sonicating the concentrated mother liquor in MeOH (5 mL) for 1 min and cooling to –27 °C for 1 h.

24. The compound exhibits the following physicochemical properties: R_f = 0.20 (hexanes/CH$_2$Cl$_2$, 7:3; Merck silica gel 60F-254 plate (visualized with 254 nm UV lamp and stained with Ce-PMA)); mp 46–47 °C (MeOH); IR (film) cm^{-1}: 2955, 2929, 2884, 2857, 1610, 1513, 1264, 1108, 1088, 1071, 1013, 913, 839, 803, 781, 762, 698; ^{1}H NMR (500 MHz, CDCl$_3$) δ: 0.23 (s, 6 H), 1.01 (s, 9 H), 4.96 (s, 2 H), 6.36 (s, 1 H), 6.92 (d, J = 8.5 Hz, 2 H), 7.30–7.39 (m, 10 H), 7.56 (d, J = 8.5 Hz, 2 H); ^{13}C NMR (126 MHz, CDCl$_3$) δ: –4.4, 18.2, 25.7, 85.2, 87.1, 104.8, 120.1, 126.3, 126.9, 128.09, 128.12, 128.50, 128.54, 128.57, 130.9, 136.6, 138.4, 156.7; MS (ESI) m/z 433.3; MS (EI) m/z (relative intensity): 326.2 (100), 297.2 (79), 269.1 (54), 241.1 (11), 220.1 (21), 179.1 (16), 149.0 (9), 91.1 (19), 73.1 (27); Anal. Calcd for C$_{27}$H$_{32}$O$_3$Si: C, 74.96; H, 7.46. Found: C, 74.69; H, 7.52.

25. Potassium carbonate (98%, powdered, 325 mesh) was purchased from Aldrich Chemical Company, Inc. and used as received without further purification.

26. Potassium fluoride (99%) was purchased from Alfa-Aesar and used as received without further purification.

27. Hydroxylamine hydrochloride (98%) was purchased from Aldrich Chemical Company, Inc. and used as received without further purification.

28. The reaction typically requires 26–37 h to go to completion and is monitored by ^{1}H NMR spectroscopy. A small aliquot of the reaction mixture is removed and treated with brine. The organic layer is separated and concentrated under reduced pressure (30 °C, 30 mmHg) to afford the NMR sample. Heating and stirring is stopped when the ratio of the signal for the protected diol (δ 4.96, 2 H) to that for the starting diol (δ 4.70, 2 H) is less than 0.05 to maximize the yield and reproducibility.

29. The salt generated in the reaction causes some emulsion during the extraction.

30. To crystallize the product, the residue is dissolved in hot EtOAc/hexanes, 3:2 (100 mL). The solution is then cooled to room temperature and left in a freezer at –27 °C for 2 h. The colorless crystals are collected by filtration (4.01 g). A second crop of crystals (0.64 g) is obtained by dissolving the concentrated mother liquor in hot EtOAc/hexanes, 3:2 (13 mL) followed by the same cooling procedure.

31. The compound exhibits the following physicochemical properties: R_f = 0.52 (hexanes/EtOAc, 1:1; Merck silica gel 60F-254 plate (visualized with 254 nm UV lamp and stained with Ce-PMA)); mp 146–147 °C (hexane/EtOAc); IR (KBr) cm^{-1}: 3499, 3393, 2895, 1452, 1385, 1198, 1178, 1045, 1013, 846, 777, 747, 705, 695; ^{1}H NMR (500 MHz, CDCl$_3$) δ: 2.92 (s, 2 H), 4.70 (s, 2 H), 7.10–7.14 (m, 4 H), 7.21–7.25 (m, 6 H); ^{13}C NMR (126 MHz, CDCl$_3$) δ: 79.1, 126.9, 127.9, 128.1, 139.8; MS (EI) m/z (relative intensity): 214.1 (< 0.5%), 107.0 (100), 79.1 (81); Anal. Calcd for C$_{14}$H$_{14}$O$_2$: C, 78.48; H, 6.59. Found: C, 78.56; H, 6.66.

Waste Disposal Information

All hazardous materials should be handled and disposed of in accordance with "Prudent Practices in the Laboratory"; National Academic Press; Washington, DC, 1995.

3. Discussion

Protection and deprotection of functional groups play a critical role in successful synthesis of multifunctionalized, complex molecules. Among a variety of protecting groups on 1,2-diols,[3] the most frequently used ones are cyclic acetals and ketals. However, since their deprotections generally require strongly acidic conditions, these protecting groups are generally incompatible with acid-labile functional groups. During the course of synthetic studies on leustroducsin B[4], we devised a novel benzylidene-type protecting group, p-(tert-butyldimethylsilyl-oxy)benzylidene group[5] (Scheme 1). This protecting group could be removed under a mild two-step sequence, desilylation of the phenolic hydroxyl group with (HF)$_3$·Et$_3$N and the subsequent treatment with weakly acidic conditions (AcOH-THF-H$_2$O). Exploitation of this protective group was the key to achieve the total synthesis of leusroducsin B bearing highly acid-sensitive functionalities.

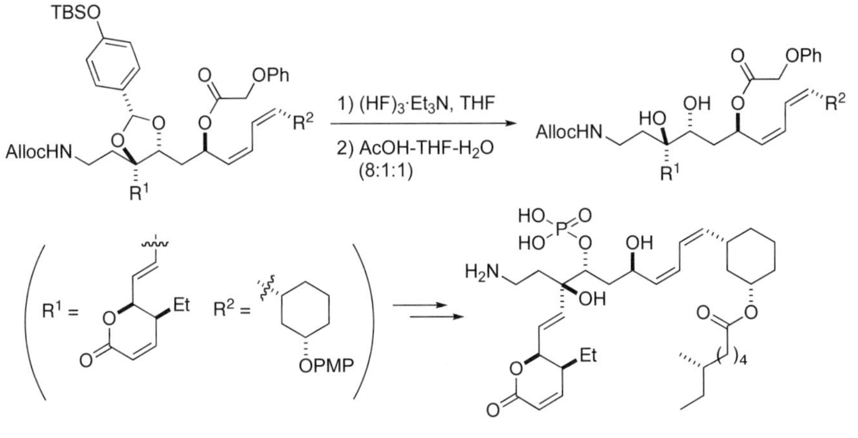

Scheme 1. Mild Acidic Deprotection of a Novel Benzylidene-Type Protecting Group.

Further investigations on the utility of this newly developed protective group revealed that this group could also be removed in a single step under mild basic conditions (Scheme 2). Thus, treatment of the *p*-hydroxybenzylidene acetal with K$_2$CO$_3$ in MeOH-H$_2$O at 70 °C gave the corresponding diol in almost quantitative yield. Furthermore, addition of NH$_2$OH·HCl and CsF dramatically facilitated the deprotection process and the single-step deprotection of the TBS ether took place smoothly. Trapping of the liberated *p*-hydroxybenzaldehyde with hydroxylamine would accelerate the deprotection process under the equilibrium conditions.

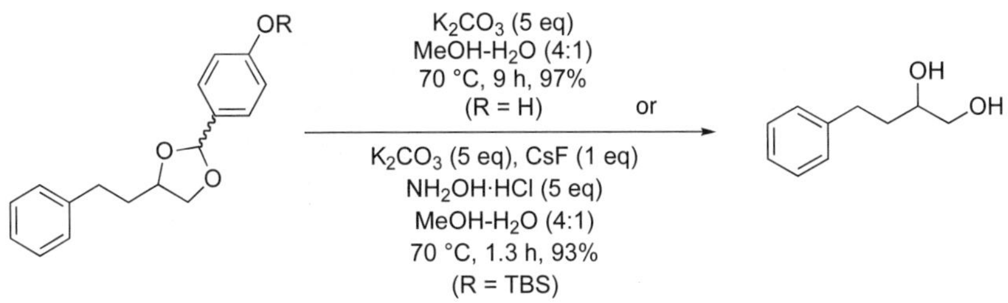

Scheme 2. Single-Step Deprotection under Basic Conditions

As shown in Table 1, acid-labile functional groups such as the THP ether and triphenymethyl group were unaffected during the deprotection process. While benzyl ester could not tolerate these basic conditions, more robust pivaloyl ester survived.

Table 1. Chemoselective Deprotection of the Benzylidene-Group under Basic Conditions[a]

(*p*-TBSO)C$_6$H$_4$ ... OTHP
2.5 h, 93%

(*p*-TBSO)C$_6$H$_4$ ... OTr
21 h, 91%

(*p*-TBSO)C$_6$H$_4$ ... OPiv
3.5 h, 92%

(*p*-TBSO)C$_6$H$_4$ ... OBz
1.5 h, 29%[b]

[a]Conditions: K$_2$CO$_3$ (5 equiv), NH$_2$OH·HCl (5 equiv), CsF (1 equiv), MeOH-H$_2$O (4:1), 70 °C. [b]Hydrolysis of benzoyl group occurred.

1. Graduate School of Pharmaceutical Sciences, University of Tokyo, 7-3-1 Hongo, Bunkyo-ku, Tokyo 113-0033, Japan.
2. Graduate School of Pharmaceutical Sciences, Tohoku University, Aramaki, Aoba-ku, Sendai 980-8578, Japan.
3. Greene, T. W.; Wuts, P. G. M. *Protective Groups in Organic Synthesis*, 4th ed.; Wiley & Sons: New York, **2006**; p 321.
4. Shimada, K.; Kaburagi, Y.; Fukuyama, T. *J. Am. Chem. Soc.* **2003**, *125*, 4048-4049.
5. Greene, T. W.; Wuts, P. G. M. *Protective Groups in Organic Synthesis*, 4th ed.; Wiley & Sons: New York, **2006**; p 339. (b) Kaburagi, Y.; Osajima, H.; Shimada, K.; Tokuyama, H.; Fukuyama, T. *Tetrahedron Lett.* **2004**, *45*, 3817-3821.

Appendix
Chemical Abstracts Nomenclature; (Registry Number)

tert-Butyldimethylsilyl chloride: Silane, chloro(1,1-dimethylethyl)dimethyl-; (18162-48-6)

138

4-Hydroxybenzaldehyde: Benzaldehyde, 4-hydroxy-; (123-08-0)

Imidazole: 1*H*-Imidazole; (288-32-4)

4-(Dimethylamino)pyridine: 4-Pyridinamine, *N,N*-dimethyl-; (1122-58-3)

Trimethyl orthoformate: Methane, trimethoxy-; (149-73-5)

Pyridinium *p*-toluenesulfonate: Benzenesulfonic acid, 4-methyl-, compd. with pyridine (1:1); (24057-28-1)

(1*R*,2*R*)-1,2-Diphenylethane-1,2-diol: 1,2-Ethanediol, 1,2-diphenyl-, (1*R*,2*R*)-; (52340-78-0)

Potassium fluoride: Potassium fluoride (KF); (7789-23-3)

Potassium carbonate: Carbonic acid, potassium salt (1:2); (584-08-7)

Hydroxylamine hydrochloride: Hydroxylamine, hydrochloride (1:1); (5470-11-1)

Tohru Fukuyama received his Ph.D. in 1977 from Harvard University with Yoshito Kishi. He remained in Kishi's group as a postdoctoral fellow until 1978 when he was appointed as Assistant Professor of Chemistry at Rice University. After seventeen years on the faculty at Rice, he returned to his home country and joined the faculty of the University of Tokyo in 1995, where he is currently Professor of Pharmaceutical Sciences. He has primarily been involved in the total synthesis of complex natural products of biological and medicinal importance. He often chooses target molecules that require development of new concepts in synthetic design and/or new methodology for their total synthesis.

Hiroyuki Osajima was born in Nishio, Aichi in 1978. He received his B.S. in 2003 and M.S. in 2005 from the University of Tokyo under the direction of Professor Tohru Fukuyama. During his M.S. studies, he worked on the development of the new benzylidene-type protecting group and synthetic studies toward pyrroloacridine alkaloid, plakinidines. Currently, he works for Tokyo CRO, Inc. as a CMC regulatory consultant.

Hideto Fujiwara was born in Iwate, Japan in 1984. He received his B.S. in 2007 from Tohoku Pharmaceutical University, where he carried out undergraduate research in the laboratories of Professor Yasuyuki Endo. In the same year, he started his doctoral studies at Graduate School of Pharmaceutical Sciences, Tohoku University under the supervision of Professor Hidetoshi Tokuyama. His current research interest is total synthesis of natural products.

Kentaro Okano was born in Tokyo in 1979. He received his B.S. in 2003 from Kyoto University, where he carried out undergraduate research under the supervision of Professor Tamejiro Hiyama. He then moved to the laboratories of Professor Tohru Fukuyama, the University of Tokyo and started his Ph.D. research on synthetic studies toward antitumor antibiotic yatakemycin by means of the copper-mediated aryl amination strategy. In 2007, he joined the faculty of Tohoku University, where he is currently an assistant professor in Professor Hidetoshi Tokuyama's group. His current research interest is natural product synthesis based on the development of new synthetic methodologies.

Hidetoshi Tokuyama was born in Yokohama in 1967. He received his Ph.D. in 1994 from Tokyo Institute of Technology under the direction of Professor Ei-ichi Nakamura. He spent one year (1994-1995) at the University of Pennsylvania as a postdoc with Professor Amos B. Smith, III. He joined the group of Professor Tohru Fukuyama at the University of Tokyo in 1995 and was appointed as Associate Professor in 2003. In 2006, he moved to Tohoku University, where he is currently Professor of Pharmaceutical Sciences. His research interest is on the development of synthetic methodologies and total synthesis of complex natural products.

Wen-Tau T. Chang obtained his BSc (Hon.) from the University of Auckland, New Zealand in 2005. Under the guidance of Assoc. Prof. Brent Copp, he investigated compounds with 4-thiazolidinone scaffold against the cell growth of *Mycobacterium tuberculosis*. In 2006, he began his graduate studies at the University of Illinois at Urbana-Champaign and joined the research group of Prof. Scott Denmark. His research interests include the transition-metal catalyzed transformation of organosilanol(ate)s.

SYNTHESIS OF 2-[3,3'-DI-(*TERT*-BUTOXYCARBONYL)-AMINODIPROPYLAMINE]-4,6,-DICHLORO-1,3,5-TRIAZINE AS A MONOMER AND 1,3,5-[*TRIS*-PIPERAZINE]-TRIAZINE AS A CORE FOR THE LARGE SCALE SYNTHESIS OF MELAMINE (TRIAZINE) DENDRIMERS

Submitted by Abdellatif Chouai, Vincent J. Venditto, and Eric E. Simanek.[1]
Checked by Ruth E. McDermott and John A. Ragan.

Org. Synth. **2009**, *86*, 141-150
Published on the Web 12/18/2008

1. Procedure

Caution! Cyanuric chloride is a lachrymator and causes burns on contact with the skin. All operations with this reagent should be carried out in a well-ventilated hood.

    A.   3,3'-Di-(tert-butoxycarbonyl)-aminodipropylamine. A 1-L, three-necked, round-bottomed flask equipped with a magnetic stirrer, a 500-mL addition funnel, a temperature probe and a static nitrogen inlet is charged with 3,3'-diaminodipropylamine (28.2 mL, 0.20 mol, 1.0 equiv) (Note 1), 300 mL of tetrahydrofuran (Note 1), and *N,N*-diisopropylethylamine (100 mL, 0.57 mol, 2.8 equiv) (Note 1). A 500-mL Erlenmeyer flask is charged with 2-(*tert*-butoxycarbonyloxyimino)-2-phenylacetonitrile (BOC-ON) (100 g, 0.41 mol, 2.0 equiv) (Note 1) and 300 mL of tetrahydrofuran. The resulting solutions are separately stirred at 0 °C for 30 min (Note 2). The BOC-ON solution is then transferred to the addition funnel and added dropwise to the solution of 3,3'-diaminodipropylamine over a 90-100 min period. After addition is complete, the solution is left to stir at 0 °C for 3 h, warmed to ambient temperature, and left to stir for an additional 20 h. The solvent is removed using a rotary evaporator at 39 °C (Note 3) and the residue is dissolved in 400 mL of dichloromethane (Note 1). The organic solution is washed with 10% NaOH (3 x 200 mL) (Notes 1 and 4), a saturated, aqueous solution of sodium chloride (1 x 300 mL) (Note 1), and dried over sodium sulfate (Note 1). Following filtration, the solvent is removed using a rotary evaporator at 32–39°C (Note 3) to afford the product as an oily material which is precipitated as an off-white solid by addition of hexane (500 mL) (Note 1) and traces of MeOH (3 mL) (Note 1). After standing in the freezer for 24 h, the solids are filtered, washed with hexane, and dried under vacuum overnight to provide the product as an off-white solid (54.2–55.9 g, 82–84 %), mp 68.1–70.0 °C (Notes 5 and 6).

    B.   2-[3,3'-Di-(tert-Butoxycarbonyl)-aminodipropylamine]-4,6-dichloro-1,3,5-triazine. A 3-L, three-necked, round-bottomed flask equipped with a mechanical stirrer, static nitrogen inlet and 1-L addition funnel is charged with cyanuric chloride (30.5 g, 0.165 mol, 1.00 equiv) (Note 1) and 300 mL of acetone (Note 1). The resulting solution is left to stir at 0 °C for 1 h (Note 7). A cooled solution of 3,3'-di-(*tert*-butoxycarbonyl)-aminodipropylamine (54.9 g, 0.166 mol, 1.01 equivalent) in acetone (686 mL) is then added dropwise to the cyanuric chloride solution over a period of 3 h (the internal temperature remained at or below 2 °C during this addition). A white suspension forms during the course of addition. Sodium bicarbonate (13.9 g, 0.166 mol, 1.00 equiv) (Note 1) in water (195 mL) is added dropwise over a period of 1 h. A yellow mixture is obtained after complete addition. The resulting solution is left to stir at 0 °C for 3 h, which resulted in the formation of a white suspension. The mixture is allowed to warm to ambient temperature and stirred for an additional 15 h. The reaction mixture is concentrated without filtration (to approximately 200 mL) on a rotary evaporator at 31–40 °C (Note 3). The resulting aqueous suspension is filtered. The solids are dissolved in 600 mL of dichloromethane (Note 1). The organic phase is washed with water (3 x 250 mL), and a saturated, aqueous solution of sodium chloride (300 mL) (Note 1). The organic layer is dried with sodium sulfate (Note 1), filtered, and the solvent is concentrated using a rotary evaporator at 40 °C (Note 3). The resulting solids are dried under vacuum (Note 5) to provide the product as an off-white solid (76.5 g, 0.160 mol, 96 %), mp 122.4-125.7 °C (Notes 8, 9 and 10).

C.   *1,3,5-[Tris-N-(tert-butoxycarbonyl)-piperazine]-triazine.* A 2-L, three-necked, round-bottomed flask equipped with a magnetic stirrer, reflux condenser, temperature probe, glass stopper and static nitrogen inlet is charged with cyanuric chloride (10.0 g, 54.2 mmol, 1.00 equiv) (Note 1) and tetrahydrofuran (1 L) (Note 1?). *N-(tert-*Butoxycarbonyl)-piperazine (34.0 g, 183 mmol, 3.38 equiv) (Note 1) is added in ~10 g portions over 17 min (during the addition, the temperature rose from 20 to 28 °C; an ambient temperature water bath was used to moderate the exotherm). White solids are formed in the reaction mixture during the addition of the piperazine. *N,N*-Diisopropylethylamine (96.2 mL, 552 mmol, 10.2 equiv) (Note 1) is added, and the reaction mixture is stirred at ambient temperature for 1 h, then heated to an internal temperature of 66 °C for 20 h at which point the reaction was judged to be complete by HPLC (Note 11). Upon cooling to ambient temperature, a white precipitate forms (Note 12). The solvent is removed using a rotary evaporator at 31 °C (Note 3). The white residue is taken up in dichloromethane (300 mL) (Note 1) and washed with water (2 x 150 mL), 10% NaHSO$_4$ (2 x 150 mL) and a saturated, aqueous solution of sodium chloride (2 x 100 mL) (Note 1). The organic layer is dried over anhydrous sodium sulfate (Note 1), filtered, and the solvent is removed using a rotary evaporator at 35–41 °C (Note 3). The resulting white solids are granulated in EtOAc (35 mL, ca. 1 mL/g) (Note 1) to yield a white crystalline material (31.0–32.0 g, 90–93 %), mp 223.4–226.1 °C (Note 13).

D.   1,3,5-[Tris-piperazine]-triazine. A 1-L, three-necked, round-bottomed flask, fitted with a magnetic stirrer, condenser, nitrogen inlet, and 250 mL addition funnel is charged with 1,3,5-[*N*-(*tert*-butoxycarbonyl)-piperazine]-triazine (27.9 g, 44.1 mmol, 1.00 equiv) and 286 mL of methanol (Note 1). The solution is left to stir at 0 °C for 30 min. A solution of 153 mL (0.918 mol, 21 equiv) of 6N hydrochloric acid (Note 1) is then added over 70 min keeping the temperature at ~ 1 °C and the resulting light yellow slurry is left to stir at 0 °C for 2 h (Note 2). The reaction slurry is then allowed to warm to ambient temperature over 3 h and then slowly heated to an internal temperature of 40 °C for 12 h (the slurry became homogeneous at 27 °C). Off-gassing was observed as the temperature increased. The volatile organic components are removed using a rotary evaporator 34–40 °C (Note 3) until only ca. 100 mL of water remains. The resulting aqueous solution is cooled to 0 °C and made alkaline (pH = 14) by addition of 237 mL (657 mmol, 15 equiv) of a 10% NaOH solution (Note 1). The resulting alkaline solution is extracted with chloroform (3 x 250 mL) (Note 1), and the organic phases are combined and dried over sodium sulfate (Note 1). The solvent is filtered and evaporated at 34 °C (Note 3) to afford the product as a white solid (14.1 g, 96 %), mp 200–208 °C (Notes 14 and 15).

2. Notes

1.   All solvents and reagents were purchased from commercial suppliers and used as received.   Acetone (99.5%), 2-(*tert*-butoxycarbonyloxyimino)-2-phenylacetonitrile (BOC-ON), chloroform (99.8%), 3,3'-diaminodipropylamine (98%), dichloromethane (99.6%), *N,N*-diisopropylethylamine (99%), methanol (99.8%), sodium chloride, sodium sulfate (anhydrous) and tetrahydrofuran (99.9%, anhydrous) were purchased from Sigma-Aldrich. Cyanuric chloride was purchased from Alfa Aesar. *N*-Boc-piperazine was purchased from AK Scientific. Hydrochloric acid and sodium bicarbonate were purchased from J. T. Baker. Sodium hydroxide was purchased from Fisher Scientific. Hexanes (98.5%) were purchased from Mallinckrodt.

2. All reactions were performed in an ambient-temperature lab (ca. 23 °C). The submitters performed all experiments in a walk-in cold room (0 °C).

3. All solvent evaporations were performed using a rotary evaporator using appropriate temperature and pressure to afford efficient concentration (THF: 75-150 mmHg; CH_2Cl_2: 200-300 mmHg; acetone: 200-300 mmHg; $CHCl_3$ and methanol: 15-75 mmHg).

4. The first and second extractions resulted in a yellow aqueous solution (color of the byproduct of BOC-ON), while the third extraction gave a clear aqueous solution.

5. All products were dried using a Büchi vacuum pump with a vacuum pressure of 76 mmHg.

6. The product has the following spectral characteristics: TLC R_f = 0.0 (silica gel 60 $F_{254,}$ EMD Chemicals, Inc. in 5:95 methanol:dichloromethane); IR (neat) cm^{-1}: 3342, 2975, 2931, 1686, 1518, 1365, 1273, 1250, 1168; ^{1}H NMR (400 MHz, CDCl$_3$) δ: 1.41 (s, 18 H), 1.60–1.66 (m, 4 H), 2.63 (t, 4 H, J = 6.6), 3.15–3.20 (br, 4 H), 5.2 (br, 2 H); ^{13}C NMR (100 MHz, CDCl$_3$) δ: 28.4(s), 29.7 (s), 38.9 (s), 47.4 (s), 78.9 (s), 156.1 (s); MS (CI), m/z 332.2 (M+H). Anal. Calcd for $C_{16}H_{33}N_3O_4$: C, 57.98; H, 10.04; N, 12.68. Found: C, 57.86; H, 9.84; N, 12.51.

7. The submitters noted that after 1 h at 0 °C, undissolved cyanuric chloride remained visible in the reaction flask (estimated at 10–20% of the initial charge). The presence of this material does not adversely affect the reaction. The checkers observed a homogeneous solution at this stage.

8. During melting point determination, some bubbling was observed above 125 °C.

9. The product has the following characteristics: TLC R_f = 0.3 (silica gel 60 $F_{254,}$ EMD Chemicals, Inc. in 5% methanol:dichloromethane); IR (neat) cm^{-1}: 3349, 2976, 1691, 1573, 1475, 1233, 1160, 847, 733; ^{1}H NMR (400 MHz, CDCl$_3$) δ: 1.41 (s, 18 H), 1.77 (tt, apparent quintet, 4 H, J = 6.6), 3.07-3.12 (br/quartet-depending on sample concentration, 4 H), 3.60 (t, 4 H, J = 7.1), 5.05 (br, 2H); ^{13}C NMR (100 MHz, CDCl$_3$) δ: 27.7 (s), 28.3 (s), 37.3 (s), 44.9 (s), 79.3 (s), 156.0 (s), 164.7 (s), 170.1 (s); MS (CI): m/z 479, 379. Anal. Calcd for $C_{19}H_{32}Cl_2N_6O_4$: C, 47.60; H, 6.73; N, 17.54; Cl, 14.79. Found: C, 47.87; H, 6.82; N, 17.48; Cl, 14.47.

10. In a second run using 53.2 g of the starting di-(BOC)-dipropylamine, 71.5 g of product was obtained (93%).

11. The reaction was monitored by HPLC on a 4.6 x 50 Zorbax C-8 column at 30 °C with a flow rate 0.7 mL per min of eluent composed of 95% of 0.5% $HClO_4$ and 5% ACN. RT cyanuric chloride = 6.68 min, RT product =11.54 min. Also TLC (silica gel, 5% methanol:dichloromethane), R_f(product) = 0.35, Silica Gel 60 F_{254}, EMD Chemicals, Inc.

12. The precipitate corresponds to the formation of the hydrochloride salt of N,N-diisopropyl ethylamine.

13. The product has the following characteristics: TLC, R_f 0.35 (silica gel 60 F_{254}, EMD Chemicals, Inc. in 5% methanol:dichloromethane); IR (neat) cm^{-1}: 1679, 1535, 1419, 1227, 998, 725; ^{1}H NMR (400 MHz, CDCl$_3$) δ: 1.46 (s, 27 H), 3.42 (m, 12 H), 3.72 (m, 12 H); ^{13}C NMR (100 MHz, CDCl$_3$) δ: 28.4, 43.0, 79.9, 154.8, 165.2; MS (CI) m/z 634.3; Anal. Calcd for $C_{30}H_{51}N_9O_6$: C, 56.85; H, 8.11; N, 19.89. Found: C, 56.69; H, 8.26; N, 19.82.

14. In a second run, 29.2 g of starting triazine provided 14.5 g of product (94%).

15. The product has the following characteristics: TLC R_f = 0.0 (silica gel 60 F_{254}, EMD Chemicals, Inc. in 10% methanol:dichloromethane); IR (neat) cm^{-1}: 3278, 2846, 1523, 1433, 1242, 1007, 806, 728; ^{1}H NMR (400 MHz, CDCl$_3$) δ: 1.62 (s, 3 H), 2.81 (t, 12 H, J = 5.0), 3.68 (t, 12 H, J = 5.0); ^{13}C NMR (100 MHz, CDCl$_3$) δ: 44.3, 46.0, 165.2; MS (CI) m/z 334.4 Anal. Calcd for $C_{15}H_{27}N_9$: C, 54.03; H, 8.16; N, 37.81. Found: C, 53.72; H, 8.32; N, 37.48.

Safety and Waste Disposal Information

All hazardous materials should be handled and disposed of in accordance with "Prudent Practices in the Laboratory"; National Academy Press; Washington, DC, 1995.

3. Discussion

s-Triazines have seen applications as precursors for polymers,[2] as scavenging resins in organic manipulations,[3] as components of host–guest assemblies,[4] as ligand scaffolds for catalysis,[5] and in medicinal chemistry.[6] The appeal of the s-triazine core is largely due to the ease of systematic substitution of the chlorine atoms with amine nucleophiles to generate a variety of structures. Consecutive substitution reactions with amine nucleophiles can proceed in a one-pot procedure: the first substitution on cyanuric chloride occurs in minutes at 0 °C; the second substitution occurs in 12–24 hours at ambient temperature; the third substitution typically occurs in 12–24 hours and requires temperatures above 80 °C.[7]

Our interest in *s*-triazines derives from their use in dendrimer synthesis.[8] Triazine dendrimers offer distinct advantages including tractable syntheses and reactivity that can be efficiently manipulated in a stepwise manner to construct pure monodisperse products with unique compositional diversity.[9] This paper describes the use of 1,3,5-triazine in the synthesis of 2-[3,3'-di-(tert-butoxycarbonyl)-aminodipropylamine]-4,6,-dichloro-1,3,5-triazine as a building block and 1,3,5-[*tris*-piperazine]-triazine as a core.

The building block, a dichloro-1,3,5-triazine substituted with a branching group presenting two Boc-protected amines, is available in 96% overall yield from the reaction of cyanuric chloride with 3,3'-di-(*tert*-butoxycarbonyl)-aminodipropylamine in acetone-water, without chromatographic purification. The latter is prepared from the reaction of 3,3'-diaminodipropylamine with BOC-ON in THF under simple experimental conditions. This building block can be regarded as AB_2B' (where A is the first chloride displaced via nucleophilic aromatic substitution, B' is the second chloride to be displaced, and B represents the BOC-protected amines). It undergoes clean, stepwise nucleophilic aromatic substitution (S_NAr) reactions to afford intermediate monochlorotriazine bearing macromolecules.

The synthesis of the core proceeds by the reaction of cyanuric chloride with *N*-(*tert*-butoxycarbonyl)-piperazine in THF at 66 °C in the presence of *N,N*-diisopropyl ethylamine at moderate reaction times (16 h) (Step C). Cleavage of the BOC-group using 6N hydrochloric acid in methanol at 0 °C for 3 h and at room temperature for 12 h followed by neutralization affords the desired product in high yield (Step D). The synthesis of higher generation dendrimers relies on iterative addition of the AB_2B' building block to core structure followed by capping and deprotection steps.[10]

1. Department of Chemistry, Texas A&M University, College Station, TX 77843. simanek@mail.chem.tamu.edu

2. (a) Hirschberg, J. H. K. K.; Ramzi, A.; Sijbesma, R. P.; Meijer, E. W.; *Macromolecules* **2003**, *36*, 1429-1432. (b) Berl, V.; Schmutz, M.; Krische, M. J.; Khoury, R. G.; Lehn, J. -M. *Chem. Eur. J.* **2002**, *8*,

1227-1244. (c) Choi, I. S.; Li, X.; Simanek, E. E.; Akaba, R.; Whitesides, G. W. *Chem. Mater.* **1999**, *11*, 684-690.

3. Marsh, A., Carlisle, S. J.; Smith, S.C . *Tetrahedron Lett.* **2001**, *42*, 493-496.

4. (a) Linton, B.; Hamilton, A. D. *Curr. Opin. Chem. Biol.* **1999**, *3*, 307-312. (c) Simanek, E. E.; Isaacs, L.; Li, X.; Wang, C. C. C.; Whitesides, G. W. *J. Org. Chem.* **1997**, *62*, 8994-9000.

5. (a) Gamez, P.; de Hoog, P.; Lutz, M.; Spek, A. L.; Reedijk, J. *Inorg. Chim. Acta* **2003**, *351*, 319-325. (b) Britovsek, G. J. P.; Gibson, V. C.; Hoarau, O. D.; Spitzmesser, S. K.; White, A. J. P.; Williams, .; D. J. *Inorg. Chem.* **2003**, *42*, 3454-3465.

6. (a) Pattarawarapan, M.; Reyes, S.; Xia, Z.; Zaccaro, M. C.; Saragovi, H. U.; Burgess, K. *J. Med. Chem.* **2003**, *46*, 3565-3567. (b) Pitts, W. J.; Guo, J.; Dhar, T. G. M.; Shen, Z.; Gu, H. H.; Watterson, S. H.; Bednarz, M. S.; Chen, B. –C.; Barrish, J. C.; Bassolino, D.; Cheney, D.; Fleener, C. A.; Rouleau, K. A.; Hollenbaugh, D. L.; Iwanowicz, E. J. *Bioorg. Med. Chem. Lett.* **2002**, *12*, 2137-2140.

7. (a) Blotny, G. *Tetrahedron* **2006**, *62*, 9507-9522 and references therein. (b) Steffensen, M. B.; Hollink, E.; Kuschel, F.; Bauer, M.; Simanek, E. E. *J. Polym. Sci., Part A: Polym. Chem.* **2006**, *44*, 3411-3433. (c) Steffensen, M. B.; Simanek, E. E. *Org. Lett.* **2003**, *5*, 2359-2361. (d) Chen, H. –T.; Neerman, M. F.; Parrish, A. R.; Simanek, E. E. *J. Am. Chem. Soc.* **2004**, *126*, 10044-10048. (e) Umali, A. P.; Simanek, E. E. *Org. Lett.* **2003**, *5*, 1245-1247. (f) Cuthbertson, W. W.; Moffatt, J. S. *J. Chem. Soc.* **1948**, 561-564.

8. (a) Hollink, E.; Simanek, E. E. *Org. Lett.* **2006**, *8*, 2293-2295. (b) Chouai, A.; Simanek, E.E.; *J. Org. Chem.* **2008**, *73*, 2357-2366.

9. (a) Lim, J.; Simanek, E.E. *Org. Lett.* **2008**, *10*, 201-204. (b) Steffensen, M.B.; Simanek, E.E. *Angew. Chem. Int. Ed.* **2004**, *43*, 5178-5180.

10. Crampton, H.; Hollink, E.; Perez, L. M.; Simanek, E. E. *New J. Chem.* **2007**, *31*, 1283-1290.

Appendix
Chemical Abstracts Nomenclature; (Registry Number)

3,3'-Diaminodipropylamine; (56-18-8)

2-(*tert*-Butoxycarbonyloxyimino)-2-phenylacetonitrile; (58632-95-4)

Cyanuric chloride; (108-77-0)

N-(*tert*-Butoxycarbonyl)-piperazine; (57260-71-6)

N,N-Diisopropyl ethylamine; (7087-68-5)

2-[3,3'-Di-(*tert*-butoxycarbonyl)-aminodipropylamine]-4,6-dichloro-1,3,5-triazine; 12-Oxa-2,6,10-triazatetradecanoic acid, 6-(4,6-dichloro-1,3,5-triazin-2-yl)-13,13-dimethyl-11-oxo-, 1,1-dimethylethyl ester; (947602-03-1)

1,3,5-[Tris-piperazine]-triazine: 1,3,5-Triazine, 2,4,6-tri-1-piperazinyl-; (19142-26-8)

Eric E. Simanek was born in 1969 in Tuscola, IL. He obtained a B.S. in Chemistry in 1991 from the University of Illinois at Urbana-Champaign while working in the laboratories of the late Dr. Kenneth L. Rinehart, Jr. After completing doctoral studies with Dr. George M. Whitesides at Harvard University in 1997, he joined Dr. Chi-Huey Wong's laboratory at The Scripps Research Institute. Since joining Texas A&M University in 1998, he has risen through the ranks to Professor of Chemistry. His interests lie in drug delivery and K-20 education.

Abdellatif Chouai was born in 1971 in Morocco. He earned a B.S. in Chemistry in 1995 from the University of Sidi Mohamed Ben Abdellah, Morocco, and a Ph.D. in Organic Chemistry from University of Houston in 2003 under Dr. Randolph P. Thummel. He joined Dr. Kim R. Dunbar's group at Texas A&M University as a postdoctoral research associate where he worked on reversible DNA biosensors complexes and bimetallic complexes as photodynamic therapy agents. He then moved to a research scientist position in 2006 in Dr. Simanek's laboratory. His research focused on an industrial scale production of triazine-based dendrimers and application in drug delivery. Currently, Dr. Chouai holds a professional development chemist position with BASF Corporation.

Vincent J. Venditto was born in 1981 in Philadelphia, PA. He earned a B.S. in Chemistry from Gettysburg College in 2003 and began working in the laboratory of Dr. Martin W. Brechbiel in the Radioimmune and Inorganic Chemistry Section of the National Cancer Institute within the NIH. After two years at the NIH, Vincent joined of Dr.Simanek's laboratory at Texas A&M University to pursue a graduate degree in chemistry. His graduate work focuses on the synthesis of triazine-based dendrimers as drug carriers for a range of therapeutic applications.

Ruth McDermott was born in Boston, MA and received a B.S. in Chemistry from the University of Massachusetts at Boston. She completed her undergraduate research under the direction of J.-P. Anselme. She is currently part of the Chemical Research and Development group at Pfizer in Groton, CT where she has been employed for 20 years. She enjoys photography and participating in short-term mission trips in the US and abroad.

LARGE SCALE, GREEN SYNTHESIS OF A GENERATION-1 MELAMINE (TRIAZINE) DENDRIMER

A.

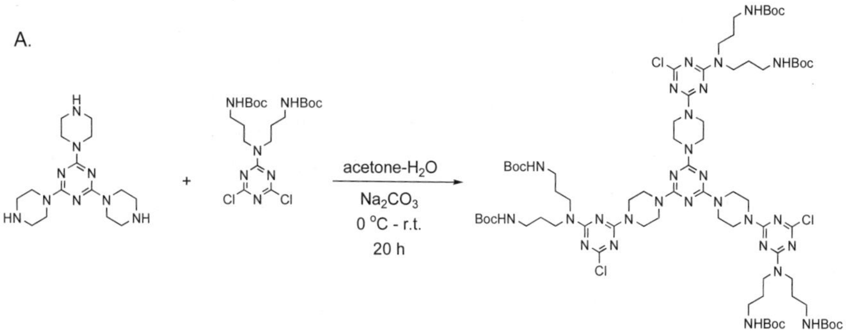

B.

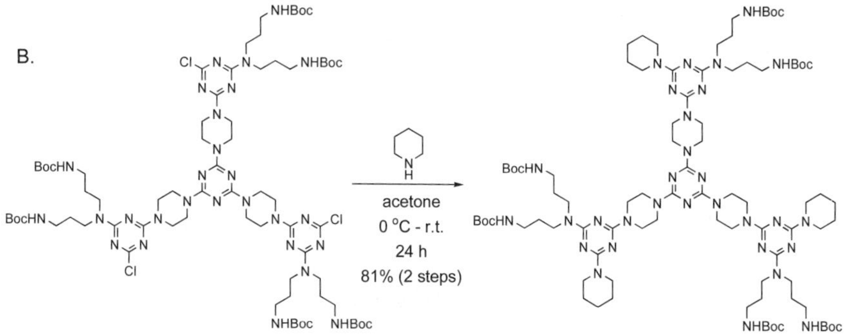

Submitted by Abdellatif Chouai, Vincent J. Venditto, and Eric E. Simanek.[1]
Checked by Brian C. Vanderplas and John A. Ragan.

1. Procedure

A.   G1-[N(CH₂CH₂CH₂NHBoc)₂]₆-Cl₃. In a 4-L, 4-necked, jacketed reaction vessel equipped with a 250-mL addition funnel, temperature probe, static N_2 and mechanical stirrer, 2-[3,3'-di-(*tert*-butoxycarbonyl)-aminodipropylamine]-4,6-dichloro-1,3,5-triazine (73.0 g, 0.152 mol, 3.5 equiv) (Note 1) is dissolved in acetone (1 L) (Note 2) and cooled to 0 °C (Note 3). Separately, a chilled solution of 1,3,5-[*tris*-piperazine]-triazine (14.5 g, 43.5 mmol, 1.0 equiv) (Note 4) in H_2O (500 mL) is prepared and treated with a solution of sodium carbonate (46.1 g, 0.435 mol, 10 equiv)

(Note 5) in 250 mL of H_2O. This solution is left to stir at 0 °C for 30 min. The resulting aqueous solution is added in a dropwise fashion to the acetone solution at 0 °C over a period of 2 h. The white suspension obtained after complete addition is left to stir at 0 °C for 2.5 h before warming gradually to 21 °C, and then stirred for an additional 20 h (Note 6). The white solid is collected by filtration on a 15 cm-diameter Büchner funnel. The reaction vessel is rinsed with 500 mL water, which is subsequently used to wash the filter cake (Notes 7 and 8). The wet solids are transferred back to the rinsed reaction vessel and dissolved in CH_2Cl_2 (1.5 L) (Note 9), washed with water (3 x 200 mL) (Note 10), a saturated, aqueous solution of sodium chloride (1 x 1.5 L) (Note 11), and then dried with 230 g sodium sulfate (Note 12). Following filtration, the solvent is removed using a rotary evaporator at 30 °C (Note 8) and dried under vacuum (Note 13) to yield an off-white crude material (76.1 g) (Note 14). This material is used in the next step without further purification (Note 15, 16, and 17).

B. *G1-[N(CH$_2$CH$_2$CH$_2$NHBoc)$_2$]$_6$-Piperidine$_3$.* In a 4-L, 4-necked, jacketed reaction vessel (Note 3) equipped with a temperature probe, static N_2 inlet, glass stopper and mechanical stirrer, G1-[N(CH$_2$CH$_2$CH$_2$NHBoc)$_2$]$_6$-Cl$_3$ (74.8 g, 43.5 mmol, 1.0 equiv) (Note 18) is suspended in acetone (3 L) (Note 2) and left to stir at 0 °C for 1 h. Piperidine (79.3 mL, 68.4 g, 803 mmol, 18.5 equiv) (Note 19) is added in a single portion and the mixture is stirred at 0 °C for 4 h. A white suspension started to form after 30 min. The mixture is warmed to 21 °C and stirred for an additional 20 h, at which time the reaction was judged to be complete by HPLC (Note 20). The resulting suspension is filtered, washed with acetone (100 mL), and air dried overnight to afford 97.3 g of a white solid (Note 21). The white solid is dissolved in CH_2Cl_2 (1000 mL) (Notes 9 and 22), transferred to a 2-L separatory funnel and washed with a 5% HCl solution (4 x 300 mL) (Notes 23 and 24), 5% NaOH solution (1 x 300 mL) (Note 25), and a saturated, aqueous solution of sodium chloride (1 x 300 mL) (Note 11). The organic phase is dried over sodium sulfate (108 g) (Note 12) and the solvent is removed on a rotary evaporator at 30 °C (Note 8) to afford an off-white solid that is dried in a vacuum oven for 96 h (Note 13) to provide 67.4 g of the title product (86% yield over two steps) (Notes 26, 27 and 28).

2. Notes

1. The building block, 2-[3,3'-di-(*tert*-butoxycarbonyl)-aminodipropylamine]-4,6-dichloro-1,3,5,-triazine, was prepared by the reaction of cyanuric chloride with 3,3'-di-(*tert*-butoxycarbonyl)-aminodipropylamine in acetone-water. This procedure is described in the preceding *Organic Syntheses* preparation.

2. Acetone was obtained from J. T. Baker and used as received.

3. All reactions were performed in a 4-L jacketed reaction vessel (see photograph below). Temperature was controlled via circulating coolant through the vessel jacket. The submitters performed all reactions in standard round-bottomed glassware in a walk-in cold room with a temperature of 0 °C. The checkers utilized the jacketed reactor for convenience, but believe that standard glassware with ice-bath cooling could be used with equal success.

4. The core, 1,3,5-[*tris*-piperazine]-triazine, was prepared in two steps from the reaction of cyanuric chloride with *N*-(*tert*-butoxycarbonyl)-piperazine in tetrahydrofuran followed by deprotection using 6N hydrochloric acid in methanol. This procedure is described in the preceding preparation.

5. Sodium carbonate was purchased from J. T. Baker.

6. Monitoring the reaction mixture by TLC (SiO$_2$, 20:1 CH$_2$Cl$_2$:CH$_3$OH) confirmed that the reaction was complete. Furthermore, a

ninhydrin test showed that all of the 1,3,5-[*tris*-piperazine]-triazine had been consumed (this material appears as a purple spot at the baseline).

7. The mother liquor was concentrated to a volume of approximately 700 mL and the precipitated solids were collected. HPLC analysis showed this second crop to be almost entirely unreacted $C_3N_3[N(CH_2CH_2CH_2NHBoc)_2]Cl_2$ and was therefore discarded.

8. All solvent evaporations were performed using a rotary evaporator at a vacuum pressure of 75–200 mmHg.

9. Dichloromethane was obtained from J. T. Baker and used as received.

10. Separations were fast with sharp phase interfaces. The first aqueous wash was hazy, subsequent washes were clear. HPLC confirmed there was essentially no product in the aqueous washes. The submitters reported a thick emulsion during the extraction; this discrepancy may be related to their collection of a second crop of solids from the mother liquors (see Note 7).

11. Sodium chloride was purchased from J. T. Baker.

12. Anhydrous sodium sulfate was purchased from Sigma-Aldrich.

13. All products were dried in a vacuum oven at 38 °C, 200 mmHg pressure with a slow nitrogen sweep.

14. TLC and HPLC analysis of this material showed the presence of the desired product.

15. A small amount of the Step A product was purified for spectral characterization using column chromatography on silica gel eluting with 10% EtOAc:CH$_2$Cl$_2$ to give first the unreacted starting material $C_3N_3[N(CH_2CH_2CH_2NHBoc)_2]Cl_2$ as a white solid followed by (50:50) EtOAc:CH$_2$Cl$_2$ to give the product as a white solid.

16. The exact melting point for this compound could not be measured. The submitters report that the product formed a viscous oil, which stuck to wall of the melting tube at 87–90 °C and then become less viscous at 212–216 °C, where it proceeded to fall to the bottom of the melting tube and boil. The checkers also observed no sharp melting point, and found the material to gradually soften to a viscous oil at 112–118 °C.

17. The product has the following characteristics: TLC R$_f$ = 0.28 (Silica Gel 60 F254, EMD Chemicals, Inc., 20:1 CH$_2$Cl$_2$:CH$_3$OH); IR (KBr pellet) cm^{-1}: 3375, 2976, 2931, 1713, 1571, 1539, 1493, 1437, 1390, 1367, 1248, 1167, 1081, 1041, 999, 983, 880, 801, 620, 465; ^{1}H NMR (400 MHz, CDCl$_3$) δ: 1.44 (s, 54 H, C(CH$_3$)$_3$), 1.75 (m, 12 H, NCH$_2$CH$_2$), 3.08 (m, 12

H, CH$_2$NHBoc), 3.57 (m, 12 H, Boc-NCH$_2$), 3.82 (m, 24 H, CH$_2$, piperazine), 4.84 (br, 3 H, NH), 5.58 (br, 3 H, NH); ^{13}C NMR (100 MHz, CDCl$_3$) δ: 28.0 (s, NCH$_2$CH$_2$), 28.1 (s, NCH$_2$CH$_2$), 28.6 (s, C(CH$_3$)$_3$), 28.7 (s, C(CH$_3$)$_3$), 37.0 (s, CH$_2$NHBoc), 38.0 (s, CH$_2$NHBoc), 42.9 (s, CH$_2$), 43.1 (s, CH$_2$), 43.6 (s, CH$_2$, piperazine), 44.1 (s, CH$_2$), 79.1 (s, C(CH$_3$)$_3$), 79.5 (s, C(CH$_3$)$_3$), 156.1 (s, C(O)), 156.4 (s, C(O)), 164.6 (s, C$_3$N$_3$), 165.2 (s, C$_3$N$_3$), 165.4 (s, C$_3$N$_3$), 169.6 (s, C$_3$N$_3$); HRMS (Thermo LTQ FT Ultra): Calcd for (M+H): 1660.8746. Found: 1660.87555. Anal calcd for C$_{72}$H$_{120}$Cl$_3$N$_{27}$O$_{12}$: C, 52.02; H, 7.28; N, 22.75; Cl, 6.40. Found: C, 52.14; H, 7.30; N, 22.63; Cl, 6.48

18. Molar amount assumed based on a 100% yield in Step A.

19. Piperidine was purchased from Acros Chemical Co., Inc. and used as received.

20. Starting material and final product are resolved in the submitter's TLC system (SiO$_2$, 5% CH$_3$OH:CH$_2$Cl$_2$), but the intermediate mono- and bis-piperidine adducts are not distinguishable from product. The following HPLC system gives baseline resolution of starting material, product, and the mono- and bis-piperidine intermediates: Halo C18 column, 4.6 x 50 mm, 2.7 μm, 50 °C, 1.5 mL/min, 280 nm UV detection. Mobile phase: 95/5 0.5% HClO$_4$/acetonitrile, gradient to 5/95 HClO$_4$/acetonitrile over 3 minutes, isocratic hold for 4 min, gradient to 95/5 0.5% HClO$_4$/acetonitrile over 7 min. R$_t$ of G1-[N(CH$_2$CH$_2$CH$_2$NHBoc)$_2$]$_6$-Cl$_3$ = 4.07 min; R$_t$ of G1-[N(CH$_2$CH$_2$CH$_2$NHBoc)$_2$]$_6$-Cl$_2$ + piperidine$_1$ = 3.91 min; R$_t$ of G1-[N(CH$_2$CH$_2$CH$_2$NHBoc)$_2$]$_6$-Cl + piperidine$_2$ = 3.76 min; R$_t$ of G1-[N(CH$_2$CH$_2$CH$_2$NHBoc)$_2$]$_6$-Piperidine$_3$ = 3.59 min.

21. TLC analysis of the crude compound showed one spot under UV-lamp; however, after a ninhydrin stain a second spot was observed at the baseline corresponding to the presence of piperidine.

22. The checkers noted that if the crude product is dried overnight in a vacuum oven at this stage (38 °C, 200 mmHg, N$_2$-sweep), the material requires approximately twice as much dichloromethane to redissolve (ca. 20 mL/g). However, if it is air-dried, the solubility described in the current procedure is observed (10 mL/g). This suggests that during drying a less-soluble, anhydrous solid form is generated.

23. Hydrochloric acid was reagent-grade and obtained from Sigma-Aldrich.

24. After four extractions with 5% HCl, a ninhydrin stain confirmed the disappearance of piperidine.

25. Sodium hydroxide was purchased from Fisher Scientific.

26. The exact melting point for this compound could not be measured. The submitters report that the product started to decompose at 206–210 °C and turned brownish yellow. Then a liquid was observed at 210–214 °C, which later began to boil at 216 °C. The checkers also observed no sharp melting point, but observed a gradual softening of the solids from 75–100 °C, followed by a gradual change to a viscous oil from 110–122 °C.

27. The product has the following characteristics: TLC R_f = 0.36 (Silica Gel 60 $F_{254,}$ EMD Chemicals, Inc., 20:1 CH_2Cl_2:CH_3OH); IR (KBr pellet) cm^{-1}: 2975, 2931, 2853, 1717, 1530, 1487, 1434, 1366, 1293, 1249, 1173, 997; ^{1}H NMR (400 MHz, CDCl$_3$) δ: 1.42 (s, 54 H, C(CH$_3$)$_3$), 1.56 (br, 12 H, C$_5$H$_{10}$N, β-H), 1.62 (br, 6 H, C$_5$H$_{10}$N, γ-H), 1.71 (br, 12 H, NCH$_2$CH$_2$), 3.06 (br, 12 H, CH$_2$NHBoc), 3.59 (br, 12 H, CH$_2$, Boc-NCH$_2$), 3.73 (br, 12 H, C$_5$H$_{10}$N, α-H), 3.80 (br, 24 H, CH$_2$, piperazine), 5.26 (br, 6 H, NH); ^{13}C NMR (100 MHz, CDCl$_3$) δ: 25.1 (C$_5$H$_{10}$N, γ -C), 26.0 (C$_5$H$_{10}$N, β-C), 27.8 (s, NCH$_2$CH$_2$), 28.7 (s, C(CH$_3$)$_3$), 37.4 (s, CH$_2$NHBoc), 41.9 (C$_5$H$_{10}$N, α-C), 43.2 (s, Boc-NCH$_2$), 43.4 (s, Boc-NCH$_2$), 44.4 (s, CH$_2$, piperazine), 79.1 (s, C(CH$_3$)$_3$), 156.2 (s, C(O)), 165.1 (s, C$_3$N$_3$), 165.5 (s, C$_3$N$_3$), 166.1 (s, C$_3$N$_3$); HRMS (Thermo LTQ FT Ultra) [M+H] calcd for C$_{87}$H$_{150}$N$_{30}$O$_{12}$: 1808.2120. Found: 1808.20932. Anal. Calcd for C$_{87}$H$_{150}$N$_{30}$O$_{12}$: C, 57.78; H, 8.36; N, 23.24. Found: C, 57.45; H, 8.10; N, 22.90.

28. In a run performed on approximately 2/3 scale (50.3 g of the dichlorotriazine and 10.0 g of *tris*-piperazine-triazine), an 87% overall yield for the two steps was obtained.

Safety and Waste Disposal Information

All hazardous materials should be handled and disposed of in accordance with "Prudent Practices in the Laboratory"; National Academy Press; Washington, DC, 1995.

3. Discussion

In the past two decades, dendrimer syntheses have attracted the attention of numerous research groups worldwide.[2] Dendrimers are monodisperse macromolecules possessing a regular treelike array of

156

branching units. Dendrimers consist of a core unit, building blocks, and a large number of functional groups at the surface. Synthesizing monodisperse macromolecules demands a high level of synthetic control, which is achieved through stepwise fashion via a convergent or a divergent approach.[3] Because of their unique size and globular shape, dendrimers have potential uses in drug delivery, energy harvesting and conversion, catalysis, and optics.[4]

A variety of new dendrimers have been reported which incorporate a wide range of functionalities including ethers,[5] amides,[6] esters,[7] and alkynes[8]. Of the many examples of dendrimers that are described today, only five are commercially available. This limitation is due to the difficulties associated with producing large quantities. Not all of these materials are single chemical entities: some are mixtures. A further obstacle resides in the usage of sophisticated building blocks, and/or expensive chemical substances, and/or excess reagents, and/or complicated purification procedures. Our research is aimed at the development of an environmentally benign process and scaleable synthesis of a first generation (G1) melamine dendrimer that circumvent many of the barriers listed above.

The synthesis is performed in two steps by means of a divergent route, utilizing a "process friendly" solvent and without column chromatography. The building block, 2-[3,3'-di-(*tert*-butoxycarbonyl)-aminodipropylamine]-4,6-dichloro-1,3,5-triazine, is treated with 1,3,5-[*tris*-piperazine]-triazine in acetone-water and in presence of potassium carbonate to afford the G1-chloride intermediate. Subsequent treatment of the G1-chloride with excess piperidine affords the amino-functionalized G1 dendrimer in 81% overall yield. Preparation of both building blocks are described in the preceding *Organic Syntheses* procedure.

We have developed a simple and efficient strategy to produce multi-gram quantities of a G1-amine terminated dendrimer. This green and scalable synthesis can be extended to higher generations of dendrimers.[9]

1. Department of Chemistry, Texas A&M University, College Station, TX 77843.

2. (a) Tomalia, D. A.; Naylor, A. M.; Goddard, W. A., III. Angew. Chem. 1990, 102, 119-157. (b) Newkome, G. R.; Moorefield, C. N.; Vögtle, F. *Dendrimers and Dendrons*; Wiley-VCH Publishers: New York, NY, 2001. (c) Fréchet, J. M. J.; Tomalia, D. A.; *Dendrimers and Other*

Dendritic Polymers Molecules; Wiley-VCH Publishers: New York, NY, 2001. (d) Inoue, K. *Prog. Polym. Sci.* **2000**, *25*, 453-571. (e) Hecht, S.; Fréchet, J. M. J. *Angew. Chem., Int. Ed.* **2001**, *40*, 74-91.

3. (a) Tomalia, D. A.; Baker, H.; Dewald, J.; Hall, M.; Kallos, G.; Martin, S.; Roeck, J.; Ryder, J.; Smith, P. *Polym. J.* **1985**, *17*, 117-132. (b) Hawker, C. J.; Fréchet, J. M. J. *J. Am. Chem. Soc.* **1990**, *112*, 7638-7647.

4. (a) Jiang, D.-L.; Aida, T. *Nature (London)* **1997**, *388*, 454-456. (b) Knapen, J. W. J.; Van der Made, A. W.; de Wilde, J. C.; Van Leeuwen, P. W. N. M.; Wijkens, P.; Grove, D. M.; Van Koten, G. *Nature (London)* **1994**, *372*, 659-663. (c) Percec, V.; Cho, C. G.; Pugh, C.; Tomazos, D. *Macromolecules* **1992**, *25*, 1164-1176. (d) Jansen, J. F. G. A.; de Brabander-Van den Berg, E. M. M.; Meijer, E. W. *Science* **1994**, *266*, 1226-1229. (e) Hawker, C. J.; Fréchet, J. M. J. *J. Chem. Soc., Perkin Trans. 1* **1992**, 2459-2469. (f) Tang, M.; Redemann, C. T.; Szoka, F. C. *Bioconjugate Chem.* **1996**, *7*, 703-714. (g) Malik, N.; Wiwattanapatapee, R.; Klopsch, R.; Lorenz, K.; Frey, H.; Weener, J. W.; Meijer, E. W.; Paulus, W.; Duncan, R. *J. Controlled Release* **2000**, *65*, 133-148.

5. Nlate, S.; Ruiz, J.; Sartor, V.; Navarro, R.; Blais, J. C.; Astruc, D. *Chem.-Eur. J.* **2000**, *6*, 2544-2553.

6. Brouwer, A. J.; Mulders, S. J. E.; Liskamp, R. M. J. *Eur. J. Org. Chem.* **2001**, 1903-1915.

7. Peng, Z.; Pan, Y.; Xu, B.; Zhang, J. *J. Am. Chem. Soc.* **2000**, *122*, 6619-6623.

8. Moore, J. S.; Xu, Z. *Macromolecules* **1991**, *24*, 5893-5894.

9. (a) Crampton, H.; Hollink, E.; Perez, L. M.; Simanek, E. E. *New J. Chem.* **2007**, *31*, 1283-1290. (b) Chouai, A.; Simanek, E.E.; *J. Org. Chem.* **2008**, *73*, 2357-2366.

Appendix
Chemical Abstracts Nomenclature (Collective Index Number); (Registry Number)

2-[3,3'-Di-(*tert*-butoxycarbonyl)-aminodipropylamine]-4,6-dichloro-1,3,5-triazine; 12-Oxa-2,6,10-triazatetradecanoic acid, 6-(4,6-dichloro-1,3,5-triazin-2-yl)-13,13-dimethyl-11-oxo-, 1,1-dimethylethyl ester; (947602-03-1)

1,3,5-[Tris-piperazine]-triazine: 1,3,5-Triazine, 2,4,6-tri-1-piperazinyl-; (19142-26-8)

G1-[N(CH$_2$CH$_2$CH$_2$NHBoc)$_2$]$_6$-Cl$_3$; (1016650-75-1)

G1-[N(CH$_2$CH$_2$CH$_2$NHBoc)$_2$]$_6$-Piperidine$_3$; (1016650-76-2)

Eric E. Simanek was born in 1969 in Tuscola, IL. He obtained a B.S. in Chemistry in 1991 from the University of Illinois at Urbana-Champaign while working in the laboratories of the late Dr. Kenneth L. Rinehart, Jr. After completing doctoral studies with Dr. George M. Whitesides at Harvard University in 1997, he joined Dr. Chi-Huey Wong's laboratory at The Scripps Research Institute. Since joining Texas A&M University in 1998, he has risen through the ranks to Professor of Chemistry. His interests lie in drug delivery and K-20 education.

Abdellatif Chouai was born in 1971 in Morocco. He earned a B.S. in Chemistry in 1995 from the University of Sidi Mohamed Ben Abdellah, Morocco, and a Ph.D. in Organic Chemistry from University of Houston in 2003 under Dr. Randolph P. Thummel. He joined Dr. Kim R. Dunbar's group at Texas A&M University as a postdoctoral research associate where he worked on reversible DNA biosensors complexes and bimetallic complexes as photodynamic therapy agents. He then moved to a research scientist position in 2006 in Dr. Simanek's laboratory. His research focused on an industrial scale production of triazine-based dendrimers and application in drug delivery. Currently, Dr. Chouai holds a professional development chemist position with BASF Corporation.

Vincent J. Venditto was born in 1981 in Philadelphia, PA. He earned a B.S. in Chemistry from Gettysburg College in 2003 and began working in the laboratory of Dr. Martin W. Brechbiel in the Radioimmune and Inorganic Chemistry Section of the National Cancer Institute within the NIH. After two years at the NIH, Vincent joined Dr. Simanek's laboratory at Texas A&M University to pursue a graduate degree in chemistry. His graduate work focuses on the synthesis of triazine-based dendrimers as drug carriers for a range of therapeutic applications.

Brian Vanderplas was born and raised in St. Louis, Missouri. Following a tour of duty in the US Army he attended Southwest Missouri State University where he received a B.S. in Chemistry in 1983. He was then employed by Sigma Chemical Company in St. Louis as a Production Chemist, and in 1986 moved to Pfizer Global Research & Development in Groton, Connecticut where he is currently a Senior Scientist in the Chemical Research & Development group. He is a 2008 recipient of the ACS Technical Achievements in Organic Chemistry Award.

160

THE DIRECT ACYL-ALKYLATION OF ARYNES.
PREPARATION OF METHYL 2-(2-ACETYLPHENYL)ACETATE.

67% yield

Submitted by David C. Ebner, Uttam K. Tambar, and Brian M. Stoltz.[1]
Checked by Morten Storgaard, Nathan D. Ide, John A. Ragan, and
Jonathan A. Ellman.

1. Procedure

Methyl 2-(2-acetylphenyl)acetate. An oven-dried (Note 1) 500-mL, three-necked, round-bottomed flask is equipped with a magnetic stir bar, and cesium fluoride (19.7 g, 130 mmol, 2.5 equiv) (Note 2) is added. The flask is fitted with a reflux condenser in the middle neck, an adaptor equipped with a thermometer in one of the side necks, and a septum in the other side neck. The condenser is equipped with a vacuum adaptor connected to a Schlenk line (nitrogen/vacuum manifold). The connected glassware is evacuated under high vacuum (0.025 mmHg) and carefully back-filled with nitrogen. This procedure is repeated twice to secure an oxygen-free atmosphere in the flask. Dry MeCN (260 mL) (Note 3) is added to the flask via a syringe through the septum. While stirring the reaction mixture, methyl acetoacetate (5.60 mL, 6.01 g, 51.8 mmol, 1.00 equiv) (Note 4) and 2-(trimethylsilyl)phenyl trifluoromethanesulfonate (15.7 mL, 19.3 g, 64.7 mmol, 1.25 equiv) (Note 5) are added through the septum via syringes. The flask is submerged in an oil bath (100 °C) and the reaction mixture is heated to reflux (internal temperature: 78–81 °C, reached after 15 min). The mixture is stirred at reflux temperature for 40 min (Note 6). Initially, the reaction mixture is a white opaque suspension, but during heating it changes color to yellow and upon reflux the color changes to orange. As the reaction progresses the opaque solution becomes transparent and the color changes back to yellow. Throughout the entire time a white precipitate is present in the flask. The reaction flask is removed from the oil bath and allowed to cool to ambient temperature (23 °C) over the course of 1 h. The flask is disconnected from the condenser and the nitrogen inlet, and the reaction mixture is diluted with saturated aqueous NaCl solution (200 mL) (Note 7). This mixture is carefully poured into a 1-L separatory funnel (Note 8). After the layers are separated, the aqueous layer is extracted with Et_2O (3 × 200 mL) (Notes 9 and 10). The combined organic layers are dried over anhydrous Na_2SO_4 (Note 11) and filtered (Note 12). Concentration of the dried organic layers is effected by rotary evaporation (35 °C, 45 mmHg) which affords an orange, viscous oil. Partial purification is achieved by flash chromatography (4.5 × 24 cm, 170 g silica gel) (Note 13) using a gradient of hexanes to 40% Et_2O in hexanes. The crude product is loaded on the column with benzene (20 mL) (Note 14). Fraction collection (50 mL fractions) is begun as the crude product is eluted first with 100 mL of hexanes (Note 15), followed by 1500 mL of 9:1 hexanes:Et_2O, 2000 mL of 4:1 hexanes:Et_2O, 1000 mL of 7:3 hexanes:Et_2O and finally 750 mL of 3:2 hexanes:Et_2O. Fractions 38–85 are concentrated by rotary evaporation (30–35 °C, 45 mmHg) to afford 7.96 g (80%) of a slightly yellow solid (Note 16). The partially purified product is further purified by bulb-to-bulb distillation (Note 17) at 124–130 °C (0.75 mmHg) (Note 18) which affords 6.63 g (67%) of the title compound as an off-white, crystalline solid (Notes 19, 20, 21, 22 and 23).

2. Notes

1. The glassware and magnetic stir bar are dried in an oven (150 °C) overnight before use and assembled while still hot and cooled to ambient temperature (23 °C) under high vacuum (0.025 mmHg).

2. Cesium fluoride (CsF) (99.9%) was purchased from the Sigma-Aldrich Chemical Company, Milwaukee, WI. The submitters used the reagent directly as received, but the checkers found that it may be important to dry CsF prior to use. The CsF is dried in a desiccator in high vacuum (0.025 mmHg) overnight at room temperature (23 °C) in the presence of P_2O_5.

3. The submitters used acetonitrile (MeCN) that was dried by passage through an activated alumina column under argon. The checkers used acetonitrile (HPLC grade, 0.2 micron filtered) from Fisher Scientific Chemicals that was distilled over CaH_2 under nitrogen atmosphere prior to use.

4.   Methyl acetoacetate (99%) was purchased from the Sigma-Aldrich Chemical Company, Milwaukee, WI. The reagent was used as received without further purification.

5.   2-(Trimethylsilyl)phenyl trifluoromethanesulfonate (97%) was purchased from the Sigma-Aldrich Chemical Company, Milwaukee, WI. The reagent was used as received without further purification. Submitters reported that the reagent can be prepared alternatively by a method of Peña and coworkers.[2]

6.   The progress of the reaction is followed by thin-layer chromatography (TLC) analysis on E. Merck silica gel 60 F254 precoated plates (0.25 mm) (used by submitters) or Dynamic Adsorbents, Inc. glass plates coated with 250 mm F-254 silica gel (used by checkers) with 1:4 EtOAc:hexanes as the eluent. The plates are visualized by UV and *p*-anisaldehyde staining (0.5 mL *p*-anisaldehyde in 50 mL glacial acetic acid and 1 mL 97% H_2SO_4). The title compound and a side product identified as methyl 2-(2-acetylphenyl)-2-phenylacetate both have very similar R_f values; 0.43 and 0.53, respectively. Both compounds are visible in UV and with *p*-anisaldehyde staining. The title compound stains reddish brown and the side product stains pale brown. The completion of the reaction is determined by disappearance of 2-(trimethylsilyl)phenyl trifluoromethanesulfonate, which has a $R_f = 0.91$ visualized by UV. This starting material cannot be visualized with *p*-anisaldehyde staining. To clarify the disappearance, cross-spotting with pure 2-(trimethylsilyl)phenyl trifluoromethanesulfonate is used since many by-products are formed nearby the diagnostic spot.

7.   Sodium chloride (NaCl), crystalline, was purchased from Fisher Scientific Chemicals.

8.   The remaining solids in the flask are not transferred to the separatory funnel to avoid clogging of the stopcock. Instead, the solids are washed with the portions of Et_2O before the solvent is transferred to the separatory funnel for extraction.

9.   Ethyl ether (Et_2O) anhydrous, stabilized, HPLC grade, was purchased from Fisher Scientific Chemicals and was used without further purification.

10.   Extraction with Et_2O forms a white, opaque emulsion that only slowly separates.

11.   Sodium sulfate (Na_2SO_4) anhydrous, powder, was purchased from EMD Chemicals Inc.

12. Removal of the drying agent was carried out by using a Wilmad Labglass (60 mL, size M) sintered glass funnel by vacuum filtration.

13. The submitters used ICN silica gel (particle size 0.032–0.063 mm) and the checkers used silica gel 60 (0.040–0.063 mm), 230–400 mesh ASTM purchased from Merck KGaA.

14. Benzene was purchased from EMD Chemicals Inc.

15. Hexanes, HPLC grade, were purchased from Fisher Scientific.

16. Purification by flash chromatography gives a mixture of the title compound and a side product, methyl 2-(2-acetylphenyl)-2-phenylacetate,[3] in an 87:13 ratio as determined by ^{1}H NMR. This side product was obtained pure by the submitters using preparative thin-layer chromatography (3:2 hexanes:Et$_2$O eluent) and exhibits the following spectroscopic properties: ^{1}H NMR (300 MHz, CDCl$_3$) δ: 2.56 (s, 3 H), 3.71 (s, 3 H), 5.76 (s, 1 H), 7.04–7.09 (m, 1 H), 7.18–7.25 (m, 2 H), 7.25–7.40 (m, 5 H), 7.73–7.77 (m, 1 H). ^{13}C NMR (75 MHz, CDCl$_3$) δ: 29.3, 52.2, 53.7, 127.0, 127.3, 128.7, 129.3, 129.6, 130.6, 131.8, 137.1, 138.1, 138.7, 173.3, 202.0. IR (thin film) υ 1735, 1681, 1254, 1201, 1161 cm^{-1}. HRMS (EI$^+$) calcd for [C$_{17}$H$_{16}$O$_3$]$^+$: m/z 268.1100, found 268.1105.

17. Bulb-to-bulb distillation was performed with a Büchi Glass Oven B-585 Kugelrohr by the submitters. The checkers used an oven from Aldrich with an internal thermometer installed. The temperature is controlled by a variable autotransformer from Staco Engery Products Co. model 3PN1010B (in: 120 V, 50/60 Hz, out: 0 – 140 V, 10 apm, 1.4 KVA). The bulbs are connected to a Trico-Folberth air pressure wiper and the entire instrument is connected to a mercury manometer from Kontes Scientific Glassware Instruments. The receiving bulb is cooled with an acetone/dry ice bath.

18. The submitters reported a boiling point at 159–165 °C (1.1 mmHg).

19. The yellow solid melts at 70–80 °C. When 100 °C is reached the receiving bulb is exchanged with a new one to avoid contamination with low-boiling impurities. Initially, the product distillate is a colorless oil, which solidifies as the distillation proceeds. The distillation stops when 59% of the title compound is recovered leaving a residue (2.00 g) containing a 50:50 mixture of the title compound and the side product. This residue is transferred from a 250-mL round-bottomed flask to a 50-mL round-bottom flask and is resubjected to distillation to afford an additional 762 mg (8%) of the title compound. The brown distillation residue (1.08 g) primarily

164

contains the side product and contains less than 5 mol% of the title compound.

20. The title compound contains 2–3% of methyl 2-(2-acetylphenyl)-2-phenylacetate, as determined by ^{1}H NMR.

21. The title product exhibits the following properties: mp 53–55 °C (lit.[4] mp 57–59 °C). MS (ES+) m/z 193 (100%, M + H$^+$). IR (neat) υ 3004, 2954, 1732, 1674, 1217, 1168 cm^{-1}. ^{1}H NMR (400 MHz, CDCl$_3$) δ: 2.54 (s, 3 H), 3.64 (s, 3 H), 3.91 (s, 2 H), 7.21 (d, J = 7.3 Hz, 1 H), 7.32 – 7.43 (m, 2 H), 7.78 (d, J = 7.6 Hz, 1 H). ^{13}C NMR (100 MHz, CDCl$_3$) δ: 28.2, 39.6, 51.3, 127.0, 129.6, 131.6, 132.2, 133.9, 136.6, 171.4, 200.6. Anal. calcd for C$_{11}$H$_{12}$O$_3$: C, 68.74; H, 6.29; found: C, 68.65; H, 6.55.

22. The submitters also suggest an alternative procedure for the final purification: After chromatography the material can be purified by crystallization. To 8.00 g of the yellow solid is added 1200 mL of pentane. The mixture is warmed to dissolve the solid. The yellow solution is allowed to cool to room temperature before cooling to –20 °C in a freezer. After 9 h, the off-white solid that has formed is collected by vacuum filtration through a Büchner funnel. The solid is washed with cold pentane (2 × 25 mL) and dried under vacuum to afford 5.12 g (51% yield) of the title compound. This material contains 2.4% of the side product by ^{1}H NMR. Attempts by the initial checkers (Ide and Ragan) to perform this recrystallization gave variable results, with one run providing pure material in modest yield (43%), and another run providing material still containing 12% of the side product (57% recovery). The checkers noted that the product crystals were very dense spheres that were firmly attached to the sides of the flask, suggesting that the material may have initially come out of solution as oil droplets adhered to the flask wall, and then subsequently crystallized. Regardless of the explanation, distillation appears to provide a more robust purification.

23. The checkers discovered that the pure product slowly decomposed upon storage at room temperature. It is therefore recommended to store it in the freezer below –18 °C.

Waste Disposal Information

All hazardous materials should be handled and disposed of in accordance with "Prudent Practices in the Laboratory"; National Academy Press; Washington, DC, 1995.

3. Discussion

While arynes have historically received attention from physical organic chemists, their use as reagents in synthetic organic chemistry is limited because of the harsh conditions needed to generate arynes, and the uncontrolled reactivity exhibited by these species. We have developed the acyl-alkylation of arynes, which is a mild and direct aryne insertion into a carbon-carbon bond.[5,6] With this reaction, two carbon-carbon bonds are formed in a single step, often with exquisite regiocontrol. The acyl-alkyation reaction is the net result of benzyne insertion into the α,β C–C single bond of the β-keto ester, presumably by a formal [2+2] cycloaddition/ fragmentation cascade.[7]

Scheme 1

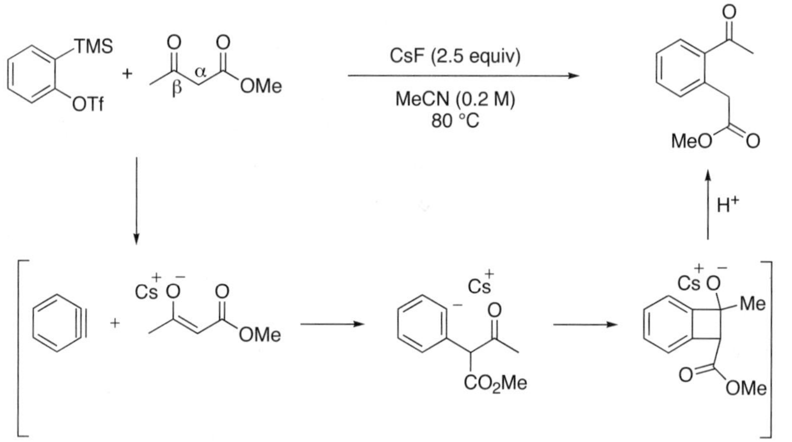

The structures that are accessed by this methodology would otherwise require multi-step sequences for their preparation. The product was independently prepared according to a literature procedure.[8] The product obtained through our methodology was identical by all spectroscopic data to the compound prepared by this alternative method.

The reaction tolerates substitution at the γ-position (Table 1, entries 2-6), including aliphatic and aromatic groups. Heteroatoms may also be incorporated into the β-keto ester side chain, albeit in slightly lower yields (Table 1, entry 5). Additionally, the ester moiety can be varied while maintaining the efficiency of the reaction. For example, β-keto esters of

166

Table 1. Acyl-Alkylation of Benzyne

entry	substrate [a]	product	yield [b]
1			67%
2			78%
3			84%
4[c]			85%
5			53%
6			99%
7			72%
8	($R = C_6H_{13}$)	($R = C_6H_{13}$)	75%

[a] 1.25 equiv of aryne precursor relative to β-keto ester. [b] Isolated yield.
[c] 2 equiv of aryne precursor relative to β-keto ester.

more complex alcohols such as menthol and cholesterol provide the desiredacyl-alkylation products in good yield (Table 1, entries 7 and 8). In general, the mild reaction conditions allow for a considerable degree of substitution on the β-keto ester subunit.

Substituted arynes can also be employed in this methodology. Methyl acetoacetate can react with arynes possessing mono-substitution at the *ortho*- and *meta*-positions (Table 2, entries 1-2) as well as disubstitution (Table 2, entry 3) to produce high yields of the corresponding acyl-alkylation products. Additionally, entries 1 and 3 demonstrate that heteroatom substituents are well tolerated.

Table 2. Acyl-Alkylation of Substituted Arynes

entry	aryne precursor [a]	product	yield [b]
1			95%
2[c]			82%[d]
3			75%

[a] 2 equiv of aryne precursor relative to β-keto ester. [b] Isolated yield. [c] 1.25 equiv of aryne precursor relative to β-keto ester. [d] Mixture of *meta*- and *para*- regioisomers (1.2 : 1).

Of particular note is the extension of this strategy toward the convergent synthesis of medium-sized carbocycles in a single step from simple starting materials. Recently this procedure for synthesizing benzannulated carbocycles has been utilized in an enantioselective synthesis of the alkaloid (+)-amurensinine.[9]

Figure 1. Medium-sized Carbocycles

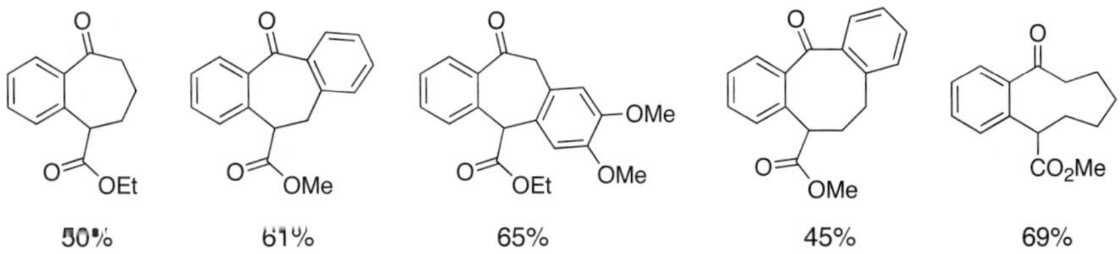

| 50% | 61% | 65% | 45% | 69% |

1. Division of Chemistry and Chemical Engineering, California Institute of Technology, Pasadena, CA 91125. Email: stoltz@caltech.edu.
2. Peña, D.; Cobas, A.; Pérez, D.; Guitián, E. *Synthesis* **2002**, 1454-1458.
3. Breslow, R.; Kivelevich, D. *J. Org. Chem.* **1961**, *26*, 679-681.
4. Halford, J. O.; Raiford, R. W., Jr.; Weissmann, B. *J. Org. Chem.* **1961**, *26*, 1898-1901.
5. Tambar, U. K.; Stoltz, B. M. *J. Am. Chem. Soc.* **2005**, *127*, 5340-5341.
6. Subsequent to our report, similar aryne insertions into C-C bonds were disclosed: (a) Yoshida, H.; Watanabe, M.; Ohshita, J.; Kunai, A. *Chem. Commun.* **2005**, 3292-3294. (b) Yoshida, H.; Watanabe, M.; Ohshita, F.; Kunai, A. *Tetrahedron Lett.* **2005**, *46*, 6729-6731.
7. This mild method for generating benzyne from *ortho*-silyl aryl triflates was initially developed by Kobayashi: Himeshima, Y; Sonoda, T.; Kobayashi, H. *Chem. Lett.* **1983**, *12*, 1211-1214.
8. Cruces, J.; Estévez, J. C.; Castedo, L.; Estévez, R. J. *Tetrahedron Lett.* **2001**, *42*, 4825-4827.
9. Tambar, U. K.; Ebner, D. C.; Stoltz, B. M. *J. Am. Chem. Soc.* **2006**, *128*, 11752-11753.

Appendix
Chemical Abstracts Nomenclature; (Registry Number)

Methyl 2-(2-acetylphenyl)acetate: Benzeneacetic acid, 2-acetyl-, methyl ester; (16535-88-9)

Cesium fluoride; (13400-13-0)

Methyl acetoacetate: Butanoic acid, 3-oxo-, methyl ester; (105-45-3)

2-(Trimethylsilyl)phenyl trifluoromethanesulfonate: Methanesulfonic acid, 1,1,1-trifluoro-, 2-(trimethylsilyl)phenyl ester; (88284-48-4)

Brian M. Stoltz was born in Philadelphia, PA in 1970 and obtained his B.S. degree from the Indiana University of Pennsylvania in Indiana, PA. After graduate work at Yale University in the labs of John L. Wood and an NIH postdoctoral fellowship at Harvard in the Corey labs, he took a position at the California Institute of Technology. A member of the Caltech faculty since 2000, he currently is the Ethel Wilson Bowles and Robert Bowles Professor of Chemistry and a KAUST GRP Investigator. His research interests lie in the development of new methodology for general applications in synthetic chemistry.

David Ebner was born in 1980 in Stillwater, Minnesota. In 2000, he received his B.S. in chemistry and B.A. in mathematics at the University of St. Thomas, working with Tom Ippoliti. He subsequently began graduate studies at the California Institute of Technology under the direction of Brian Stoltz. After completing his Ph.D. on the palladium-catalyzed enantioselective oxidation of secondary alcohols in 2008, he began postdoctoral research in the laboratories of Erik Sorensen as an NIH Postdoctoral Fellow. His research interests include the development of novel synthetic methodology and the total synthesis of natural products.

Uttam Krishan Tambar was born on November 22, 1978 in Barnsley, England. In 2000, he obtained his A.B. in Chemistry and Physics at Harvard University, where he conducted research with Cynthia Friend and Stuart Schreiber. Uttam performed his Ph.D. studies in the laboratory of Brian Stoltz at the California Institute of Technology where he developed convergent methodologies for the synthesis of biologically active natural products. Since 2006, Uttam has been conducting research with James Leighton at Columbia University, where he has developed a general enantioselective aza-Diels-Alder reaction with acyclic dienes and an enantioselective [3+2] cycloaddition for the synthesis of heterocycles.

Morten Storgaard was born in Denmark in 1980. He graduated from Technical University of Denmark in 2006 with a M.Sc. degree in chemistry and in 2007 he continued as a Ph.D. student under the supervision of professor David Tanner and Dr. Bernd Peschke from Novo Nordisk. His research has mainly been focusing on palladium catalyzed coupling reactions towards the synthesis of biologically active compounds. In the summer and fall of 2008 he visited the group of Jonathan A. Ellman at University of California, Berkeley, working on the rhodium-catalyzed enantioselective synthesis of amines.

Nathan D. Ide was born in 1979 in Grand Haven, MI. In 2001, he obtained his B.S. in chemistry from Hope College in Holland, MI. While at Hope College, he worked with Stephen K. Taylor on the enzymatic resolution of γ- and δ-hydroxyamides. In 2001, he joined the research group of David Y. Gin at the University of Illinois in Urbana-Champaign, IL. His research efforts focused on the synthesis/reactivity of aziridine-containing peptides and the total synthesis (-)-crambidine. After obtaining his Ph.D. in 2006, he joined the Chemical Research & Development group at Pfizer in Groton, CT.

CONVENIENT PREPARATION OF 3-ETHOXYCARBONYL BENZOFURANS FROM SALICYLALDEHYDES AND ETHYL DIAZOACETATE

Submitted by Matthew E. Dudley, M. Monzur Morshed, and M. Mahmun Hossain.[1]
Checked by David Hughes.

1. Procedure

Caution: Ethyl diazoacetate (EDA) and other diazo compounds are potentially explosive and therefore must be handled with caution. They are also toxic and prone to cause development of specific sensitivity. A well-ventilated hood should be used for the entire procedure.

5-Chloro-3-ethoxycarbonylbenzofuran. A 250-mL, three-necked, round-bottomed flask is fitted with a reflux condenser, thermometer, and a pennyhead stopper (Note 1). A 3-cm oval stir bar is added to the flask. 5-Chlorosalicylaldehyde (5.00 g, 31.9 mmol) (Note 2) is added to the flask along with 11 mL CH_2Cl_2 (Note 3) to form a thick white slurry, then 54% $HBF_4 \cdot OEt_2$ (0.50 g, 3.2 mmol, 0.1 equiv) (Note 4) is added by syringe directly to the stirring solution through the neck containing the pennyhead stopper, whereby a color change from a white slurry to an orange slurry is noted. The pennyhead stopper is then replaced with a clean and dry 100-mL pressure-equalizing addition funnel with the Teflon stopcock closed (Note 5). Ethyl diazoacetate (90% pure) (5.79 g, 45.7 mmol, 1.43 equiv) is added to the addition funnel and diluted to 33 mL with dichloromethane (Notes 5-7). The EDA solution is then added to the reaction mixture at a rate of 5-7 mL per minute, resulting in steady evolution of nitrogen (total addition time was 5-7 minutes) and an increase in temperature from the initial 23 °C to 38 °C with a gentle reflux (Notes 5, 6 and 8). Once addition is complete gas evolution ceases within a minute and the reaction is assayed for completion by ^1H NMR (Note 9). A heating mantle is placed under the round-bottomed flask. The equalizing addition funnel is replaced with a pennyhead stopper.

172

At this point the solution is dark yellow in color. A simple distillation apparatus is assembled in place of the original reflux condenser using a distillation adapter, Claisen head adapter, and a 100-mL receiving flask. The reaction mixture is concentrated by distilling CH_2Cl_2 (30-33 mL) and free EDA to a maximum pot temperature of 58–61 °C leaving a clear yellow viscous oil. After removal of the heating mantle the solution is cooled to room temperature using a water bath, then 4.0 g of 95-98% H_2SO_4 (39.2 mmol) (Note 10) is added by syringe directly to the stirring viscous oil through the neck containing the pennyhead stopper over a period of 1-2 minutes. The addition is only slightly exothermic, with a temperature rise of about 5 °C using a water-cooling bath, and a color change from a clear yellow oil to a cloudy yellow-orange suspension is noted. After 10 minutes, the second step is complete and ready for neutralization. With water bath cooling and rapid agitation, the acidic reaction mixture is neutralized by slow addition of 100 mL saturated $NaHCO_3$ solution (Note 11). The empty addition funnel is removed and replaced with a pennyhead stopper. The water bath is removed and replaced with a heating mantle. The aqueous mixture is azeotropically distilled, while maintaining high speed stirring in order to remove any trace of residual CH_2Cl_2 to a maximum pot temperature of 90 °C. Upon reaching 90 °C, the heating mantle is removed and the aqueous mixture is allowed to cool to room temperature while continuing to maintain rapid stirring (Note 12). After crystallizing overnight, the crude yellow solid product is filtered using a Büchner funnel and dried to constant weight. The yield of crude material (~90% pure by NMR) is 5.99–6.08 g (84–86% uncorrected for purity).

The crude material is purified by recrystallization. To a 250-mL, three-necked, round-bottomed flask equipped with a thermometer, stopper, and 100 mL addition funnel are added a 3-cm oval stir bar, 30 mL of anhydrous EtOH, in which 0.3 g of NaOH has been dissolved, and the crude product (5.97 g) (Note 13). The mixture is stirred with warming to 35 °C in order to dissolve the product, then 70 mL of deionized water is added dropwise from the addition funnel over 15 min. The product begins to crystallize after 10 mL water is added. On completion of the water addition, the mixture is stirred for 5 min at room temperature, then filtered using a Büchner funnel and filter paper. The flask is washed with 2 x 15 mL of deionized water and the washes are then used to rinse the crystalline product on the Büchner funnel. The crystalline product is allowed to dry on the

Büchner funnel for 30 min to 1 h to provide 4.84 g (68%) of product as a tan, free-flowing solid (Note 14).

2. Notes

1.   A mechanical stirrer and a four-necked flask can be substituted for the magnetic stir bar and three-necked flask apparatus with round-bottomed flasks smaller than 500 mL. The submitters strongly suggest use of a mechanical stirrer over magnetic stirring in order to maintain vigorous stirring and agitation as noted throughout the procedure.

2.   5-Chlorosalicylaldehyde (98%) was purchased from Alfa Aesar and was used as received. The checkers used 5-chlorosalicylaldehyde purchased from Aldrich. NMR suggested a purity of about 96%. Yields are uncorrected for purity.

3.   CH$_2$Cl$_2$ (99.5%) was purchased from VWR International and distilled from P$_2$O$_5$. The checkers used dichloromethane (99.5%) from Sigma-Aldrich and used without purification. The water content was measured by Karl-Fischer titration and found to be <50 ppm.

4.   HBF$_4$·OEt$_2$ (tetrafluoroboric acid, 54% w/w in diethyl ether) solution was purchased from Aldrich Chemical Company as a clear colorless or clear yellow solution. Solutions that are dark brown or purple in color are unacceptable. The checkers added tetrafluoroboric acid based on weight; the tetrafluoroboric acid was drawn into a weighed syringe to the appropriate weight, discharged into the flask, and re-weighed to determine the accurate amount added.

5.   A Teflon stopcock is recommended for ethyl diazoacetate (EDA) as greased glass stopcocks tend to leak in the presence of the EDA solution. For reactions run with magnetic stirring, the addition funnel was placed on the middle neck such that the EDA could drop directly into the vortex of the stirring solution. The checkers used ethyl diazoacetate purchased from Aldrich. The EDA was assayed via NMR using toluene as an internal standard, as follows. Approximately 50 mg each of EDA and toluene (99.5% purity) were accurately weighed into a 5-mm NMR tube, diluted with about 0.8 mL of CDCl$_3$, and analyzed using ^{1}H NMR with a 5 second delay to ensure full relaxation. The CH$_3$ protons were integrated and compared to the integration of the CH$_2$ and CH$_3$ groups of EDA. A fresh bottle from Aldrich was assayed at 90 wt% with 9% dichloromethane. An older bottle assayed at 78 wt % with 15 wt% dichloromethane.

6. The Merck Index cautions against using EDA with sulfuric acid.[2] In our hands, we have had no issues or incidents regarding use of tetrafluoroboric acid and EDA. Information is available concerning the safety of EDA on an industrial scale.[3]

7. The submitters report that the previously published preparation of EDA[4] can also be used. By omitting the final distillation of dichloromethane, the requisite 7-10 mol% EDA solution is obtained.

8. In cases of larger scale reactions, overwhelming of the condenser by refluxing CH_2Cl_2 may be prevented by using a room temperature water bath to control excessive temperatures. The water bath is unnecessary for a 5 g scale reaction. The EDA solution should be dripped directly into the rapidly stirring reaction mixture and not on the walls of the flask.

9. The completion of reaction was determined by ^1H NMR. A 50 mg sample of the batch was diluted with 0.7 mL $CDCl_3$ and analyzed by NMR. The phenol and aldehyde resonances (at 10.9 ppm and 9.9 ppm) were integrated and compared to the combined integration of all resonances from 6 to 8 ppm (3 aromatic protons for starting material, intermediates, and products). If more than 5% unreacted aldehyde remained, an appropriate amount of EDA was added to complete the reaction.

10. ACS reagent grade (95.0-98.0%) H_2SO_4 was purchased from Fischer Chemical Company and was used as received. The checkers added sulfuric acid by weight; the sulfuric acid was drawn into a weighed syringe to the appropriate weight, discharged into the flask, and re-weighed to determine the accurate amount added.

11. ACS reagent grade $NaHCO_3$ (99.7-100 %) was purchased from Aldrich Chemical Company and was used as received. The initial addition of saturated bicarbonate should be dropwise due to a rapid exotherm; after the first 5 mL are added, the exotherm subsides as the mixture is cooled by CO_2 evolution. The final 95 mL can be added over 5 min.

12. The entire azeotrope procedure lasts approximately 15 min and generally less than 1 mL is distilled. The majority of the CH_2Cl_2 distills at about 72 °C and the azeotrope is stopped once water is seen distilling into the receiver (90 °C). Upon cooling to room temperature the pH of the aqueous mixture was 8-9 as measured with pH paper. This part of the procedure is a trituration and requires rapid stirrer speed. The oily melt must be efficiently dispersed throughout the aqueous solution on cooling to room temperature or it will oil out and eventually form an impure gummy solid. In addition, the mixture must slowly cool to room temperature by equilibrating

with the surrounding air. Forced cooling (i.e., ice baths, water jackets, etc.) should be avoided. In many cases, it may take several hours for complete trituration and eventual crystal formation to occur. It is suggested to let this process go overnight under the highest possible stir speed. The checkers found that if gumming occurred, the mixture could be re-heated to 80 °C and allowed to re-cool. This provided solids in all cases. As an alternate to crystallization from the aqueous solution, the checkers also used the following extractive procedure. The final distillation at 90 °C is omitted, and the aqueous mixture after neutralization is extracted with 2 x 50 mL ethyl acetate. The EtOAc extracts are combined, washed with 40 mL brine, filtered through a bed of sodium sulfate, and concentrated by rotary evaporation (30 mmHg, bath temp 40 °C). The resulting material crystallized on standing and was recrystallized from EtOH/water as described in the text to provide 5.26 g of product (74% yield).

13. NaOH pellets were dissolved in ethanol using sonication. The recrystallization process can be repeated if needed for desired purity, although loss of yield is approximately 15% for each recrystallization. ACS reagent grade ethanol, absolute 200 proof (≥99.5%) and ACS reagent grade NaOH pellets, (≥97.0%) were used as received from Aldrich chemical.

14. Physical and spectroscopic properties of 5-chloro-3-ethoxycarbonylbenzofuran are as follows: mp 60.5–61.5 °C; ^{1}H NMR (CDCl$_3$, 400 MHz) δ: 1.43 (t, J = 7.2 Hz, 3 H), 4.42 (q, J = 7.2 Hz, 2 H), 7.32 (dd, J = 2.1, 8.8 Hz, 1 H), 7.44 (d, J = 8.8 Hz, 1 H), 8.03 (d, J = 2.1 Hz, 1 H), 8.26 (s, 1 H); ^{13}C NMR (CDCl$_3$, 100 MHz) δ: 14.4, 60.8, 112.7, 114.6, 121.8, 125.6, 126.0, 130.0, 152.0, 153.9, 162.9; HRMS (EI) m/z calcd for C$_{11}$H$_9$O$_3$Cl: 224.02407; found: 224.0241. Anal. Calcd for C$_{11}$H$_9$O$_3$Cl: C, 58.82; H, 4.04. Found: C, 58.54; H, 4.01.

Safety and Waste Disposal Information

All hazardous materials should be handled and disposed of in accordance with "Prudent Practices in the Laboratory"; National Academy Press; Washington, DC, 1995.

3. Discussion

3-Substituted benzofuran derivatives are medicinally and biologically important compounds.[5] Many of the benzofuran syntheses to date result in the formation of 2-substituted or 2,3-disubstituted benzofurans, although 3-substituted benzofurans are not often reported. Nonetheless, at least three other synthetic methods have appeared recently. All three methods involve multi-step palladium-mediated intramolecular cyclizations from *o*-halo phenols or *o*-dihalobenzenes, which have limited commercial availability and high cost. One method involves carbonylative cyclization of complex *o*-alkynylphenols.[6a,b,c] A second involves cyclization by enolate *O*-arylation.[6d] A third method, recently reported by Malona, et al involves an improved synthesis of 3-ethoxycarbonyl benzofuran (entry 1) in 74% overall yield from iodophenol. The two-step method uses a Michael addition to ethyl propiolate to give *E*-3-(2-iodophenoxy)-2-propenoic acid ethyl ester in 88% yield followed by an intramolecular Heck reaction (84% yield).[6e] The synthetic procedure given here is higher yielding than the other three aforementioned synthetic methods. Furthermore, this procedure uses the relatively inexpensive and commercially available salicylaldehydes, in lieu of the more expensive *o*-halo phenols or *o*-dihalobenzenes.[7] The reaction has been shown to proceed through a unique hemiacetal which subsequently dehydrates to the benzofuran product as detailed in Scheme 1.[7] Upon protonation of the carbonyl oxygen, the activated salicylaldehyde undergoes nucleophilic attack by EDA, followed by an aryl migration with concomitant loss of N_2.[8] The resulting aryl propanal tautomerizes rapidly to form a 3-hydroxyaryl acrylate, which again isomerizes to a stable cyclic hemiacetal in the presence of acid.[7]

Scheme 1

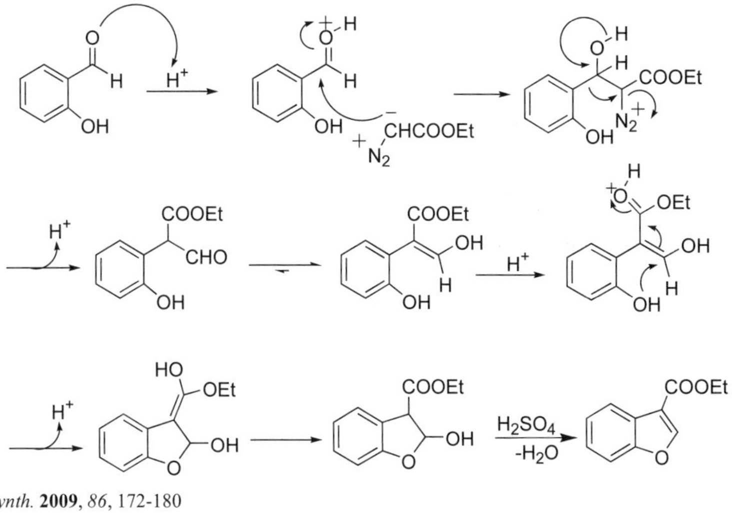

The following procedure can be generalized for any of the 3-substituted benzofurans represented in Table 1. No inert N_2 atmosphere is required. The entire two-step, one-pot reaction can be accomplished in less than 2 hours. The procedure yields product of analytical purity after recrystallization. In cases where EDA is undertitrated, an excess of 5-chlorosalicylaldehyde starting material will remain. Conversely, overtitration of EDA results in formation of diethyl diglycolate impurity. Both of these impurities are easily observed by ^1H NMR. The ^1H NMR chemical shifts for the phenolic and aldehydic protons of 5-chlorosalicylaldehyde are: (CDCl$_3$, 300 MHz) δ: 11.01 (s, 1 H), 9.83 (s, 1H) respectively. The ^1H NMR chemical shifts for the methylene and –CH$_2$ ester protons of diethyl diglycolate are: (CDCl$_3$, 300 MHz) δ: 4.24 (s, 4H), 4.23 (q, 4H) respectively. Both 5-chlorosalicylaldehyde and diethyl diglycolate are acidic compared to the product and can therefore be removed by employing a recrystallization procedure from ethanolic NaOH solution.

Table 1. Isolated Yields of 3-Ethoxycarbonyl Benzofurans

Entry	Aldehyde	Acid Ester	Yield (%)
1			91
2			100
3			97
4			100

1. Department of Chemistry and Biochemistry, University of Wisconsin-Milwaukee, Milwaukee, 3210 N Cramer St. WI 53201. E-mail: mahmun@uwm.edu

2. O'Neil, M.J.; Ed. "The Merck Index: An Encyclopedia of Chemicals, Drugs, and Biologicals (Merck Index, 14 edition)" Merck; 2006. p 2564.

3. a) Thayer A., M., *Chem. Eng. News* **2005**, *83*, 43. b) Bolm, C.; Kasyan, A.; Drauz, K.; Gunther, K.; Raabe, G. *Angew. Chem. Int. Ed.* **2000**, *39*, 2288. c) Stinson, S.; C. *Chem. Eng. News*, **2000**, *78*, 63. d) Clark, J. D.; Heise, J. D.; Shah, A. S.; Peterson, J. C.; Chou, S. K.; Levine, J.; Karakas, A. M.; Ma, Y.; Ng, K. Y.; Patelis, L.; Springer, J, R.; Stano, D. R.; Wettach, R. H.; Dutra, G. A. *Org. Proc. Res. Dev.* **2004**, *8*, 176.

4. Searle, N.E. *Org. Syn., Coll. Vol. 4*, Rabjohn, N.; Ed.; John Wiley and Sons: New York 1963 p.424.

5. a) Weissberger, A.; Taylor, E.C.; Eds. *The Chemistry of Heterocyclic Compounds, Vol 29: Benzofurans*, John Wiley and Sons: New York, 1974. b) Boyle, E.A.; Morgan F.R.; Markwell, R.E.; Smith, S.A.; Thomson, M.J.; Ward, R.W.; Wyman, P.A. *J. Med. Chem.* **1986**, *29*, 894-898. c) Pieters, L.; Van Dyck, S.; Gao, M.; Bai, R.; Hamel, E.; Vlietinck, A.; Lemiere, G. *J. Med. Chem.* **1999**, *42*, 5475. d) Ma, C.-Y.; Liu, W.K.; Che C.-T. *J. Natural Products*, **2002**, *65*, 206. e) Baxendale, I.R.; Griffiths-Jones, C.M.; Ley, S.V.; Tranmer, G.K. *Synlett* **2006**, *3*, 427.

6. a) Nan, Y; Miao, H.; Yang, Z. *Org. Lett.* **2000**, *2*, 297-299. b) Liao, Y.; Reitman, M.; Zhang, Y.; Fathi, R.; Yang, Z. *Org. Lett.* **2002**, *4*, 2607. c) Lu, K.; Luo, T.; Xiang, Z.; You, Z.; Fathi, R.; Chen, J.; Yang, Z. *J. Comb. Chem.* **2005**, *7*, 958 d) Willis, M.C.; Taylor, D.; Gillmore, A.T. *Org. Lett.* **2004**, *6*, 4755. e) Malona, J.A.; Colbourne, J.M.; Frontier, A.J. *Org. Lett.* **2006**, *8*, 5661-5664.

7. Dudley, M. E.; Morshed, M. M.; Hossain, M. M. *Synthesis.* **2006**, 1711-1714.

8. Dudley, M. E.; Morshed, M. M.; Brennan, C. L.; Islam, M. S.; Ahmad, M. S.; Atuu, M. R.; Branstetter, B.; Hossain, M. M. *J. Org. Chem.* **2004**, *69*, 7599

Appendix
Chemical Abstracts Nomenclature; (Registry Number)

3-Benzofurancarboxylic acid, 5-chloro-, ethyl ester; (899795-65-4)

EDA: Acetic acid, 2-diazo-, ethyl ester; (623-73-4)

Tetrafluoroboric acid diethyl etherate; (67969-82-8)

5-Chlorosalicylaldehyde: Benzaldehyde, 5-chloro-2-hydroxy-; (635-93-8)

M. Mahmun Hossain received his M.Sc. degree in chemistry from Dhaka University, Bangladesh. In 1985, he received his Ph.D. from the University of South Carolina. After about 3 years of postdoctoral study with Professor Jack Halpern at the University of Chicago, he joined the Department of Chemistry at the University of Wisconsin-Milwaukee as an Assistant Professor. In 1994, he was promoted to Associate Professor, and received a research award from the UWM Foundation for his outstanding research and creativity. He is an author of nearly 60 publications and has presented more than 100 seminars, posters, and papers at various meetings, universities, and colleges.

Matt Dudley received a B.A. degree in German in 1994 from the University of Denver and an ACS-approved B.S. degree in both Chemistry and Biology in 1998 from the University of Wisconsin-Oshkosh. After an internship award at Aldrich Chemical Company, he joined Aldrich in 1998 as an Associate Chemist and was promoted to Chemist in 1999, where he received three awards for large scale and cGMP processes, and safety. He joined the Hossain group at the University of Wisconsin-Milwaukee in 2002 under fellowship. He received the 2006 Outstanding Teaching Assistant Award and is studying benzofurans and asymmetric-aryl quaternary carbon centers.

Monzur Morshed received his B.Sc. (Honors) and M.Sc. in Chemistry from Jahangirnagar University, Bangladesh. He also completed an M.S. degree in Wood Science from Mississippi State University (MSU). He began Ph.D. studies in Organic Chemistry at the University of Wisconsin-Milwaukee in 2002. Since then he has been an active member in Professor Mahmun Hossain's group and has received multiple Chancellor's Fellowship Awards. He was awarded an ACS Milwaukee Section Travel Grant in 2007. He was also recently honored with the University's Outstanding Teaching Assistant Award for 2007.

PREPARATION OF (S)-tert-ButylPHOX
(Oxazole, 4-(1,1-dimethylethyl)-2-[2-(diphenylphosphino)phenyl]-4,5-dihydro- (4S)-)

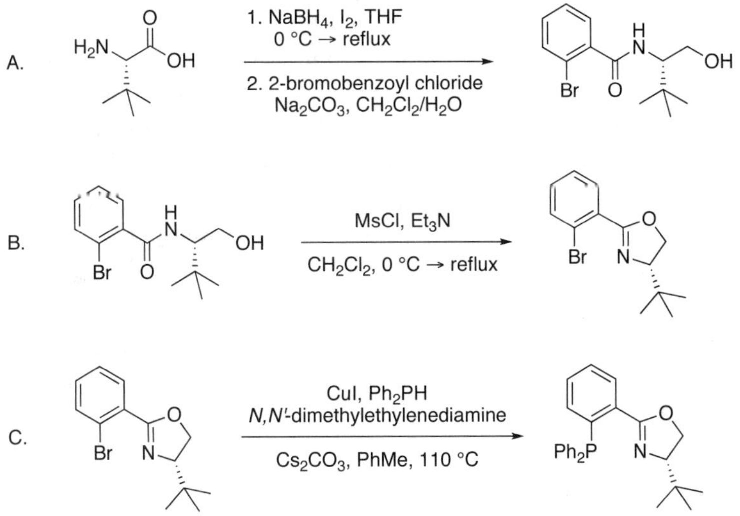

A.

1. NaBH₄, I₂, THF
   0 °C → reflux

2. 2-bromobenzoyl chloride
   Na₂CO₃, CH₂Cl₂/H₂O

B.

MsCl, Et₃N

CH₂Cl₂, 0 °C → reflux

C.

CuI, Ph₂PH
N,N'-dimethylethylenediamine

Cs₂CO₃, PhMe, 110 °C

Submitted by Michael R. Krout, Justin T. Mohr, and Brian M. Stoltz.*[1]
Checked by Andreas Schumacher and Andreas Pfaltz.

1. Procedure

Caution! This procedure should be carried out in an efficient fume hood due to the evolution of hydrogen gas during the reaction.

A. 2-Bromo-N-[(1S)-1-(hydroxymethyl)-2,2-dimethylpropyl]-benzamide. An oven-dried, 500-mL, 3-neck flask equipped with a 3.0 cm × 1.4 cm, egg-shaped, teflon-coated magnetic stirring bar, pressure-equalizing addition funnel, an internal thermometer, and a reflux condenser (central neck) equipped with a two-tap Schlenk adapter connected to a bubbler and an argon/vacuum manifold (Note 1) is assembled hot and cooled under a stream of argon. The flask is charged with (L)-*tert*-leucine (5.00 g, 38.1 mmol, 1.00 equiv, 99% ee) (Note 2) and tetrahydrofuran (100 mL, 0.38 M) (Note 3) under a positive pressure of argon. The resulting suspension is cooled to approximately 4 °C in an ice-water bath and sodium borohydride

(3.46 g, 91.5 mmol, 2.40 equiv) (Note 2) is added in one portion (Note 4). The addition funnel is charged with a solution of iodine (9.67 g, 38.1 mmol, 1.00 equiv) (Note 2) in tetrahydrofuran (25 mL) (Note 3) via syringe and added dropwise to the suspension over 30 min. After complete addition, the bath is removed, the addition funnel and the thermometer are removed and replaced by glass stoppers and the reaction is warmed to reflux (80 °C oil bath temperature). After 18 h the reaction is allowed to cool to ambient temperature and methanol (50 mL) (Note 3) is added slowly resulting in an almost clear solution (Note 5). After stirring for 30 min the solution is quantitatively transferred to a 500-mL, 1-necked flask with methanol (ca. 50 mL) and concentrated on a rotary evaporator under reduced pressure (40 °C, ca. 53 mmHg) to a white semi-solid. The resulting material is dissolved in 20% aqueous potassium hydroxide (75 mL) and stirred for 5 h at ambient temperature with a 3.0 cm × 1.4 cm, egg shaped, teflon-coated magnetic stirring bar. The aqueous phase is extracted with dichloromethane (6 × 60 mL) and the combined organic extracts are dried over sodium sulfate (ca. 7 g), filtered, and concentrated on a rotary evaporator under reduced pressure (40 °C, 38 mmHg) and dried under vacuum (0.13 mmHg) to yield 4.42–4.45 g (37.7–37.9 mmol, 99% yield) of crude (S)-tert-leucinol as a colorless oil (Note 6). This material is used in the following step without purification.

A 500-mL flask containing a 3.0 cm × 1.4 cm, egg shaped, teflon-coated magnetic stirring bar is charged with crude (S)-tert-leucinol (4.42 g, 37.7 mmol, 1.00 equiv), dichloromethane (125 mL, 0.30 M) (Note 3) and then a solution of sodium carbonate (11.98 g, 113.1 mmol, 3.00 equiv) (Note 2) in distilled water (95 mL) (Note 3) is added at ambient temperature. The biphasic mixture is stirred vigorously to emulsify and neat 2-bromobenzoyl chloride (5.67 mL, 43.3 mmol, 1.15 equiv) (Note 2) is added dropwise via syringe over approximately 15 min. The reaction flask is capped with a two-tap Schlenk adapter connected to a bubbler (Note 1) and stirred for 10 h, after which time the layers are partitioned in a 500 mL separatory funnel and the aqueous phase is extracted with dichloromethane (4 × 50 mL). The combined organic extracts are stirred with 1 N potassium hydroxide solution in methanol (19 mL) in a 500 mL Erlenmeyer flask with a 3.0 cm × 1.4 cm, egg-shaped, teflon-coated magnetic stirring bar for 30 min at ambient temperature and then acidified to neutral pH with 1 N hydrochloric acid (ca. 16 mL). Water (25 mL) is added, the phases are partitioned in a 1-L separatory funnel, and the aqueous phase is extracted with dichloromethane (4 × 35 mL). The combined organic extracts are washed with saturated brine

Org. Synth. **2009**, *86*, 181-193

(75 mL), dried over sodium sulfate (3.00 g), filtered, and concentrated on a rotary evaporator under reduced pressure (40 °C, ca. 11 mmHg) to an off-white solid. The crude white solid is dissolved in a minimal amount of hot acetone (ca. 10 mL) (Note 3) and hexanes (Note 3) are added until a cloudy solution is obtained (ca. 45 mL). The crystals formed upon cooling and aging for 3 hours at 0 °C are collected, washed with hexanes, and dried under vacuum to afford 2-bromo-*N*-[(1*S*)-1-(hydroxymethyl)-2,2-dimethylpropyl]-benzamide (9.23–9.61 g, 30.8–32.0 mmol) as white blocks (Note 7). The filtrate is concentrated and recrystallized in a similar manner (with acetone (ca. 2 mL) and hexanes (10 mL)) to provide additional product (0.62–1.13 g, 2.1–3.76 mmol) as white blocks (Note 8), for a combined yield of 9.85–10.74 g (32.8–35.77 mmol, 86–94 % yield over two steps).

B. 2-(2-Bromophenyl)-4-(1,1-dimethylethyl)-4,5-dihydro-(4S)-oxazole. An oven-dried, 500-mL, 3-necked flask equipped with a 3.0 cm × 1.4 cm, egg-shaped, teflon-coated magnetic stirring bar, an internal thermometer, a glass stopper and a reflux condenser (central neck) equipped with a two-tap Schlenk adapter connected to a bubbler and an argon/vacuum manifold (Note 1) is assembled hot and cooled under a stream of argon. The flask is charged with 2-bromo-*N*-[(1*S*)-1-(hydroxymethyl)-2,2-dimethylpropyl]-benzamide (9.85 g, 32.8 mmol, 1.00 equiv), dichloromethane (170 mL, 0.19 M) (Note 3), and triethylamine (11.0 mL, 78.6 mmol, 2.40 equiv) (Note 2) under a positive pressure of argon. The resulting colorless solution is cooled to approximately 4 °C in an ice-water bath and neat methanesulfonyl chloride (2.92 mL, 37.7 mmol, 1.15 equiv) (Note 2) is added dropwise via syringe over 3 min, at which point the solution turns slightly yellow. The reaction is warmed to reflux (50 °C oil bath temperature) while monitoring conversion by TLC (Note 9). Upon completed cyclization, the reaction is allowed to cool to ambient temperature and 60 mL of saturated aqueous sodium bicarbonate is added with vigorous stirring for 5 min. The layers are partitioned in a 1-L separatory funnel, the aqueous phase is extracted with dichloromethane (2 × 35 mL), the combined organic phases are washed with saturated brine (75 mL), dried over anhydrous magnesium sulfate (1.50 g), filtered, and concentrated on a rotary evaporator under reduced pressure (40 °C, 23 mmHg) to afford a red-brown semi-solid. The residue is dissolved in a minimal amount of dichloromethane (ca. 35 mL), dry-loaded onto silica gel (8 g), and purified by silica gel chromatography (Note 10) to afford 8.85–8.86 g (31.4 mmol, 96% yield) of 2-(2-bromophenyl)-4-(1,1-dimethyl-ethyl)-4,5-dihydro-(4*S*)-

oxazole as a pale yellow oil. This material solidifies when placed in a –20 °C freezer and is preferred in this state for the subsequent reaction (Note 11).

C. 4-(1,1-dimethylethyl)-2-[2-(diphenylphosphino)phenyl]-4,5-dihydro-(4S)-oxazole ((S)-tert-ButylPHOX). A 150-mL Schlenk flask equipped with a glass valve, a glass stopper, and a 1.7 cm × 0.7 cm, egg-shaped, teflon-coated magnetic stirring bar is dried with a heat gun under vacuum and cooled under argon atmosphere. The glass stopper is removed under a positive pressure of argon and the flask is charged with copper(I) iodide (19.0 mg, 0.10 mmol, 0.005 equiv) (Note 2), diphenylphosphine (4.35 mL, 25.0 mmol, 1.25 equiv) (Note 2), N,N'-dimethylethylenediamine (53 μL, 0.50 mmol, 0.025 equiv) (Note 2) and toluene (20 mL) (Note 3). The flask is sealed with the glass stopper and the colorless contents are stirred at ambient temperature for 20 min. The glass stopper is then removed under a positive pressure of argon and the flask is charged with 2-(2-bromophenyl)-4-(1,1-dimethylethyl)-4,5-dihydro-(4S)-oxazole (5.64 g, 20.0 mmol, 1.00 equiv), cesium carbonate (9.78 g, 30.0 mmol, 1.50 equiv) (Note 2), and toluene (20 mL, 0.50 M total) to wash the neck and walls of the flask. The flask is equipped with a reflux condenser with a two-tap Schlenk adapter connected to a bubbler and an argon/vacuum manifold (Note 1). The now yellow heterogeneous reaction is placed in a 110 °C oil bath and vigorously stirred under argon atmosphere (Note 12). Following consumption of starting material (Note 13), the reaction is allowed to cool to ambient temperature, filtered through a pad of celite, and the filter cake is washed with dichloromethane (2 × 40 mL) (Note 14). The filtrate is concentrated on a rotary evaporator under reduced pressure (40 °C, 15 mmHg) to a pale yellow semi-solid. The residue is dissolved in a minimal amount of dichloromethane (ca. 40 mL) (Note 15), dry-loaded onto silica gel (10 g), and purified by silica gel chromatography eluting with 24:1 hexanes/diethyl ether until excess Ph₂PH elutes, then with a 9:1 dichloromethane/diethyl ether mixture until the desired product elutes (Note 16). The combined fractions are concentrated on a rotary evaporator under reduced pressure (40 °C, 14 mmHg) to a viscous, pale yellow oil and layered with acetonitrile (ca. 5 mL) to facilitate crystallization (Notes 3 and 17). The flask is swirled while crystals form within seconds (Note 18). After approximately 15 minutes, the flask is placed under high vacuum to remove volatiles to afford 6.81 g (17.6 mmol, 88% yield) of (S)-*tert*-ButylPHOX as white blocks (Note 19).

2. Notes

1.   A two-tap Schlenk adapter connected to a bubbler and an argon/vacuum manifold is illustrated in Yu, J.; Truc, V.; Riebel, P.; Hierl, E. and Mudryk, B. *Org. Synth.* **2008**, *85*, 64–71.

2.   Submitters and checkers purchased (L)-*tert*-leucine (99%, 99% ee), sodium borohydride (98%), cesium carbonate (99%), and *N,N'*-dimethylethylenediamine (99%) from Aldrich and used as received. (2)-Bromobenzoyl chloride (98%) and methanesulfonyl chloride (99.5%) were purchased from Acros and used as received. Copper iodide (98%) was purchased from Strem and used as received. Submitters and checkers purchased triethylamine (99.5%) from Aldrich and distilled it from calcium hydride prior to use. Submitters and checkers purchased diphenylphosphine (99%) from Strem and transferred it through a cannula to a dry Schlenk tube under nitrogen to prolong reagent life. The submitters purchased iodine (≥99%) and sodium carbonate (99%) from Aldrich and used as received. The checkers purchased iodine (puriss. p. a.) and potassium hydroxide (puriss. p. a.) from Riedel-de-Haën (puriss. p. a.) and sodium carbonate (99.5%) from Merck and used as received.

3.   Submitters distilled tetrahydrofuran from sodium 9-fluorenone ketyl[2] prior to use. Submitters and checkers used methylene chloride, toluene, and acetonitrile purified by passage through an activated alumina column under argon.[3] The submitters purchased reagent grade acetone from EMD, and hexanes and methanol (both ACS grade) were purchased from Fisher and used as received. The submitters used distilled water purified with a Barnstead NANOpure Infinity UV/UF system. The checkers used tetrahydrofuran (VWR, HPLC-grade) dried using a Pure-Solve™ system. Reagent grade acetone was purchased from VWR, methanol (Baker analyzed) was purchased from J.T. Baker and hexanes were distilled.

4.   The evolution of hydrogen gas during the addition of sodium borohydride is minor due to the adequate size of the reaction flask and the surface area of cooling. This is readily vented through the oil bubbler.

5.   The initial reaction quench with methanol proceeds with vigorous gas evolution. Methanol should be added dropwise until the intensity of gas evolution abates.

6.   The reduction product, (*S*)-*tert*-leucinol, can be purified by distillation,[4] but this was not necessary for this application. The material showed the following characterization data: ^{1}H NMR (400 MHz, CDCl$_3$) δ:

0.89 (s, 9 H), 2.49 (dd, J = 10.2, 3.9 Hz, 1 H), 3.19 (t, J = 10.2 Hz, 1 H), 3.70 (dd, J = 10.2, 3.9 Hz, 1 H). This material may also be purchased from commercial sources, but is less expensive in the amino acid form. Similar amino acid reductions have appeared in *Organic Syntheses*.[5]

7.   2-Bromo-*N*-[(1*S*)-1-(hydroxymethyl)-2,2-dimethylpropyl]-benzamide showed the following characterization data: mp 117–118 °C from acetone/hexanes; R_f = 0.14 (2:1 hexanes/acetone); ^{1}H NMR (500 MHz, CDCl$_3$) δ: 1.03 (s, 9 H), 2.40 (br dd, J = 5.0 Hz, 1 H), 3.64–3.70 (m, 1 H), 3.91-3.97 (m, 1 H), 4.04–4.08 (m, 1 H), 6.20 (br d, J = 8.4 Hz, 1 H), 7.27 (ddd, J = 7.9, 7.9, 1.6 Hz, 1 H), 7.35 (dd, J = 7.5, 7.5 Hz, 1 H), 7.55 (dd, J = 7.6, 1.5 Hz, 1 H), 7.59 (d, J = 8.0 Hz, 1 H); ^{13}C NMR (126 MHz, CDCl$_3$) δ: 27.1, 33.8, 60.3, 63.0, 119.0, 127.6, 129.7, 131.3, 133.3, 137.9, 168.6; IR (ATR) 3223, 3065, 2961, 1627, 1544 cm^{-1}; HRMS (FAB+) m/z calc'd for C$_{13}$H$_{19}$NO$_2$Br [M+H]$^+$: 300.0599, found 300.0590; MS (FAB) m/z (relative intensity): 300 (100%), 185 (49%), 77 (21%); [α]$_D^{20}$ +17.3 (c 2.38, methanol); Anal. calc'd. for C$_{13}$H$_{18}$NO$_2$Br: C, 52.01; H, 6.04; N, 4.67. Found: C, 52.01; H, 5.95; N, 4.51.

8.   In some cases the resulting filtrate was purified by silica gel flash chromatography, eluting with a 3:1 → 2:1 hexanes/acetone gradient to afford an additional 2–6% of an off-white amorphous solid that is spectroscopically identical to the crystalline material.

9.   Reaction progress can be monitored by TLC analysis (the checkers used Polygram®SIL/UV$_{254}$-TLC-plates from Macherey-Nagel) using 2:1 ethyl acetate/hexanes as the eluent with UV visualization (R_f amide = 0.28, R_f mesylate = 0.43, R_f bromooxazoline = 0.64). Mesylate formation is typically complete upon final addition of methanesulfonyl chloride, whereas cyclization to the bromooxazoline typically requires ca. 5 h at 50 °C to complete.

10.   Flash chromatography column dimensions: 3 cm diameter × 20 cm height of silica gel (checkers used "Silica Gel 60" (0.040-0.063 mm) from Merck), eluting with 200 mL of 9:1 hexanes/ethyl acetate, then 450 mL of 6:1 hexanes/ethyl acetate, collecting ca. 20–25 mL fractions. Fraction purity can be assayed by TLC (the checkers used Polygram®SIL/UV$_{254}$-TLC-plates from Macherey-Nagel) analysis using 4:1 hexanes/ethyl acetate with UV visualization. This method of purification removes color from the crude material and a minor impurity at R_f = 0.33.

11.   2-(2-Bromophenyl)-4-(1,1-dimethylethyl)-4,5-dihydro-(4*S*)-oxazole showed the following characterization data: mp 47–48 °C; R_f = 0.27

(4:1 hexanes/ethyl acetate); ^{1}H NMR (500 MHz, CDCl$_3$) δ: 1.00 (s, 9 H), 4.11 (dd, J = 10.2, 8.0 Hz, 1 H), 4.26 (dd, J = 8.3, 8.3 Hz, 1 H), 4.38 (dd, J = 10.2, 8.7 Hz, 1 H), 7.27 (ddd, J = 7.7, 7.6, 1.9 Hz, 1 H), 7.33 (ddd, J = 7.5, 7.5, 1.1 Hz, 1 H), 7.63 (dd, J = 8.0, 0.9 Hz, 1 H), 7.66 (dd, J = 7.6, 1.3 Hz, 1 H); ^{13}C NMR (126 MHz, CDCl$_3$) δ: 26.0, 34.1, 69.0, 76.8, 121.9, 127.1, 130.3, 131.3, 131.5, 133.7, 162.8; IR (ATR) 2958, 1660, 1476, 1358, 1095, 1022, 959 cm^{-1}; HRMS (FAB+) m/z calc'd for C$_{13}$H$_{17}$NOBr [M+H]$^+$: 282.0493, found 282.0488; MS (FAB) m/z (relative intensity): 282 (100%), 224 (10%), 183 (17%), 77 (12%); $[\alpha]_D^{20}$ −48.9 (c 3.77, hexane); Anal. calc'd. for C$_{13}$H$_{16}$NOBr: C, 55.33; H, 5.72; N, 4.96. Found: C, 55.37; H, 5.70; N, 4.84.

12. Submitters sealed the teflon valve and placed the sealed flask in a 110 °C oil bath protected with a blast shield. Reactions performed with minimal stirring or that cease to stir result in incomplete conversion. The preferred stirring rate of the coupling reaction is ca. 700 setting (ca. 700 rpm) on an IKAmag RET basic stir/hot plate (a range between 500–800 rpm is sufficient). Additionally, the color of the reaction becomes an intense yellow within 5–10 minutes of heating. The color of the inorganic base then dominates as it turns to light gray, and finally to a dark maroon/purple color after several hours.

13. The reaction typically requires 21 h to reach complete conversion. Reaction progress can be monitored by TLC analysis (the checkers used Polygram®SIL/UV$_{254}$-TLC-plates from Macherey-Nagel) using 4:1 hexanes/diethyl ether as the eluent (developed twice) with UV visualization (R$_f$ bromooxazoline = 0.17, R$_f$ reduced oxazoline = 0.27, R$_f$ tert-ButylPHOX = 0.33, R$_f$ Ph$_2$PH = 0.51).

14. Fritted glass funnel (Por. 3, pore size 15–40 µm), 4 cm diameter × 5 cm height) filled with 16 g of celite.

15. Submitters dissolved the pale yellow semi-solid in a minimal amount of dichloromethane (ca. 40 mL) and diethyl ether (ca. 50 mL).

16. Flash chromatography column dimensions: 5 cm diameter × 16 cm height of silica gel, (checkers used "Silica Gel 60" (0.040-0.063 mm) from Merck). Checkers eluted with 500 mL of 24:1 hexanes/diethyl ether, then 400 mL of 9:1 dichloromethane/diethyl ether, collecting ca. 50 mL fractions. Fraction purity can be assayed by TLC (the checkers used Polygram®SIL/UV$_{254}$-TLC-plates from Macherey-Nagel) analysis using 4:1 hexanes/diethyl ether with UV visualization. The mixture of products may

contain reduced arene, starting bromooxazoline, and desired (S)-tert-ButylPHOX, depending on the extent of reaction (Note 12).

17. As the percentage of desired (S)-tert-ButylPHOX in the crude mixture increases, the oil readily solidifies upon concentration under reduced pressure. To decrease the time required to induce crystallization, this oil can then be dissolved in diethyl ether and further concentrated. Additionally, acetonitrile efficiently promotes crystallization of (S)-tert-ButylPHOX in concentrated solutions.

18. If the reaction is pushed to completion, the material obtained from this simple purification is typically quite pure (no impurities were detected by ^{1}H NMR analysis of the crude oil). If the purity is unsatisfactory, this crystalline material can be recrystallized with hot acetonitrile. A typical recrystallization is performed as follows: in an experiment run on 20.0 mmol scale, 7.033 g (18.15 mmol) of crude product was dissolved in a minimal amount (ca. 8–10 mL) of boiling acetonitrile and allowed to cool slowly to ambient temperature. The crystals are then filtered and washed with ca. 15–25 mL of hexanes, then dried under high vacuum to yield 6.613 g (17.07 mmol, 85.3% yield) of white blocks. This material is analytically pure by ^{1}H NMR and all other spectroscopic data (see Note 19).

19. In a run carried out on half-scale, 3.49 g of (S)-tert-ButylPHOX was obtained (90% yield). (S)-tert-ButylPHOX showed the following characterization data; mp 113–114 °C from acetonitrile; R_f = 0.33 (4:1 hexanes/diethyl ether); ^{31}P NMR (162 MHz, CDCl$_3$) δ: –5.49 (s); ^{1}H NMR (400 MHz, CDCl$_3$) δ: 0.73 (s, 9 H), 3.88 (dd, J = 10.2, 8.2 Hz, 1 H), 4.01 (dd, J = 8.3 Hz, 8.3 Hz, 1 H), 4.08 (dd, J = 10.2, 8.5 Hz, 1 H), 6.87 (ddd, J = 7.7, 4.0, 0.8 Hz, 1 H), 7.33–7.21 (m, 11 H), 7.36 (apparent dt, J = 7.6, 1.3 Hz, 1 H), 7.94 (ddd, J = 7.7, 3.7, 0.9 Hz, 1 H); ^{13}C NMR (101 MHz, CDCl$_3$) δ: 25.8, 33.6, 68.3, 76.7 (overlaps with CHCl$_3$-rest-signal, detected in DEPT135-experiment), 128.0, 128.3 (d, J_{CP} = 5.9 Hz), 128.2 (2 lines), 128.5 (d, J_{CP} = 9.7 Hz), 129.8 (d, J_{CP} = 3.1 Hz), 130.3, 132.0 (d, J_{CP} = 19.7 Hz), 133.6 (d, J_{CP} = 20.2 Hz), 134.1, 134.3 (d, J_{CP} = 21.0 Hz), 138.3 (d, J_{CP} = 9.7 Hz), 138.5 (d, J_{CP} = 12.6 Hz), 138.8 (d, J_{CP} = 25.5 Hz), 162.7 (d, J_{CP} = 2.7 Hz); IR (ATR) 3069, 2955, 2897, 2866, 1653, 1583, 1475, 1433, 1354, 1090, 1024, 955, 742, 669, 580, 511 cm^{-1}; HRMS (FAB+) m/z calc'd for C$_{25}$H$_{27}$NOP [M+H]$^+$: 388.1830, found 388.1831; MS (EI) m/z (relative intensity): 388 (1%), 372 (9%), 330 (55%), 302 (100), 228 (6), 183 (12); $[\alpha]_D^{20}$ –75.2 (c 0.925, CHCl$_3$); Anal. calc'd. for C$_{25}$H$_{26}$NOP: C, 77.50; H, 6.76; N, 3.62. Found: C, 77.22; H, 6.82; N, 3.57. The enantiomeric excess of

(S)-*tert*-ButylPHOX can be determined by analytical supercritical fluid chromatography; this was performed using a Berger Analytix SFC (Thar Technologies) equipped with a Chiralcel® OJ-H column (4.6 mm x 25 cm) obtained from Daicel Chemical Industries, Ltd. and a diode array detector. The assay conditions are 10% ethanol, 35 °C, 2 mL/min flow rate, with visualization at 210 nm (optimal), retention times: (R) enantiomer = 4.67 min, (S) enantiomer = 5.17 min. The minor (R) enantiomer can not be detected from SFC analyses of ligand prepared from the reported procedure, and is therefore >99% ee. This crystalline material is stable indefinitely at ambient temperatures in a closed container under an atmosphere of nitrogen or argon.

Safety and Waste Disposal Information

All hazardous materials should be handled and disposed of in accordance with "Prudent Practices in the Laboratory"; National Academy Press; Washington, DC, 1995.

3. Discussion

This synthesis of (S)-*tert*-ButylPHOX (4-(1,1-dimethylethyl)-2-[2-(diphenylphosphino)phenyl]-4,5-dihydro-(4S)-oxazole) is a modification of our previously reported procedure.[6] Several improvements have been implemented that facilitate large-scale preparation of this ligand. Recrystallization of 2-bromo-N-[(1S)-1-(hydroxymethyl)-2,2-dimethyl-propyl]-benzamide obviates the previous need for flash column chromatography. Subsequent oxazoline formation is now accomplished via mesylate displacement with improved efficiency and yield. The use of methanesulfonyl chloride enables rapid mesylate formation at milder temperatures, and aqueous reaction workup is favored over the previous method, where incomplete hydrolysis of *p*-toluenesulfonyl chloride complicated purification. The copper(I) iodide-catalyzed phosphine coupling[7] has been optimized to maximize the efficiency of the reaction by minimizing the use of catalyst and diamine ligand, as well as reducing the quantities of phosphine, cesium carbonate, and solvent. Finally, a procedure to purify (S)-*tert*-ButylPHOX is described, using a simple silica gel plug, followed by crystallization with acetonitrile (or recrystallization when necessary), to afford the ligand as a white crystalline solid in four steps from (L)-*tert*-leucine in excellent overall yield (71.9–80.4% over four steps).

The phosphinooxazoline[8] (*S*)-*tert*-ButylPHOX is a chiral P/N-ligand useful for an array of organometallic transformations, including alkylations,[8,9] desymmetrizations of *meso*-anhydrides,[10] Heck reactions,[11] hetero-Diels–Alder cycloadditions,[12] Meerwein–Eschenmoser Claisen rearrangements,[13] and hydrogenations.[14] Our laboratory has recently described its use as a uniquely effective ligand for the palladium-catalyzed

Table 1. PHOX derivatives prepared via this protocol.[6b]

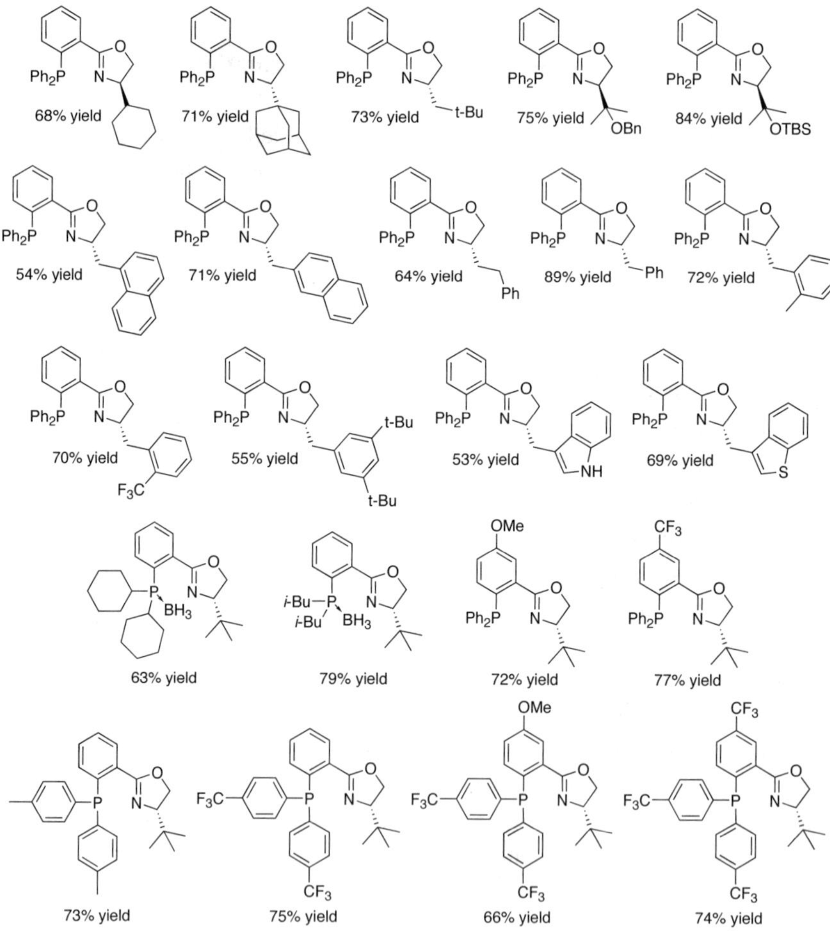

asymmetric decarboxylative allylation[6,15] and protonation[16] of prochiral ketone enolates. This synthesis of (S)-*tert*-ButylPHOX highlights improvements of a general and efficient strategy to access PHOX ligands of varied structure and electronics in substantial quantities (Table 1).[6b]

1. Department of Chemistry and Chemical Engineering, M/C 164-30, California Institute of Technology, Pasadena, California, 91125; E-mail: stoltz@caltech.edu.
2. Kamaura, M.; Inanaga, J. *Tetrahedron Lett.* **1999**, *40*, 7347–7350.
3. Pangborn, A. B.; Giardello, M. A.; Grubbs, R. H.; Rosen, R. K.; Timmers, F. J. *Organometallics* **1996**, *15*, 1518–1520.
4. McKennon, M. J.; Meyers, A. I. *J. Org. Chem.* **1993**, *58*, 3568–3571.
5. (a) Dickman, D. A.; Meyers, A. I.; Smith, G. A.; Gawley, R. E. *Org. Synth. Coll. Vol. VII*, **1990**, 530–533. (b) Gage, J. R.; Evans, D. A. *Org. Synth., Coll. Vol. VIII*, **1993**, 528–531.
6. (a) Behenna, D. C.; Stoltz, B. M. *J. Am. Chem. Soc.* **2004**, *126*, 15044–15045. (b) Tani, K.; Behenna, D. C.; McFadden, R. M.; Stoltz, B. M. *Org. Lett.* **2007**, *9*, 2529–2531.
7. Gelman, D.; Jiang, L.; Buchwald, S. L. *Org. Lett.* **2003**, *5*, 2315–2318.
8. (a) von Matt, P.; Pfaltz, A. *Angew. Chem., Int. Ed. Engl.* **1993**, *32*, 566–568. (b) Sprinz, J.; Helmchen, G. *Tetrahedron Lett.* **1993**, *34*, 1769–1772. (c) Dawson, G. J.; Frost, C. G.; Williams, J. M. J.; Coote, S. J. *Tetrahedron Lett.* **1993**, *34*, 3149–3150.
9. (a) Helmchen, G.; Pfaltz, A. *Acc. Chem. Res.* **2000**, *33*, 336–345. (b) Weiβ, T. D.; Helmchen, G.; Kazmaier, U. *Chem. Commun.* **2002**, *12*, 1270–1271.
10. Cook, M. J.; Rovis, T. *J. Am. Chem. Soc.* **2007**, *129*, 9302–9303.
11. Loiseleur, O.; Hayashi, M.; Schmees, N.; Pfaltz, A. *Synthesis* **1997**, *11*, 1338–1345.
12. Yao, S.; Saaby, S.; Hazell, R. G.; Jørgensen, K. A. *Chem.–Eur. J.* **2000**, *6*, 2435–2448.
13. Linton, E. C.; Kozlowski, M. C. *J. Am. Chem. Soc.* **2008**, *130*, 16162–16163.
14. Legault, C. Y.; Charette, A. B. *J. Am. Chem. Soc.* **2005**, *127*, 8966–8967.
15. (a) Mohr, J. T.; Behenna, D. C.; Harned, A. M.; Stoltz, B. M. *Angew. Chem., Int. Ed.* **2005**, *44*, 6924–6927. (b) Seto, M.; Roizen, J. L.; Stoltz,

B. M. *Angew. Chem., Inte. Ed.* **2008**, *47*, 6873–6876. (c) Mohr, J. T.; Krout, M. R.; Stoltz, B. M. *Org. Synth.* **2009**, *86*, [please update with appropriate reference].

16. (a) Mohr, J. T.; Nishimata, T.; Behenna, D. C.; Stoltz, B. M. *J. Am. Chem. Soc.* **2006**, *128*, 11348–11349. (b) Marinescu, S. C.; Nishimata, T. N.; Mohr, J. T.; Stoltz, B. M. *Org. Lett.* **2008**, *10*, 1039–1042.

Appendix
Chemical Abstracts Nomenclature; (Registry Number)

(L)-*tert*-Leucine: L-Valine, 3-methyl-; (20859-02-3)

(S)-*tert*-Leucinol: 1-Butanol, 2-amino-3,3-dimethyl-, (2S)-; (112245-13-3)

Sodium borohydride: Borate(1-), tetrahydro-, sodium (1:1); (16940-66-2)

2-Bromobenzoyl chloride: Benzoyl chloride, 2-bromo-; (7154-66-7)

Methansulfonyl chloride; (124-63-0)

Copper iodide; (1335-23-5)

Diphenylphosphine: Phosphine, diphenyl-; (829-85-6)

N,N'-dimethylethylenediamine: 1,2-Ethanediamine, N1,N2-dimethyl-; (110-70-3)

Cesium carbonate: Carbonic acid, cesium salt (1:2); (534-17-8)

(S)-*tert*-ButylPHOX: Oxazole, 4-(1,1-dimethylethyl)-2-[2-(diphenylphosphino)phenyl]-4,5-dihydro-, (4S)-; (148461-16-9)

Brian M. Stoltz was born in Philadelphia, PA in 1970 and obtained his B.S. degree from the Indiana University of Pennsylvania in Indiana, PA. After graduate work at Yale University in the labs of John L. Wood and an NIH postdoctoral fellowship at Harvard in the Corey labs he took a position at the California Institute of Technology. A member of the Caltech faculty since 2000, he currently is the Ethel Wilson Bowles and Robert Bowles Professor of Chemistry and a KAUST GRP Investigator. His research interests lie in the development of new methodology for general applications in synthetic chemistry.

192

Michael R. Krout received his B.S. degree in biochemistry from the Indiana University of Pennsylvania in 2002. He then worked in the medicinal chemistry department at Merck Research Laboratories in West Point, PA, where he was involved in the development of non-steroidal selective androgen receptor modulators aimed toward the treatment of osteoporosis. In the fall of 2003, he joined the lab of Professor Brian Stoltz at Caltech where he has worked toward his Ph.D. as a Lilly fellow. His research interests include the development of catalytic, asymmetric methods and their utility in natural product total synthesis.

Justin T. Mohr received his A.B. degree in chemistry in 2003 from Dartmouth College where he conducted research with Professor Gordon W. Gribble. He joined the laboratories of Professor Brian M. Stoltz at Caltech in 2003 where he has pursued Ph.D. studies as a Lilly fellow. His research interests include the development of enantioselective reactions and applications in natural product total synthesis.

Andreas Schumacher was born in Binningen (Switzerland) in 1983. He studied chemistry at the University of Basel (Switzerland) where he obtained his B.Sc. in Chemistry in 2006 and his M.Sc. in Chemistry in 2008. He started his Ph.D, studies in 2008 under the supervision of Prof. Andreas Pfaltz at the University of Basel and is currently working in the field of Ir-catalyzed enantioselective hydrogenation.

PREPARATION OF (S)-2-ALLYL-2-METHYLCYCLOHEXANONE
(Cyclohexanone, 2-methyl-2-(2-propen-1-yl)-, (2S)-)

Submitted by Justin T. Mohr, Michael R. Krout, and Brian M. Stoltz.*[1]
Checked by Christian Ebner and Andreas Pfaltz.

1. Procedure

Caution! This procedure should be carried out in an efficient fume hood due to the evolution of hydrogen gas during the reaction. Appropriate precautions should be taken to avoid inhalation or direct contact with iodomethane or allyl alcohol. The former is a known carcinogen, and the latter is a potent toxin due to its in vivo metabolism to acrolein, a known carcinogen.

A. *1-Methyl-2-oxo-cyclohexanecarboxylic acid 2-propenyl ester.* A 500-mL, single-necked, round-bottomed flask equipped with a large magnetic stir bar (38 x 8 mm) is charged with 50.0 g of pimelic acid (313 mmol, 1.00 equiv), 156 mL of toluene, and 63.9 mL of allyl alcohol (939 mmol, 3.00 equiv) (Note 1). The mixture is stirred vigorously to create a uniform suspension, and 297 mg of *p*-toluenesulfonic acid monohydrate (1.57 mmol, 0.005 equiv) is added. A Dean–Stark trap and a water-cooled condenser with a two-tap Schlenk adapter connected to a bubbler and an argon/vacuum manifold (Note 2) are affixed to the flask, and the resulting

Org. Synth. **2009**, *86*, 194-211
Published on the Web 2/10/2009

suspension is heated to reflux (120 °C oil bath temperature). The mixture in the flask soon became homogeneous. After 16 h at reflux, approximately 11 mL of water had accrued in the Dean–Stark trap. The vessel is cooled to ambient temperature and the solution is transferred to a separatory funnel (500 mL). The organic solution is washed successively with saturated aqueous sodium bicarbonate (3 x 15 mL) and brine (2 x 15 mL) and then dried over anhydrous magnesium sulfate (6 g). After filtration through cotton, the organic solution is concentrated by rotary evaporation under vacuum (60 °C, 15 mmHg) and then the last traces of solvent are removed under high vacuum (0.15 mmHg) to yield 72.3–74.6 g of diallyl pimelate (301–311 mmol, 96–99% yield) as a slightly yellow-colored, free-flowing liquid. GC analysis indicated >99% purity (Note 3).

A flame-dried, three-necked, 1-L flask equipped with a glass stopper, a water-cooled reflux condenser with a two-tap Schlenk adapter connected to a bubbler and an argon/vacuum manifold (Note 2), a rubber septum, and a magnetic stir bar is charged with 13.2 g of 60% sodium hydride (331 mmol, 1.10 equiv) and tetrahydrofuran (250 mL) (Note 1). The flask is immersed in a water bath (22 °C) and a solution of 72.2 g of crude diallyl pimelate (301 mmol, 1.00 equiv) in 50 mL of tetrahydrofuran is added in a steady stream via cannula during the course of 5 min. Some moderate bubbling (hydrogen evolution) of the reaction mixture is observed during the addition. Following the addition, the suspension is heated to 40 °C and then stirred for 10 h (Note 4) whereupon the reaction mixture turned to a clear, yellowish solution. Once the starting material is consumed (based on TLC, Note 5), 24.3 mL of neat iodomethane (391 mmol, 1.30 equiv) (Note 1) is added to the mixture, which became a white suspension. After an additional 15 h at 40 °C, the mixture is cooled to ambient temperature (22 °C) and water (60 mL) is added carefully via syringe over the course of 9 min to obtain a clear, yellowish solution. The mixture is transferred to a 1-L, single-necked, round-bottomed flask, the THF is removed by rotary evaporation under vacuum (40 °C, 150 mmHg) (Note 6), and the remaining solution is transferred to a separatory funnel (500 mL) and diluted with ethyl acetate (100 mL). The phases are separated and the aqueous phase is extracted with ethyl acetate (3 x 75 mL). The combined organic extracts are washed with brine (1 x 50 mL) and dried over anhydrous magnesium sulfate (5 g). After filtration through cotton, the organic solution is concentrated by rotary evaporation under vacuum (40 °C, 75 mmHg) to yield a yellow liquid. The material is purified by short-path distillation until the first drop of the distillate turned yellow to give 52.9–53.0 g (270 mmol, 90% yield) (Note 7) of a clear, colorless liquid boiling from 69–72 °C/0.08 mmHg. GC analysis found >99% product purity (Note 8).

B. (2S)-2-Methyl-2-(2-propen-1-yl)-cyclohexanone. A flame-dried, 50-mL, conical flask equipped with a rubber septum is charged with a portion of 1-methyl-2-oxo-cyclohexanecarboxylic acid 2-propenyl ester, placed under vacuum (0.06 mmHg) for 60 min to remove any dissolved gases, and then backfilled with argon. A 1-L, three-necked, round-bottomed flask is equipped with a stir bar, two rubber septa, and a two-tap Schlenk adapter connected to a bubbler and an argon/vacuum manifold (Note 2). The apparatus is flame-dried under vacuum and backfilled with dry argon (three cycles). After cooling the flask to ambient temperature, 435 mL of anhydrous tetrahydrofuran (Notes 1 and 9) is added and the flask is immersed in a 30 °C water bath. A twelve-inch needle is inserted through one of the septa and used to bubble dry argon gas through the liquid for 30 min. The needle is removed and then 1.02 g of tris(dibenzylideneacetone)dipalladium(0) (Pd$_2$(dba)$_3$, 1.11 mmol, 0.0125 equiv) and 1.03 g of (S)-tert-ButylPHOX (2.67 mmol, 0.030 equiv) (Note 1) are added. The mixture immediately became opaque and took on a golden-brown color. This mixture is stirred at 30 °C for 30 min (Note 10). Subsequently, neat 1-methyl-2-oxo-cyclohexanecarboxylic acid 2-propenyl ester (17.5 g, 89.03 mmol, 1.00 equiv) from the conical flask is added via syringe in a dropwise fashion to the catalyst mixture over the course of 10 min.

When the transfer is complete, the syringe is rinsed successively with two 5 mL portions of anhydrous tetrahydrofuran into the reaction mixture. Upon addition of the substrate to the catalyst mixture, the color changed to olive green. The mixture is maintained at 30–32 °C for 22–23 h (Note 11), when TLC indicated complete consumption of the starting material (Note 12). The olive green-colored mixture is then passed through a pad of silica gel (5 cm diameter x 5 cm height) and rinsed with diethyl ether (200 mL). The bright yellow filtrate is concentrated by rotary evaporation under vacuum (150 mmHg, 40 °C) (Note 13). The liquid is then transferred to a 50-mL round-bottomed flask and distilled through a short path apparatus into a receiving flask immersed in an ice water bath to provide 11.5–12.8 g (75.7–84.2 mmol, 85–95% yield) of (S)-2-allyl-2-methylcyclohexanone as a clear, colorless liquid boiling from 91–93 °C/16 mmHg that is analytically pure based on standard techniques (Note 14). Analysis of this material by

GC on a chiral stationary phase found 86–87% enantiomeric excess (Note 15). In a reaction that gave 85% yield after distillation, additional product was obtained by subjecting the material remaining in the distillation pot to flash chromatography on silica gel (Notes 16 and 17), which provided an additional 1.14 g of product (7.50 mmol, 8% yield), also of 86% ee, for a combined yield of 12.64 g (83.2 mmol, 93% yield) (Note 18).

C. *Enrichment of (2S)-2-methyl-2-(2-propen-1-yl)-cyclohexanone via (2E)-2-[(2S)-2-methyl-2-(2-propenyl)cyclohexylidene]-hydrazinecarboxamide.* Into a 250-mL, pear-shaped flask is added 5.58 g of sodium acetate (68.0 mmol, 1.00 equiv), 8.38 g of semicarbazide hydrochloride (74.8 mmol, 1.10 equiv), 75 mL of purified water (Note 1), and a large magnetic stir bar. The solution is stirred until all of the solids dissolved. At this point, 10.34 g of neat 2-allyl-2-methylcyclohexanone (68.0 mmol, 1.00 equiv) is added via syringe. When the addition is complete, a two-tap Schlenk adapter connected to a bubbler and an argon/vacuum manifold (Note 2) is attached and the mixture is heated to 60 °C for 14 h (Note 19). The thick slurry is vacuum filtered (water aspirator) directly through filter paper on a porcelain Büchner funnel and rinsed with water (2 x 20 mL). The white solid is dried for 30 min on the funnel, transferred to a 250-mL round-bottomed flask, which then is immersed in a 50 °C water bath. The white solid is dried under vacuum (0.3 mmHg) until a constant mass of 12.2 g (58.3 mmol, 86% yield) is achieved (about 8 h). At this point, the semicarbazone is found to have 90–91% ee (measured by reverting to the ketone, Note 20).

A stir bar is added to the flask and the solids are suspended in 150 mL of toluene with mixing at approximately 400 rpm. After a water-cooled reflux condenser is attached to the flask, the mixture is then heated to 110 °C (bath temperature) in an oil bath. After a few minutes at this temperature, the solids dissolve completely to afford a clear colorless solution (Note 21). Heating is discontinued and the stirred mixture is allowed to cool to ambient temperature (20 °C) overnight while still immersed in the oil bath (Note 22). The cooled heterogeneous mixture is vacuum filtered (water aspirator) through filter paper on a porcelain Büchner funnel. The solids are rinsed with toluene (2 x 10 mL) and then dried on the filter for 15 min (Note 23). The solids are transferred to a 250-mL pear-shaped flask and dried under vacuum (0.3 mmHg) until a constant mass of 10.8–10.9 g (51.7–52.2 mmol, 76–77% yield, 89–90% recovery) is observed. This material is found to have

98–99% enantiomeric excess (measured by reverting to the ketone, Note 20).

A 250-mL pear-shaped flask containing a magnetic stir bar and 10.5 g of semicarbazone (50.2 mmol) and 40 mL of diethyl ether is stirred to suspend the solids. To the suspension is added 20 mL of 3 N aqueous hydrochloric acid (Note 1). No appreciable heat evolution is observed. The mixture is stirred vigorously for 3 h at ambient temperature (20 °C), at which time all of the solids had disappeared and two clear colorless phases are observed. The biphasic mixture is transferred to a 100-mL separatory funnel and the phases are separated. The aqueous phase is extracted with diethyl ether (3 x 10 mL). The combined organic layers are then washed successively with saturated sodium bicarbonate (2 x 5 mL), water (1 x 5 mL), and brine (2 x 5 mL). The organic phase is dried over anhydrous magnesium sulfate (1 g) and then filtered through cotton and concentrated by rotary evaporation under vacuum (150 mmHg, first at 20 °C, then at 40 °C to remove the last traces of solvent) to provide 7.62–7.63 g (50.1–50.2 mmol, >99% yield) of (S)-2-allyl-2-methylcyclohexanone of 98% ee (Notes 14 and 15). GC analysis demonstrates the product is formed in >99% product purity (Note 24).

2. Notes

1. Pimelic acid (≥99%, Fluka), allyl alcohol (≥99%, Sigma-Aldrich), p-toluenesulfonic acid monohydrate (ACS reagent, ≥98.5%, Sigma-Aldrich), toluene (Baker ultra resi-analyzed, J.T.Baker), solid sodium bicarbonate (tech grade, Brenntag Schweizerhall AG), magnesium sulfate (tech. grade, Brenntag Schweizerhall AG), sodium hydride (60% dispersion in mineral oil, Acros), iodomethane (Reagent Plus, 99%, Sigma-Aldrich), tris(dibenzylideneacetone)dipalladium ($Pd_2(dba)_3$, Strem), sodium acetate (puriss. p.a., ACS reagent, anhydrous, ≥99.0% (NT), Fluka), semicarbazide hydrochloride (99%, Alfa Aesar), and hydrochloric acid (36–38 wt%, J.T.Baker), were purchased and used as received. Checkers purchased purified water (for HPLC, Fluka), submitters used water purified with a Barnstead NANOpure Infinity UV/UF system. Ethyl acetate (tech. grade, Brenntag Schweizerhall AG) was distilled prior to use, diethyl ether (tech. grade, Brenntag Schweizerhall AG) was distilled and passed through an activated alumina column under nitrogen prior to use,[2] tetrahydrofuran (HPLC grade, Fisher) was distilled from sodium 9-fluorenone ketyl[3] or

passed through an activated alumina column under argon prior to use. The ligand (S)-tert-ButylPHOX was prepared using our accompanying procedure in *Organic Syntheses*.[4,5]

2. A two-tap Schlenk adapter connected to a bubbler and an argon/vacuum manifold is illustrated in Yu, J.; Truc, V.; Riebel, P.; Hierl, E.; Mudryk, B. *Org. Synth.* **2008**, *85*, 64–71.

3. The esterification product, diallyl pimelate, may be distilled (bp 134–135 °C/0.2 mmHg), but this is not necessary for this application. Distillation of a separate sample of diallyl pimelate led to significant loss of material to unidentified polymeric byproducts formed in the distillation flask, and distillation is therefore not recommended. Product purity was measured by GC using a CE Instruments GC 8000 Top equipped with a Restek Rtx-1701 column (30.0 m x 0.25 mm) and a flame ionization detector using a method of 100 °C isothermal for 5 min, then ramp 13 °C/min to 240 °C, then 240 °C isothermal for 5 min with 60 kPa He carrier gas flow. The retention time for the product was 17.85 min. No further signals were observed by the checkers, and therefore a product purity of 98% was assigned with >99% yield. Submitters reported observation of a predominant but unidentified impurity with slightly shorter retention time than the product. The product exhibited the following characteristics: ^{1}H NMR (400 MHz, CDCl$_3$) δ: 1.33–1.40 (m, 2 H), 1.66 (apparent quintet, J = 7.7 Hz, 4 H), 2.34 (t, J = 7.6 Hz, 4 H), 4.57 (apparent dt, J = 5.7, 1.4 Hz, 4 H), 5.23 (apparent dq, J = 10.4, 1.3 Hz, 2 H), 5.30 (apparent dq, J = 17.2, 1.5 Hz, 2 H), 5.91 (ddt, J = 17.2, 10.4, 5.7 Hz, 2 H); ^{13}C NMR (101 MHz, CDCl$_3$) δ: 24.7, 28.7, 34.1, 65.1, 118.3, 132.4, 173.3; IR (neat film, NaCl) 3086, 3025, 2942, 2866, 1733, 1648, 1456, 1421, 1378, 1272, 1173, 1086, 991, 932, 734 cm^{-1}; MS (FAB, NBA) m/z (%) 242 (11), 241 (100, [M+H]$^+$), 183 (85), 137 (31), 136 (14), 125 (53), 77 (10), 69 (12), 41 (59), 39 (12); HRMS (EI) m/z calc'd for C$_{13}$H$_{20}$O$_4$ [M]$^+$: 240.1362, found 240.1355; TLC (Hex/EtOAc = 4:1) R$_f$ = 0.46. Anal calcd for C$_{13}$H$_{20}$O$_4$: C 64.98, H 8.39, found C 65.28, H 8.38.

4. Submitters reported 7 h at 22 °C and an additional 4 h at 40 °C until all starting material was consumed. After this time the checkers did not observe full conversion by TLC analysis using the TLC method described in Note 5.

5. The progression of the cyclization may be monitored by TLC analysis using 20% ethyl acetate in hexanes as eluent with KMnO$_4$ staining (submitters used *p*-anisaldehyde staining): R$_f$ diallyl pimelate = 0.46, R$_f$

cyclized intermediate = 0.58–0.77 (broad, also UV active), R_f alkylation product = 0.56. The detection of diallyl pimelate is often obscured by the cyclized intermediate.

6. Following the submitters' procedure, THF was not removed before diluting with ethyl acetate. In the checkers' hands no phase separation took place under these conditions.

7. Submitters reported 67% yield and 92% purity.

8. Using the GC method described in Note 3, 1-methyl-2-oxo-cyclohexanecarboxylic acid 2-propenyl ester has a retention time of 14.77 min. The distilled material contains a small amount (<1% by GC) of uncyclized pimelate and <1% of an unidentified byproduct (retention time of 14.61 min). Submitters report observation of 6% of uncyclized diallyl pimelate and 2% of unidentified byproduct, which does not significantly affect the subsequent step. Out of this mixture analytically pure material may be obtained by flash chromatography on silica gel using a gradient of 1.5 → 4% diethyl ether in hexanes as eluent. GC response factors between 1-methyl-2-oxo-cyclohexanecarboxylic acid 2-propenyl ester and diallyl pimelate were determined with purified products to confirm these ratios, however assuming a 1:1 response factor gave the same ratios. The product showed the following characterization data: ^{1}H NMR (400 MHz, CDCl$_3$) δ: 1.30 (s, 3 H), 1.43–1.50 (m, 1 H), 1.59–1.78 (m, 3 H), 1.98–2.05 (m, 1 H), 2.42–2.54 (m, 3 H), 4.58–4.66 (m, 2 H), 5.24 (dd, J = 10.4, 0.8 Hz, 1 H), 5.31 (dd, J = 17.2, 1.4 Hz, 1 H), 5.83–5.93 (m, 1 H); ^{13}C NMR (101 MHz, CDCl$_3$) δ: 21.4, 22.7, 27.6, 38.3, 40.8, 57.3, 65.9, 119.0, 131.6, 172.9, 208.2; IR (neat film, NaCl) 3442, 3082, 2939, 2866, 1719, 1648, 1452, 1377, 1336, 1301, 1259, 1212, 1160, 1121, 1084, 1062, 1038, 977, 936, 854, 816, 767, 668, 599 cm^{-1}; MS (EI, 70 eV) m/z (%) 196 (26, [M]$^+$), 168 (18), 139 (12), 138 (23), 137 (26), 127 (44), 111 (27), 110 (14), 109 (48), 83 (30), 82 (23), 81 (100), 69 (34), 67 (16), 55 (56), 43 (22), 41 (85), 39 (24); HRMS (EI) m/z calc'd for C$_{11}$H$_{16}$O$_3$ [M]$^+$: 196.1099, found 196.1096; TLC (Hex/EtOAc = 4:1) R_f = 0.56. Anal calcd for C$_{11}$H$_{16}$O$_3$: C 67.32, H 8.22, found C 67.17, H 8.15.

9. The substrate concentration (0.2 M) described herein yields product of slightly lower enantiomeric excess (about 1% lower) than the previously reported, optimized conditions (0.033 M in substrate). For smaller scale where overall quantity of solvent is less important, the lower substrate concentration is recommended.

10. The complexation time prior to adding substrate is important to the overall reaction. Shorter or longer complexation times led to lower product yield and incomplete substrate conversion.

11. Submitters reported 26 h reaction time.

12. Although the reaction produces an equivalent of carbon dioxide, the evolution of this byproduct is not visually apparent during the reaction. The reaction is readily evaluated by TLC analysis using 10% diethyl ether in pentane as eluent with $KMnO_4$ staining (submitters used p-anisaldehyde staining): R_f dibenzylideneacetone = 0.24 (also UV active), R_f β-ketoester = 0.33, R_f product = 0.46.

13. Care should be taken to ensure that the moderately volatile product is not lost during concentration of the filtrate. However, if a substantial amount of solvent remains, distillation of the product does not occur smoothly. At 150 mmHg and 40 °C, tetrahydrofuran and diethyl ether are easily removed and product is not lost.

14. The distilled material showed the following analytical data: ^1H NMR (400 MHz, CDCl$_3$) δ: 1.06 (s, 3 H), 1.54–1.61 (m, 1 H), 1.65–1.90 (m, 5 H), 2.23 (apparent ddt, J = 13.9, 7.3, 0.9 Hz, 1 H), 2.33–2.40 (m, 3 H), 5.01–5.06 (m, 2 H), 5.69 (apparent ddt, J = 16.6, 11.1, 7.4 Hz, 1 H); ^{13}C NMR (101 MHz, CDCl$_3$) δ: 21.2, 22.8, 27.5, 38.7, 38.9, 42.1, 48.6, 118.0, 133.9, 215.5; IR (neat film, NaCl) 3393, 3076, 2933, 2864, 1706, 1451, 1124, 995, 913 cm^{-1}; MS (EI, 70 eV) m/z (%) 152 (31, [M]$^+$), 137 (36), 123 (29), 109 (60), 108 (27), 95 (33), 94 (21), 93 (69), 83 (49), 82 (16), 81 (31), 79 (21), 69 (14), 68 (17), 67 (66), 55 (100), 53 (13), 41 (50), 39 (25); HRMS (EI) m/z calc'd for $C_{10}H_{16}O$ [M]$^+$: 152.1201, found 152.1204; TLC (Pentane/Et$_2$O = 9:1) R_f = 0.46. Anal calcd for $C_{10}H_{16}O$: C 78.90, H 10.59, found C 78.86, H 10.48; optical rotation following enrichment (Part C): $[\alpha]_D^{21.0}$ –47.0 (c 2.30, dichloromethane, 98% ee).

15. GC analyses were performed with a Fisons Instruments HRGC Mega2 series equipped with a Chiraldex G-TA column (30.0 m x 0.25 mm) and a flame ionization detector. The assay conditions for 2-allyl-2-methylcyclohexanone are 100 °C isothermal, 60 kPa H$_2$ carrier gas flow, retention times: major (S) enantiomer = 14.15 min, minor (R) enantiomer = 17.09 min. The absolute configuration was established by X-ray crystallographic analysis of a semicarbazone derivative bearing a substituent with known absolute configuration.[6]

16. Column chromatography: 5 cm diameter x 10 cm height, eluting with 10% diethyl ether in pentane, 100 mL forerun, collecting 30 mL

fractions. Product appeared in fractions 9–20. See Note 12 for TLC conditions. For smaller scale preparations, it is often convenient to perform chromatography directly rather than distilling the product.

17. In the reaction that gave 95% yield after distillation, TLC (see Note 12) of the distillation residue showed only traces of product. Therefore no flash chromatography was performed.

18. Submitters reported 76% yield after distillation and an additional 11% from flash chromatography for an overall yield of 87%.

19. Semicarbazone formation begins before the addition of ketone is complete, although conversion at room temperature is sluggish.

20. To ensure an accurate ee value, the powder was mixed thoroughly prior to measurement. The enantiomeric excess was determined by suspending a small amount of semicarbazone (approximately 10 mg) in a biphasic mixture of diethyl ether (1 mL) and 2 N aqueous hydrochloric acid (1 mL) at ambient temperature. After 30 min of stirring, all of the solids had dissolved and the organic layer was separated, dried briefly over anhydrous magnesium sulfate, filtered through cotton, and the filtrate concentrated by rotary evaporation. The residue was then dissolved in *tert*-butyl methyl ether and analyzed by GC (see Note 15 for separation conditions). The semicarbazone was homogeneous according to the proton and carbon NMR spectra, and appears to be a single geometric isomer. However a correct elemental analysis could not be achieved. The following properties were observed: mp 190–191 °C (toluene, 98% ee); ^{1}H NMR (400 MHz, CDCl$_3$) δ: 1.09 (s, 3 H), 1.41–1.48 (m, 1 H), 1.53–1.71 (m, 5 H), 2.14–2.25 (m, 2 H), 2.32–2.39 (m, 2 H), 5.00 (apparent d, J = 3.5 Hz, 1 H), 5.03 (s, 1 H), 5.68–5.9 (m, 1 H), 8.29 (s, 1 H); ^{13}C NMR (101 MHz, CDCl$_3$) δ: 21.2, 22.9, 24.7, 26.1, 38.7, 41.6, 43.1, 117.3, 134.9, 157.3, 158.7; IR (neat film, NaCl) 3465, 3243, 3198, 3074, 2967, 2860, 1695, 1665, 1567, 1477, 1374, 1111, 1078, 991, 909 cm^{-1}; MS (EI, 70 eV) m/z (%) 209 (35, [M]$^+$), 194 (44), 168 (15), 165 (100), 151 (33), 150 (70), 149 (23), 148 (10), 135 (48), 134 (28), 125 (95), 108 (36), 107 (15), 98 (35), 96 (21), 95 (18), 93 (33), 91 (18), 82 (12), 81 (63), 80 (14), 79 (30), 77 (12), 67 (42), 55 (30), 53 (17), 44 (11), 41 (48), 39 (14); HRMS (CI, CH$_4$) m/z calc'd for C$_{11}$H$_{20}$N$_3$O [M + H]$^+$: 210.1606, found 210.1599; [α]$_D^{21.0}$ −50.5 (c 1.91, methanol, 98% ee).

21. At the reported concentration, the hot toluene solution is not saturated. The additional solvent helps maintain efficient stirring as the crystallization progresses and the viscosity of the mixture increases. The

additional solvent does not significantly affect the efficiency of product recovery.

22. Stirring during the crystallization process is very important to the efficiency of the ee improvement. For example, two separate 300 mg portions of semicarbazone with 89% ee were recrystallized from hot toluene (about 3 mL) with and without stirring. Although product recovery was comparable for either procedure (81% and 80%, respectively), the unstirred crystallization provided semicarbazone of 93% ee while the stirred crystallization provided semicarbazone of 96% ee. Either procedure yields the product as very fine needles.

23. Concentration of the filtrate by rotary evaporation provided an additional 1.24–1.32 g (10–11% recovery) of semicarbazone. GC analysis of the corresponding ketone found 20–26% ee for this material (see Note 20).

24. Using the GC method described in Note 3, 2-allyl-2-methylcyclohexanone has a retention time of 11.43 min.

Safety and Waste Disposal Information

All hazardous materials should be handled and disposed of in accordance with "Prudent Practices in the Laboratory"; National Academy Press; Washington, DC, 1995.

3. Discussion

The Dieckmann cyclization protocol employed here is a modification of a similar procedures developed by Tsuji and coworkers[7] and Fuchs and coworkers.[8] This improved procedure allows preparation of racemic allyl β-keto ester substrates in two steps with a single purification. Importantly, the single-pot cyclization/alkylation is an improvement over our previously reported method that required solvent exchange.[9] The Dieckmann protocol is useful for the preparation of a number of substituted allyl β-keto esters by varying the electrophile. Possible substituents include alkyl, benzyl, substituted benzyl, and alkenyl.[9] Other more sensitive substituents may be introduced by quenching the intermediate β-keto ester enolate with aqueous acid and then alkylating the resulting β-keto ester under more mild conditions (e.g., K_2CO_3, acetone, 50 °C).[9] In this manner, the β-keto ester enolate may undergo conjugate addition, aldol, or fluorination reactions with appropriate electrophilic components.[9] The β-keto ester substrates are useful not only for enantioselective decarboxylative allylation, but also for enantioselective decarboxylative protonation reactions generating α-tertiary cycloalkanones.[10] Alternative methods for synthesis of β-keto ester substrates include acylation of ketones with diallyl carbonates,[9] allyl cyanoformates,[9,11] allyl chloroformates,[12] or allyl 1H-imidazole-1-carboxylates.[13]

The enantioselective decarboxylative allylation method from allyl β-keto esters,[9,14] based on non-enantioselective transformations pioneered by Tsuji and Saegusa,[15] represents a substantial advance in asymmetric allylation since prior methods[16] required that the putative prochiral enolate intermediate[17] be stabilized by an electron-withdrawing group (e.g., esters or aryl groups) or contain only a single acidic site.[18,19] To highlight the previous deficiency in the literature, 2-allyl-2-methylcyclohexanone had not been prepared in high enantiomeric excess prior to our work since few alternative synthetic methods are available.[20] Related enantioselective transformations for the conversion of allyl enol carbonates and silyl enol ethers to α-quaternary cycloalkanones, also based on earlier work by Tsuji,[21] have been developed by our group[6,22] and others.[23] However, β-keto ester substrates are often preferable due to the straightforward synthesis and ease of substrate handling. The procedure reported herein has been optimized for large-scale preparation and features lower catalyst loading and higher substrate concentration than our previously reported work. These changes have minimal impact on the efficiency and selectivity observed in the reaction. Improvements to purification include conditions for distillation of the product and an improved protocol for conversion to the corresponding semicarbazone derivative. Conditions for recrystallization of the semicarbazone derivative are also reported, and provide access to highly enantioenriched 2-allyl-2-methylcyclohexanone.

The scope of this transformation[9] and the related transformation of allyl enol carbonates and enol silanes[6,22] has been demonstrated to include alkyl, alkenyl, aryl, ethereal, siloxy, halogen,[24] ketone, ester, and nitrile substituents. Additionally, the ring may be appended, unsaturated, enlarged, or substituted with heteroatoms. The delivered allyl group may be substituted at the internal position. Cascade allylation has also been performed to generate two quaternary stereocenters. Good levels of enantioselectivity are observed throughout these variations and products may be obtained in 55–99% yield and 80–94% ee (Table 1).[6,9,22]

Table 1. Ketones prepared via enantioselective decarboxylative allylation.[6,9,22]

89% yield 88% ee	80% yield 91% ee	96% yield 92% ee	55% yield 82% ee	83% yield 91% ee	86% yield 81% ee

| | | R = H 99% yield 85% ee | R = CF₃ 99% yield 82% ee | R = OCH₃ 80% yield 86% ee | 87% yield 92% ee | 87% yield 91% ee |

87% yield 88% ee — 94% yield 86% ee

80% yield 86% ee — 77% yield 90% ee — 73% yield 86% ee — 85% yield 92% ee — 82% yield 87% ee

97% yield 88% ee — 96% yield 90% ee — 90% yield 85% ee — 97% yield 91% ee — 81% yield 87% ee — 90% yield 79% ee

R = H 97% yield 92% ee — R = OMe 94% yield 91% ee — 91% yield 92% ee — 86% yield 87% ee — 79% yield 93% ee — 83% yield 92% ee

59% yield 89% ee — 93% yield 88% ee — 59% yield 92% ee — 73% yield 94% ee — 76% yield 92% ee, (4:1 d.r.)

The non-enantioselective Tsuji allylation reaction has been used sparingly in total synthesis efforts.[25] Since the development of asymmetric variants, however, enantioselective decarboxylative allylation has functioned as a key asymmetric step in the synthesis of the natural products (+)-

dichroanone,[26] (+)-elatol,[27] (+)-laurencenone B,[27] (–)-cyanthiwigin F,[28] (+)-carissone,[29] and (+)-cassiol[30] as well as in an approach to the natural product zoanthenol[31] (Table 2). Other useful transformations of the product (S)-2-allyl-2-methyl cyclohexanone include elaboration to various [6.5]- and [6.6]-fused bicycles and oxidation to a caprolactone derivative (Table 3a).[6] Spirocyclic systems are accessible by employing Grubbs' olefin metathesis catalysts[32] with α,ω-dienes[18,22,27] (Table 3b). Dioxanone products may be cleaved to access acyclic keto diols and α-hydroxy esters (Table 3c).

Table 2. Synthetic targets accessed via enantioselective decarboxylative allylation.

(+)-Dichroanone   (+)-Elatol   (+)-Laurencenone B   (–)-Cyanthiwigin F

(+)-Carissone   (+)-Cassiol   ABC ring system of Zoanthenol

Table 3. (a) Derivatives of 2-allyl-2-methylcyclohexanone.[6] (b) Spirocycles accessible via ring-closing metathesis.[18,22,27] (c) Cleavage of dioxanones to access acyclic products.[22]

1. Department of Chemistry and Chemical Engineering, California Institute of Technology, Pasadena, California, 91125; E-mail: stoltz@caltech.edu.
2. Pangborn, A. B.; Giardello, M. A.; Grubbs, R. H.; Rosen, R. K.; Timmers, F. J. *Organometallics* **1996**, *15*, 1518–1520.
3. Kamaura, M.; Inanaga, J. *Tetrahedron Lett.* **1999**, *40*, 7347–7350.
4. Krout, M. R.; Mohr, J. T.; Stoltz, B. M. *Org. Synth.* **2009**, *86*, 181-193.
5. Tani, K.; Behenna, D. C.; McFadden, R. M.; Stoltz, B. M. *Org. Lett.* **2007**, *9*, 2529–2531.
6. Behenna, D. C.; Stoltz, B. M. *J. Am. Chem. Soc.* **2004**, *126*, 15044–15045.
7. Tsuji, J.; Nisar, M.; Shimizu, I.; Minami, I. *Synthesis* **1984**, *12*, 1009.
8. Pariza, R. J.; Kuo, F.; Fuchs, P. L. *Synth. Commun.* **1983**, *13*, 243–254.
9. Mohr, J. T.; Behenna, D. C.; Harned, A. M.; Stoltz, B. M. *Angew. Chem., Int. Ed.* **2005**, *44*, 6924–6927.

10. (a) Mohr, J. T.; Nishimata, T.; Behenna, D. C.; Stoltz, B. M. *J. Am. Chem. Soc.* **2006**, *128*, 11348–11349. (b) Marinescu, S. C.; Nishimata, T. N.; Mohr, J. T.; Stoltz, B. M. *Org. Lett.* **2008**, *10*, 1039–1042.

11. (a) Mander, L. N.; Sethi, S. P. *Tetrahedron Lett.* **1983**, *24*, 5425–5428. (b) Donnelly, D. M. X.; Finet, J.-P.; Rattigan, B. A. *J. Chem. Soc., Perkin Trans. 1* **1993**, 1729–1735.

12. Trost, B. M.; Bream, R. N.; Xu, J. *Angew. Chem., Int. Ed.* **2006**, *45*, 3109–3112.

13. Trost, B. M.; Xu, J. *J. Org. Chem.* **2007**, *72*, 9372–9375.

14. (a) Nakamura, M.; Hajra, K.; Endo, K.; Nakamura, E. *Angew. Chem., Int. Ed.* **2005**, *44*, 7248–7251. (b) For a related method using a bis(phosphine) ligand, see: ref 12.

15. (a) Shimizu, I.; Yamada, T.; Tsuji, J. *Tetrahedron Lett.* **1980**, *21*, 3199–3202. (b) Tsuda, T.; Chujo, Y.; Nishi, S.-i.; Tawara, K; Saegusa, T. *J. Am. Chem. Soc.* **1980**, *102*, 6381–6384.

16. (a) Hayashi, T.; Kanehira, K.; Hagihara, T.; Kumada, M. *J. Org. Chem.* **1988**, *53*, 113–120. (b) Sawamura, M.; Nagata, H.; Sakamoto, H.; Ito, Y. *J. Am. Chem. Soc.* **1992**, *114*, 2586–2592. (c) Sawamura, M.; Sudoh, M.; Ito, Y. *J. Am. Chem. Soc.* **1996**, *118*, 3309–3310. (d) Trost, B. M.; Radinov, R.; Grenzer, E. M. *J. Am. Chem. Soc.* **1997**, *119*, 7879–7880. (e) Trost, B. M.; Ariza, X. *Angew. Chem., Int. Ed.* **1997**, *36*, 2635–2637. (f) Kuwano, R.; Ito, Y. *J. Am. Chem. Soc.* **1999**, *121*, 3236–3237. (g) Trost, B. M.; Schroeder, G. M. *J. Am. Chem. Soc.* **1999**, *121*, 6759–6760. (h) You, S.-L.; Hou, X.-L.; Dai, L.-X.; Cao, B.-X.; Sun, J. *Chem. Commun.* **2000**, 1933–1934. (i) You, S.-L.; Hou, X.-L.; Dai, L.-X.; Zhu, X.-Z. *Org. Lett.* **2001**, *3*, 149–151. (j) Trost, B. M.; Schroeder, G. M.; Kristensen, J. *Angew. Chem., Int. Ed.* **2002**, *41*, 3492–3495. (k) Kuwano, R.; Uchida, K.; Ito, Y. *Org. Lett.* **2003**, *5*, 2177–2179. (l) Trost, B. M.; Schroeder, G. M. *Chem.–Eur. J.* **2005**, *11*, 174–184.

17. For a computational investigation of the mechanism of the allylation reaction, see: Keith, J. A.; Behenna, D. C.; Mohr, J. T.; Ma, S.; Marinescu, S. C.; Oxgaard, J.; Stoltz, B. M.; Goddard, W. A., III *J. Am. Chem. Soc.* **2007**, *129*, 11876–11877.

18. For a review of the development of these asymmetric allylation methods, see: Mohr, J. T.; Stoltz, B. M. *Chem.–Asian J.* **2007**, *2*, 1476–1491 and references therein.

19. For recent reviews of allylic alkylation of ketone enolates, see: (a) Braun, M.; Meier, T. *Angew. Chem., Int. Ed.* **2006**, *45*, 6952–6955. (b)

You, S.-L.; Dai, L.-X. *Angew. Chem., Int. Ed.* **2006**, *45*, 5246–5248. (c) Braun, M.; Meier, T. *Synlett* **2006**, 661–676. (d) Kazmaier, U. *Curr. Org. Chem.* **2003**, *7*, 317–328.

20. For notable alternative catalytic methods for the generation of α-quaternary cycloalkanones, see: (a) Yamashita, Y.; Odashima, K.; Koga, K. *Tetrahedron Lett.* **1999**, *40*, 2803–2806. (b) Doyle, A. G.; Jacobsen, E. N. *J. Am. Chem. Soc.* **2005**, *127*, 62–63. (c) Doyle, A. G.; Jacobsen, E. N. *Angew. Chem., Int. Ed.* **2007**, *46*, 3701–3705.

21. (a) Tsuji, J.; Minami, I.; Shimizu, I. *Chem. Lett.* **1983**, 1325–1326. (b) Tsuji, J.; Minami, I.; Shimizu, I. *Tetrahedron Lett* **1983**, *24*, 1793–1796.

22. Seto, M.; Roizen, J. L.; Stoltz, B. M. *Angew. Chem., Int. Ed.* **2008**, *47*, 6873–6876.

23. (a) Trost, B. M.; Xu, J. *J. Am. Chem. Soc.* **2005**, *127*, 2846–2847. (b) Trost, B. M.; Xu, J. *J. Am. Chem. Soc.* **2005**, *127*, 17180–17181. (c) Trost, B. M.; Xu, J.; Reichle, M. *J. Am. Chem. Soc.* **2007**, *129*, 282–283. (d) Schulz, S. R.; Blechert, S. *Angew. Chem., Int. Ed.* **2007**, *129*, 3966–3970.

24. For examples of the utility of the allylation reaction to generate tertiary fluoride stereocenters, see: (a) Ref 9. (b) Ref 14a. (c) Burger, E. C.; Barron, B. R.; Tunge, J. A. *Synlett* **2006**, 2824–2826. (d) Bélanger, É.; Cantin, K.; Messe, O.; Tremblay, M.; Paquin, J.-F. *J. Am. Chem. Soc.* **2007**, *129*, 1034–1035.

25. (a) Ohmori, N. *J. Chem. Soc., Perkin Trans. 1* **2002**, 755–767. (b) Nicolaou, K. C.; Vassilikogiannakis, G.; Mägerlein, W.; Kranich, R. *Angew. Chem., Int. Ed.* **2001**, *40*, 2482–2486. (c) Herrinton, P. M.; Klotz, K. L.; Hartley, W. M. *J. Org. Chem.* **1993**, *58*, 678–682. (d) Burns, A. C.; Forsyth, C. J. *Org. Lett.* **2008**, *10*, 97–100.

26. McFadden, R. M.; Stoltz, B. M. *J. Am. Chem. Soc.* **2006**, *128*, 7738–7739.

27. White, D. E.; Stewart, I. C.; Grubbs, R. H.; Stoltz, B. M. *J. Am. Chem. Soc.* **2008**, *130*, 810–811.

28. Enquist, J. A., Jr.; Stoltz, B. M. *Nature* **2008**, *453*, 1228–1231.

29. Levine, S. R.; Krout, M. R.; Stoltz, B. M. *Org. Lett.* **2009**, *11*, 289–292.

30. Petrova, K. V.; Mohr, J. T.; Stoltz, B. M. *Org. Lett.* **2009**, *11*, 293–295.

31. Behenna, D. C.; Stockdill, J. L.; Stoltz, B. M. *Angew. Chem., Int. Ed.* **2007**, *46*, 4077–4080.

32. (a) Scholl, M.; Ding, S.; Lee, C. W.; Grubbs, R. H. *Org. Lett.* **1999**, *1*, 953–956. (b) Stewart, I. C.; Ung, T.; Pletnev, A. A.; Berlin, J. M.; Grubbs, R. H.; Schrodi, Y. *Org. Lett.* **2007**, *9*, 1589–1592.

Appendix
Chemical Abstracts Nomenclature; (Registry Number)

Pimelic acid: Heptanedioic acid; (111-16-0)

Allyl alcohol: 2-Propen-1-ol; (107-18-6)

p-Toluenesulfonic acid monohydrate: Benzenesulfonic acid, 4-methyl-, hydrate (1:1); (6192-52-5)

Diallyl pimelate: Pimelic acid, diallyl ester; (91906-66-0)

Sodium hydride; (7646-69-7)

Iodomethane: Methane, iodo-; (74-88-4)

Allyl 1-methyl-2-oxocyclohexanecarboxylate: Cyclohexanecarboxylic acid, 1-methyl-2-oxo-, 2-propenyl ester; (7770-41-4)

Tris(dibenzylideneacetone) dipalladium(0): Palladium, tris[μ-[(1,2-η:4,5-η)-(1*E*,4*E*)-1,5-diphenyl-1,4-pentadien-3-one]]di-; (51364-51-3)

(*S*)-*tert*-ButylPHOX: Oxazole, 4-(1,1-dimethylethyl)-2-[2-(diphenylphosphino)phenyl]-4,5-dihydro-, (4*S*)-; (148461-16-9)

(*S*)-2-Allyl-2-methylcyclohexanone: Cyclohexanone, 2-methyl-2-(2-propen-1-yl)-, (2*S*)-; (812639-07-9)

Sodium acetate: Acetic acid, sodium salt (1:1); (127-09-3)

Semicarbazide hydrochloride: Hydrazinecarboxamide, hydrochloride (1:1); (563-41-7)

(*S*)-2-(2-Allyl-2-methylcyclohexylidene)hydrazinecarboxamide: Hydrazinecarboxamide, 2-[(2*S*)-2-methyl-2-(2-propenyl)cyclohexylidene]-, (2*E*)-; (812639-25-1)

Brian M. Stoltz was born in Philadelphia, PA in 1970 and obtained his B.S. degree from the Indiana University of Pennsylvania in Indiana, PA. After graduate work at Yale University in the labs of John L. Wood and an NIH postdoctoral fellowship at Harvard in the Corey labs he took a position at the California Institute of Technology. A member of the Caltech faculty since 2000, he currently is the Ethel Wilson Bowles and Robert Bowles Professor of Chemistry and a KAUST GRP Investigator. His research interests lie in the development of new methodology for general applications in synthetic chemistry.

Justin T. Mohr received his A.B. degree in chemistry in 2003 from Dartmouth College where he conducted research with Professor Gordon W. Gribble. He joined the laboratories of Professor Brian M. Stoltz at Caltech in 2003 where he has pursued Ph.D. studies as a Lilly fellow. His research interests include the development of enantioselective reactions and applications in natural product total synthesis.

Michael R. Krout received his B.S. degree in biochemistry from the Indiana University of Pennsylvania in 2002. He then worked in the medicinal chemistry department at Merck Research Laboratories in West Point, PA, where he was involved in the development of non-steroidal selective androgen receptor modulators aimed toward the treatment of osteoporosis. In the fall of 2003, he joined the lab of Professor Brian Stoltz at Caltech where he has worked toward his Ph.D. as a Lilly fellow. His research interests include the development of catalytic, asymmetric methods and their utility in natural product total synthesis.

Christian Ebner was born in Mönchengladbach (Germany) in 1984 and did his chemistry studies at the University of Basel (Switzerland). In 2008 he obtained his M.Sc. degree under the supervision of Prof. Andreas Pfaltz, whose group he joined as a Ph.D. student in May 2008. His current work deals with the synthesis of new ligands and the development of a new screening method for palladium-catalyzed reactions.

PHOSPHINE-CATALYZED [4+2] ANNULATION: SYNTHESIS OF ETHYL 6-PHENYL-1-TOSYL-1,2,5,6-TETRAHYDROPYRIDINE-3-CARBOXYLATE

Submitted by Kui Lu and Ohyun Kwon.[1]
Checked by Kay M. Brummond and Matthew M. Davis.

1. Procedure

A. Ethyl 2-methylbuta-2,3-dienoate (1). A flame-dried, 500-mL, two-necked, round-bottomed flask equipped with an egg-shaped magnetic stir bar, a rubber septum, and an argon inlet is charged with ethyl 2-(triphenylphosphoranylidene)propionate (27.18 g, 75.0 mmol) (Note 1), triethylamine (10.5 mL, 75.3 mmol) (Note 6), anhydrous dichloromethane (125 mL) (Note 7), and pentane (125 mL) (Note 8) and the mixture is stirred under argon. Acetyl chloride (5.3 mL, 75 mmol) (Note 9) is added via syringe over a period of 1 h using a syringe pump; during the addition the flask is placed in a 25 °C water bath to control the heat evolved during the reaction. During the addition the reaction turns cloudy, then a yellow-orange precipitate forms. Stirring is continued for an additional 6 h after which time the heterogeneous mixture is filtered through a Büchner funnel connected to a vacuum from a water aspirator. The precipitate is washed with pentane (50 mL) and the filtrate is concentrated to ca. 50 mL (Note 10) at 0 °C on a rotary evaporator (20–25 mmHg). Pentane (150 mL) and an egg-shaped

212

Org. Synth. **2009**, *86*, 212-224
Published on the Web 2/18/2009

magnetic stir bar are added to the residue-containing flask and the mixture is stirred vigorously for 30 min. The white precipitate (triphenylphosphine oxide) is removed by filtering the mixture through a Büchner funnel and washing the precipitate with pentane (50 mL). The filtrate is concentrated to ca. 30 mL at 0 °C on a rotary evaporator (20–25 mmHg). The residue is filtered again to remove the precipitate (triphenylphosphine oxide) and the precipitate is rinsed with pentane (10 mL). The filtrate is concentrated to ca. 20 mL as described above (Note 11). The residue is transferred to a 25-mL one-necked, round-bottomed flask along with an egg-shaped magnetic stir bar and purified by distillation (54-55 °C, 25–26 mmHg) (Note 12) to afford 5.73 g (61%) of **1** (Note 13) as a colorless oil.

B. *(E)-N-Benzylidene-4-methylbenzenesulfonamide (2)*. A flame-dried, 500-mL, one-necked, round-bottomed flask equipped with an egg-shaped magnetic stir bar is charged with *p*-toluenesulfonamide (10.25 g, 59.9 mmol) (Note 14), benzaldehyde (6.7 mL, 66 mmol) (Note 15), benzene (300 mL) (Note 16) and $BF_3 \cdot Et_2O$ (0.75 mL, 6.0 mmol) (Note 17). The flask is equipped with a Dean–Stark trap, which is attached to a reflux condenser and a gas adapter under argon and the solution is refluxed (oil bath temperature: 90–95 °C) (Note 18) for 14 h (Note 19). The yellow solution is cooled to room temperature and concentrated using a rotary evaporator (20–25 mmHg) at 35 °C to afford a brown solid. Ethyl acetate (50 mL) (Note 4) and hexane (100 mL) (Note 20) are added to the flask; a reflux condenser and gas adapter are attached and the solution is refluxed under Ar (oil bath temperature: 85–90 °C). Upon dissolution of the solid (ca. 5 min) the stir bar is removed, the solution is allowed to cool to room temperature and then cooled to –20 °C in a freezer overnight. The resulting solid is collected by filtering through a Büchner funnel connected to a vacuum and washing with hexane (100 mL) to provide 14.27 g (92%) of **2** (Note 21) as an off-white solid.

C. *Ethyl 6-phenyl-1-tosyl-1,2,5,6-tetrahydropyridine-3-carboxylate (3)*. A flame-dried, 1-L, one-necked, round-bottomed flask equipped with an egg-shaped magnetic stirring bar and a rubber septum is charged with **2** (7.78 g, 30.0 mmol) and tri-*n*-butylphosphine (1.48 mL, 5.93 mmol) (Note 22). The flask is purged with argon and anhydrous dichloromethane (600 mL) (Note 7) is added via cannula under argon; the mixture is stirred for 5 min. Ethyl 2-methylbuta-2,3-dienoate (**1**) (4.61 g, 36.5 mmol) is added to the flask via syringe over a period of 5 min. After stirring the light-yellow solution at room temperature for 13 h (Note 23), the reaction mixture is concentrated by rotary evaporation (20–25 mmHg) at 35 °C to afford a yellow residue, which is transferred to a 100-mL, one-necked, pear-shaped flask equipped with an egg-shaped magnetic stirring bar. Chloroform (20 mL) (Note 24) and hexane (20 mL) (Note 20) are added to the flask, a reflux condenser is attached, and the solution is heated to reflux under Ar (oil bath temperature: 85–90 °C). When the solid is dissolved, the stir bar is removed, the solution is allowed to cool to room temperature and then cooled to – 20 °C in a freezer overnight. The resulting solid is collected by filtering the slurry through a Büchner funnel connected to a vacuum. The solid is washed with hexane (100 mL) to provide 8.77 g (76% yield) of **3** (Note 25) as a yellow solid. The filtrate is concentrated by rotary evaporation (20–25 mmHg) at 35 °C to give an orange residue, which, when purified using silica gel flash column chromatography (Note 26), provides an additional 2.13 g (18% yield) of **3** as a white solid.

2. Notes

1.   Ethyl 2-(triphenylphosphoranylidene)propionate was purchased by the checkers from Alfa Aesar under the name (1-ethoxycarbonylethylidene)-triphenylphosphorane (97%) and was used as received. The submitters prepared this reagent from ethyl 2-bromopropanoate (Note 2) and triphenylphosphine (Note 3) using the following procedure (note that this procedure was not checked): A 500-mL, one-necked, round-bottomed flask equipped with an egg-shaped magnetic stirring bar is charged with triphenylphosphine (52.46 g, 200.0 mmol), ethyl acetate (130 mL) (Note 4), and ethyl 2-bromopropanoate (26.03 mL, 200.0 mmol). A reflux condenser is attached to the flask and the solution is heated under reflux (oil bath temperature: 75–80 °C) under argon for 24 h. The resulting white precipitate is collected through suction filtration on a Büchner funnel and washed with ethyl acetate (100 mL). The salt is dissolved in dichloromethane (500 mL) (Note 5) and the solution is transferred to a 1-L separation funnel. Aqueous sodium hydroxide solution (2 M, 200 mL) is added and, after vigorously shaking, the organic and aqueous layers are separated. The aqueous phase is extracted with dichloromethane (100 mL). The combined organic phases are washed with brine, dried over anhydrous Na_2SO_4, filtered, and concentrated at 30 °C under rotary evaporation (20–25 mmHg); drying under vacuum (0.1–0.2   mmHg)   affords   54.51   g   (75.2%)   of   ethyl 2-(triphenylphosphoranylidene)propionate   as   a   light-yellow   solid,   mp

156–158 °C.

2. Ethyl 2-bromopropionate (99%) was purchased from Aldrich Chemical Company, Inc., and was used as received.

3. Triphenylphosphine (99%) was purchased from Acros Chemical Company, Inc., and was used as received.

4. Ethyl acetate (99.9%) was purchased from Fisher Scientific and was used as received.

5. Dichloromethane (99.9%) was purchased from Fisher Scientific and was used as received.

6. Triethylamine (99.5%) was purchased from Aldrich Chemical Company, Inc. and was used as received.

7. Dichloromethane, supplied by Fisher Scientific, was purified by passing over activated alumina using the Sol-Tek ST-002 solvent purification system. The submitters distilled dichloromethane from calcium hydride.

8. Pentane (99.7%) was purchased from Fisher Scientific and was used as received.

9. Acetyl chloride (98%) was purchased from Aldrich Chemical Company, Inc. and was used as received.

10. A precipitate formed toward the end of the rotary evaporation process. Over-concentration of the solvent causes some loss of the product.

11. A small amount of precipitate (triphenylphophine oxide) was observed in the residue.

12. Due to the volatility of ethyl 2-methylbuta-2,3-dienoate (1) the condenser was cooled using a recirculating pump placed in an ice/water bath. Additionally the pressure was lowered gradually from 60 mmHg to 25 mmHg to minimize bumping.

13. The physical properties are as follows: IR (thin film) 3063, 2984, 1969, 1944, 1712, 1369, 1278, 1222, 1123, 1027, 853, 778, 615 cm^{-1}; ^{1}H NMR (500 MHz, CDCl$_3$) δ: 1.29 (t, J = 7.0 Hz, 3 H), 1.88 (t, J = 3.0 Hz, 3 H), 4.21 (q, J = 7.0 Hz, 2 H), 5.07 (q, J = 3.0 Hz, 2 H); ^{13}C NMR (125 MHz, CDCl$_3$) δ: 14.2, 14.7, 61.0, 77.8, 95.5, 167.6, 214.0. MS (EI) m/z (%) 126 (14) [M]$^+$, 98 (47%), 86 (47%), 84 (75), 53 (100); HR-EI calcd for C$_7$H$_{10}$O$_2$ 126.0681, found 126.0680. Anal. Calcd for C$_7$H$_{10}$O$_2$: C, 66.65; H, 7.99. Found: C, 66.08; H, 8.04. The product did not analyze correctly for carbon (0.57% difference from theory for the checkers, 1.08% difference for the submitters). The product is homogeneous in all other respects.

14. p-Toluenesulfonamide (98%) was purchased from Aldrich

Chemical Company, Inc., and was used as received. The material contained ≤ 2% *o*-toluenesulfonamide by ¹H NMR. The absence of observable multiplets between 8.04-8.01 ppm and 7.51-7.46 ppm and singlet at 2.70 indicated the material contained less than 2% *o*-toluenesulfonamide.

15. Benzaldehyde (99.5%) was purchased from Aldrich Chemical Company, Inc., and was used as received.

16. Benzene, supplied by EMD Chemicals, Inc., was distilled from calcium hydride.

17. Boron trifluoride-diethyl etherate (36%) was purchased from Aldrich Chemical Company, Inc., and was used as received.

18. Over time a white solid collected in the Dean-Stark trap and at the bottom of the reflux condenser. The solid was analyzed by ¹⁹F NMR (282 MHz, DMSO) and the spectrum shows a single resonance at –148.2 ppm.

19. The submitters report that the product decomposes slightly on E. Merck silica gel thin layer chromatography plates (60F-254), so ¹H NMR spectroscopic analysis of the crude reaction mixture is used to monitor the reaction. Concentrating a portion (ca. 0.5 mL) of the crude reaction mixture at 35 °C using rotary evaporation affords a residue that is dissolved in CDCl₃ (purchased from Cambridge Isotope Laboratories, Inc.). Disappearance of the signal for the NH₂ protons (ca. 5.0 ppm) in the ¹H NMR spectrum indicates that the reaction is complete.

20. Hexane (99.9%) was purchased from Fisher Scientific and was used as received.

21. The physical properties are as follows: mp 112–113 °C; IR (thin film) 1598, 1572, 1450, 1321, 1157, 1088, 782, 757 cm⁻¹; ¹H NMR (500 MHz, CDCl₃) δ: 2.45 (s, 3 H), 7.36 (d, *J* = 8.5 Hz, 2 H), 7.50 (t, *J* = 8.0 Hz, 2 H), 7.62 (tt, *J* = 7.5 and 1.5 Hz, 1 H), 7.89–7.94 (m, 4 H), 9.04 (s, 1 H); ¹³C NMR (125 MHz, CDCl₃) δ: 21.7, 128.1, 129.1, 129.8, 131.3, 132.3, 134.9, 135.1, 144.6, 170.1; MS (ESI) *m/z* 282 [M + Na]⁺; HR-ESIMS calcd for C₁₄H₁₃NO₂SNa [M + Na]⁺ 282.0565, found 282.0552. Anal. Calcd for C₁₄H₁₃NO₂S: C, 64.84; H, 5.05; N, 5.40. Found: C, 64.79; H, 5.01; N, 5.43.

22. Tri-*n*-butylphosphine (97%) was purchased from Aldrich Chemical Company, Inc., and was used as received.

23. The disappearance of starting material **2** is monitored using TLC (Note 19): R_f = 0.37 (EtOAc/hexane, 1:4) for **2**. The spots are visualized under UV light and *p*-anisaldehyde stain.

24. Chloroform, supplied by EMD Chemical, Inc., was distilled from calcium chloride.

25. The physical properties are as follows: mp 149–150 °C; IR (thin film) 2981, 1709, 1659, 1598, 1495, 1449, 1340, 1264, 1161, 1100, 958, 727 cm^{-1}; ^{1}H NMR (500 MHz, CDCl$_3$) δ: 1.29 (t, J = 7.0 Hz, 3 H), 2.43 (s, 3 H), 2.58–2.50 (m, 1 H), 2.69 (ddd, J = 19.0, 5.5, 2.5 Hz, 1 H), 3.50–3.43 (m, 1 H), 4.19 (q, J = 7.0 Hz, 2 H), 4.50 (d, J = 19.0 Hz, 1 H), 5.37 (d, J = 7.0 Hz, 1 H), 7.05–7.04 (m, 1 H), 7.32–7.27 (m, 7 H), 7.71 (d, J = 8.5 Hz, 2 H); ^{13}C NMR (100 MHz, CDCl$_3$) δ: 14.1, 21.5, 26.9, 39.5, 52.1, 60.8, 127.0, 127.2, 127.6, 127.8, 128.6, 129.7, 136.1, 137.4, 138.1, 143.4, 164.6. MS (ESI) m/z 408 [M + Na]$^+$; HR-ESIMS calcd for C$_{21}$H$_{24}$NO$_4$SNa [M + Na]$^+$ 408.1245, found 408.1208. Anal. Calcd for C$_{21}$H$_{23}$NO$_4$S: C, 65.43; H, 6.01, N, 3.63. Found: C, 64.99; H, 5.92; N, 3.61.

26. Flash column chromatography is performed using a 4-cm-wide, 25-cm-high column packed with E. Merck silica gel 60 (230–400 mesh, 100 g). The column is packed by slurrying the silica with ethyl acetate, dichloromethane, and hexane (1:1:10), loading the residue with dichloromethane, and eluting with ethyl acetate, dichloromethane, and hexane (1:1:10). The collected fractions are analyzed using TLC (Note 19), eluting with ethyl acetate, dichloromethane and hexane (1:1:10; R_f = 0.21 for **3**). The spots are visualized using UV light and p-anisaldehyde stain. Concentration of the product-containing fractions using rotary evaporation (20–25 mmHg) at 35 °C and then drying under vacuum (0.06 mmHg) provides **3**.

Safety and Waste Disposal Information

All hazardous materials should be handled and disposed of in accordance with "Prudent Practices in the Laboratory"; National Academy Press; Washington, DC, 1995.

3. Discussion

During the past decade, catalysis employing nucleophilic organic molecules, such as N-heterocyclic carbenes, amines, and phosphines, has garnered tremendous attention. This interest is due in part to the spectacular expansion of the scope of nucleophilic catalysis beyond its classical examples, such as acyloin condensations and the Morita–Baylis–Hillman (MBH) reactions. In particular, the use of activated allenes, instead of alkenes, as reactants in MBH reactions in the presence of nucleophilic tertiary phosphines has led to the development of many processes for the construction of carbocycles and heterocycles.[2] Typically, phosphine-catalyzed MBH reactions of allenes are operationally simple and do not require aqueous work-up because the two reactants and the phosphine catalyst are all organic; the products may be purified through flash column chromatography after evaporation of the reaction solvent. Among these reactions, Lu's [3+2] cycloaddition, in which the allene reacts with an alkene or imine to provide a cyclopentene or pyrroline, is particularly noteworthy.[3] As the first example of phosphine-catalyzed allene cycloaddition, Lu's reaction has been applied in the total syntheses of several natural products,[4] and a highly enantioselective variant has been established.[5] Our research group has also been engaged in the development of nucleophilic phosphine-catalyzed reactions of allenoates, and this method has served to demonstrate the ever-expanding versatility of allene/phosphine catalysis.[6] One of the most broadly useful reactions is the phosphine-catalyzed [4+2] annulation of α-alkyl allenoates and imines to form tetrahydropyridines.[7] This reaction, like Lu's [3+2] reaction, has been applied in natural product syntheses,[8,9] and an enantioselective variant of the reaction employing Gladiali's phosphepine has provided products with up to 99% ee.[9] Recently, we disclosed an all-carbon variant of the [4+2] reaction, which we used to synthesize cyclohexenes from α-alkyl allenoates and activated olefins.[10]

Using the procedure described herein, highly functionalized tetrahydropyridines are easily synthesized in high yield from N-tosylimines and 2-methyl-2,3-butadienoates in the presence of catalytic amounts of PBu$_3$.[7] Table 1 lists several examples of these products. All of the non-acidic aryl N-tosylimines tested, except for one bearing a nitro substituent on the aryl group, provided products in yields exceeding 90% (Table 1, entries 1–9). The salicyl and 2-pyrrolyl N-tosylimines did not provide their expected products (Table 1, entries 10 and 12), presumably because their acidic protons quenched the key β-phosphonium dienolate zwitterionic intermediate 4 in the catalytic cycle described in Scheme 1. This problem is easily overcome when the phenol and pyrrole moieties are protected as silyl ethers and BOC carbamates, respectively (Table 1, entries 11 and 13). Naphthyl and other heteroaryl imines are also viable substrates (Table 1, entries 14–16). Among the alkyl N-tosylimines examined, only the non-enolizable N-tosylpivalaldimine 2o afforded the desired product (Table 1, entries 17 and 18).

Table 1. Synthesis of Tetrahydropyridines **3** From Ethyl
2-Methyl-2,3-butadienoate and *N*-Tosylaldimines[a]

entry	R	product	yield (%)[b]
1	Ph (**2**)	**3**	94[c]
2	4-MeOC$_6$H$_4$ (**2a**)	**3a**	99
3	4-MeC$_6$H$_4$ (**2b**)	**3b**	95
4	3-ClC$_6$H$_4$ (**2c**)	**3c**	96
5	2-ClC$_6$H$_4$ (**2d**)	**3d**	93
6	4-FC$_6$H$_4$ (**2e**)	**3e**	95
7	4-NCC$_6$H$_4$ (**2f**)	**3f**	98
8	2-F$_3$CC$_6$H$_4$ (**2g**)	**3g**	98
9	4-O$_2$NC$_6$H$_4$ (**2h**)	**3h**	86
10	2-HOC$_6$H$_4$ (**2m**)	**3i**	0
11	2-TBSOC$_6$H$_4$ (**2n**)	**3j**	93
12	2-pyrrolyl (**2l**)	**3k**	0
13	*N*-Boc-2-pyrrolyl (**2m**)	**3l**	99
14	1-naphthyl (**2i**)	**3m**	96
15	2-furyl (**2j**)	**3n**	97
16	4-pyridyl (**2k**)	**3o**	92[d]
17	*tert*-butyl (**2o**)	**3p**	86[e]
18	*n*-propyl (**2p**)	**3q**	0[f]

[a] 1.0 mmol scale. [b] Isolated yields. [c] 30 mmol scale. [d] 30 mol% PBu$_3$ was used. [e] 3 equiv of Na$_2$CO$_3$ was added. [f] The imine decomposed to the corresponding aldehyde and *p*-toulenesulfonamide.

In the proposed mechanism, the nucleophilic phosphine adds to ethyl 2-methyl-2,3-butadienoate (**1**), forming the resonance-stabilized zwitterions **4**. Because of steric congestion at the α-carbon atom, the zwitterion **4** undergoes addition to the imine at its γ-carbon atom, producing the phosphonium amide **5**. A proton transfer converts the amide anion in **5** to the vinylogous ylide **6**; another proton transfer transforms the ylide **6** into the allylic phosphonium amide **7**. The ready conjugate addition of the amide anion to the α,β-unsaturated ester and subsequent β-elimination of the phosphine provides the cyclized product **3**. Based on pK_a values, we suspected that the proton transfer from the sulfonamide **5** to the ylide **6** would be the rate-determining step.[11] Therefore, increasing the acidity of the β′-carbon atom appears to accelerate the overall reaction. In fact, the reaction between 2-(4-cyanobenzyl)-2,3-butadienoate (**8b**) and *N*-tosylbenzaldimine (**2**) reached completion within 30 min (cf. 9 h for the corresponding reaction of 2-methyl-2,3-butadienoate), producing the expected product in 99% yield with high diastereoselectivity (dr 98:2), favoring the *syn* product.

Scheme 1. Proposed mechanism

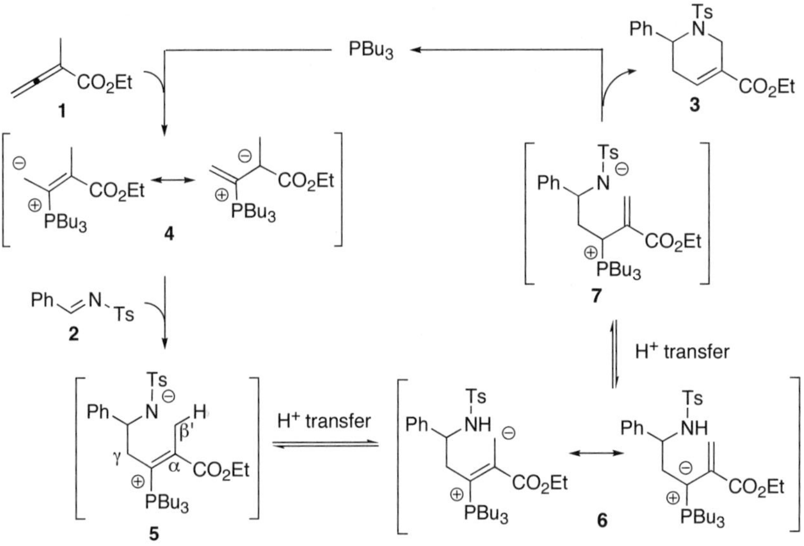

Table 2 lists examples of successful [4+2] reactions performed using 2-arylmethyl-2,3-butadienoates and *N*-tosylimines as substrates. With *N*-tosyl benzaldimine as the imine reaction partner, the yields and diastereoselectivities were excellent (Table 2, entries 1–4), except for the system in which 2-*ortho*-tolylmethyl-2,3-butadienoate was used (dr 88:12; Table 2, entry 5). A variety of electron-withdrawing and -donating substituents are tolerated in both the imine and allenoate substrates (Table 2, entries 6–11). A nitro substituent diminishes the reaction efficiency slightly (Table 2, entry 7). An *ortho* substituent in the imine substrate also results in diminished diastereoselectivities (Table 2, entries 8 and 11) and/or a lower product yield (entry 8).

220

Table 2. Synthesis of Tetrahydropyridines **9** From Ethyl
2-Arylmethyl-2,3-butadienaoates and *N*-Tosylaldimines[a]

entry	R	R´	product	yield (%)[b]	dr[c]
1[d]	Ph (**2**)	Ph (**8a**)	**9a**	92	98:2
2	Ph (**2**)	4-NCC$_6$H$_4$ (**8b**)	**9b**	99	98:2
3	Ph (**2**)	3-MeOC$_6$H$_4$ (**8c**)	**9c**	99	98:2
4	Ph (**2**)	2-FC$_6$H$_4$ (**8d**)	**9d**	99	97:3
5	Ph (**2**)	2-MeC$_6$H$_4$ (**8e**)	**9e**	82	88:12
6	4-MeOC$_6$H$_4$ (**2a**)	Ph (**8a**)	**9f**	99	97:3
7	4-O$_2$NC$_6$H$_4$ (**2h**)	Ph (**8a**)	**9g**	90	95:5
8	2-F$_3$CC$_6$H$_4$ (**2g**)	4-NCC$_6$H$_4$ (**8b**)	**9h**	80	90:10
9	3-ClC$_6$H$_4$ (**2c**)	4-NCC$_6$H$_4$ (**8b**)	**9i**	99	98:2
10	4-MeC$_6$H$_4$ (**2b**)	3-MeOC$_6$H$_4$ (**8c**)	**9j**	99	98:2
11	2-ClC$_6$H$_4$ (**2d**)	3-MeOC$_6$H$_4$ (**8c**)	**9k**	96	83:17

[a] 1.0 mmol scale. [b] Isolated yields. [c] Diastereoisomeric ratio, determined using ^{1}H NMR spectroscopy. [d] 30 mmol scale.

In summary, the formation of multisubstituted tetrahydropyridines from readily available sulfonylimines and α-alkylallenoates in the presence of catalytic amounts of tributylphosphine is an operationally simple and mild reaction.

1. Department of Chemistry and Biochemistry, University of California, Los Angeles, 607 Charles E. Young Drive East, Los Angeles, California 90095-1569.

2. For reviews on phosphine-catalyzed reactions, see: (a) Lu, X.; Zhang, C.; Xu, Z. *Acc. Chem. Res.* **2001**, *34*, 535. (b) Valentine, D. H.; Hillhouse, J. H. *Synthesis* **2003**, *3*, 317. (c) Methot, J. L.; Roush, W. R. *Adv. Synth. Catal.* **2004**, *346*, 1035. (d) Lu, X.; Du, Y.; Lu, C. *Pure Appl. Chem.* **2005**, *77*, 1985. (e) Nair, V.; Menon, R. S.; Sreekanth, A.

R.; Abhilash, N.; Biji, A. T. *Acc. Chem. Res.* **2006**, *39*, 520. (f) Denmark, S. E.; Beutner, G. L. *Angew. Chem., Int. Ed.* **2008**, *47*, 1560. (g) Ye, L.-W.; Zhou, J.; Tang, Y. *Chem. Soc. Rev.* **2008**, *37*, 1140.

3.   (a) Zhang, C.; Lu, X. *J. Org. Chem.* **1995**, *60*, 2906. (b) Xu, Z.; Lu, X. *Tetrahedron Lett.* **1999**, *40*, 549. (c) Du, Y.; Lu, X.; Yu, Y. *J. Org. Chem.* **2002**, *67*, 8901. (d) Zhu, X.-F.; Henry, C. E.; Kwon, O. *Tetrahedron* **2005**, *61*, 6276. (e) Lu, X.; Lu, Z.; Zhang, X. *Tetrahedron* **2006**, *62*, 457. (f) Xia, Y.; Liang, Y.; Chen, Y.; Wang, M.; Jiao, L.; Huang, F.; Liu, S; Li, Y.; Yu, Z.-X. *J. Am. Chem. Soc.* **2007**, *129*, 3470. (g) Mercier, E.; Fonovic, B.; Henry, C.; Kwon, O.; Dudding, T. *Tetrahedron Lett.* **2007**, *48*, 3617.

4.   (a) Du, Y.; Lu, X. *J. Org. Chem.* **2003**, *68*, 6463. (b) Wang, J.-C.; Krische, M. J. *Angew. Chem., Int. Ed.* **2003**, *42*, 5855. (c) Pham, T. Q.; Pyne, S. G.; Skelton, B. W.; White, A. H. *J. Org. Chem.* **2005**, *70*, 6369.

5.   (a) Zhu, G.; Chen, Z.; Jiang, Q.; Xiao, D.; Cao, P.; Zhang, X. *J. Am. Chem. Soc.* **1997**, *119*, 3836. (b) Wilson, J. E.; Fu, G. C. *Angew. Chem., Int. Ed.* **2006**, *45*, 1426. (c) Cowen, B. J.; Miller, S. J. *J. Am. Chem. Soc.* **2007**, *129*, 10988. (d) Fang, Y.-Q.; Jacobsen, E. N. *J. Am. Chem. Soc.* **2008**, *130*, 5660. (e) Voituriez, A.; Panossian, A.; Fleury-Bregeot, N.; Retailleau, P.; Marinetti, A. *J. Am. Chem. Soc.* **2008**, *130*, 14030.

6.   (a) Zhu, X.-F.; Henry, C. E.; Wang, J.; Dudding, T.; Kwon, O. *Org. Lett.* **2005**, *7*, 1387. (b) Zhu, X.-F.; Schaffner, A.-P.; Li, R. C.; Kwon, O. *Org. Lett.* **2005**, *7*, 2977. (c) Dudding, T.; Kwon, O.; Mercier, E. *Org. Lett.* **2006**, *8*, 3643. (d) Castellano, S.; Fiji, H. D. G.; Kinderman, S. S.; Watanabe, M.; de Leon, P.; Tamanoi, F.; Kwon, O. *J. Am. Chem. Soc.* **2007**, *129*, 5843. (e) Zhu, X.-F.; Henry, C. E.; Kwon, O. *J. Am. Chem. Soc.* **2007**, *129*, 6722. (f) Henry, C. E.; Kwon, O. *Org. Lett.* **2007**, *9*, 3069. (g) Creech, G. S.; Kwon, O. *Org. Lett.* **2008**, *10*, 429. (h) Creech, G. S.; Zhu, X.-F.; Fonovic, B.; Dudding, T.; Kwon, O. *Tetrahedron* **2008**, *64*, 6935. For phosphine-catalyzed reactions of acetylenes, see: (i) Sriramurthy, V.; Barcan, G. A.; Kwon, O. *J. Am. Chem. Soc.* **2007**, *129*, 12928.

7.   Zhu, X.-F.; Lan, J.; Kwon, O. *J. Am. Chem. Soc.* **2003**, *125*, 4716.

8.   Tran, Y. S.; Kwon, O. *Org. Lett.* **2005**, *7*, 4289.

9.   Wurz, R. P.; Fu, G. C. *J. Am. Chem. Soc.* **2005**, *127*, 12234.

10.   Tran, Y. S.; Kwon, O. *J. Am. Chem. Soc.* **2007**, *129*, 12632.

11.   The values of pK_a for benzenesulfonamide and the

methyltriphenylphosphonium ion in DMSO are 16.1 and 22.4, respectively. These values were obtained from the "Bordwell pK_a Table" at http://www.chem.wisc.edu/areas/reich/pkatable; the original references are: (a) Bordwell, F. G.; Fried, H. E.; Hughes, D. L.; Lynch, T.-Y.; Satish, A. V.; Whang, Y. E. *J. Org. Chem.* **1990**, *55*, 3330. (b) Zhang, X.-M.; Bordwell, F. G. *J. Am. Chem. Soc.* **1994**, *116*, 968.

Appendix
Chemical Abstracts Nomenclature; (Registry Number)

Ethyl 2-(triphenylphosphoranylidene)propionate: Propanoic acid, 2-(triphenylphosphoranylidene)-, ethyl ester; (5717-37-3)

Ethyl 2-methylbuta-2,3-dienoate: 2,3-Butadienoic acid, 2-methyl-, ethyl ester; (5717-41-9)

(E)-N-Benzylidene-4-methylbenzenesulfonamide: Benzenesulfonamide, 4-methyl-*N*-(phenylmethylene)-, [*N(E)*]-: (51608-60-7)

p-Toluenesulfonamide: Benzenesulfonamide, 4-methyl-; (70-55-3)

Benzaldehyde; (100-52-7)

BF$_3$·Et$_2$O; (109-63-7)

Ethyl 6-phenyl-1-tosyl-1,2,5,6-tetrahydropyridine-3-carboxylate: 3-Pyridinecarboxylic acid, 1,2,5,6-tetrahydro-1-[(4-methylphenyl)-sulfonyl]-6-phenyl-, ethyl ester; (528853-66-9)

Tri-*n*-butylphosphine: Phosphine, tributyl-; (998-40-3)

Ohyun Kwon was born in South Korea in 1968. She received her BS and MS in chemistry (with Eun Lee) from Seoul National University in 1991 and 1993, respectively. She came to the US in 1993 and obtained her Ph.D. (with Samuel J. Danishefsky) from Columbia University in 1998. After a postdoctoral stint in the laboratory of Stuart L. Schreiber at Harvard University, Ohyun Kwon started her independent career as an Assistant Professor at University of California, Los Angeles, in 2001. Her research involves the development of phosphine-catalyzed reactions and their applications in natural product synthesis and chemical biology.

 Kui Lu was born in Jiangxi Province, China, in 1981. He received his BS degree from China Agricultural University in 2002, and then a Ph.D. degree in organic chemistry from Peking University, China, in 2007 under the supervision of Professors Zhen Yang and Jiahua Chen. He is currently working as a postdoctoral fellow in Professor Kwon's group at the University of California, Los Angeles, focusing on the development of new chiral phosphines and phosphine-catalyzed reactions.

 Matthew Davis was born in 1981 in Park Forest, Illinois. In 2004 he received his B.S. degree in chemistry from Hope College in Holland, Michigan. He is currently pursuing graduate studies at the University of Pittsburgh, under the guidance of Prof. Kay Brummond. His research currently focuses on expanding the scope of the Rh(I)-catalyzed cyclocarbonylation reaction of allene-ynes.

SYNTHESIS OF POLYYNES BY IN SITU DESILYLATIVE BROMINATION AND PALLADIUM-CATALYZED COUPLING: (7-(BENZYLOXY)HEPTA-1,3,5-TRIYNYL)TRIISOPROPYLSILANE

A.

B.

C.

D.

Submitted by Soonho Hwang, Hee Ryong Kang, and Sanghee Kim.[1]
Checked by Olesya Haze and Rick L. Danheiser.

1. Procedure

A. ((3-Bromoprop-2-ynyloxy)methyl)benzene (1). A 500-mL, three-necked, round-bottomed flask (Note 1) equipped with a magnetic stir bar, two glass stoppers, and an argon inlet adapter is charged with benzyl propargyl ether (11.2 g, 76.6 mmol, 1.0 equiv) (Note 2), 150 mL of acetone (Note 3), and *N*-bromosuccinimide (15.0 g, 84.3 mmol, 1.1 equiv) (Note 2). To the resulting yellow solution, silver(I) nitrate (1.30 g, 7.65 mmol, 0.1 equiv) (Note 2) is added, and the cloudy-grey reaction mixture is stirred at 24 °C for 30 min (Notes 4, 5). The reaction mixture is transferred to a 1-L separatory funnel containing 100 mL of saturated aq $Na_2S_2O_3$ solution and diluted with 200 mL of Et_2O and 200 mL of pentane (Note 6). The organic layer is separated and washed with two 200 mL portions of brine, dried over $MgSO_4$ (5 g), filtered, and concentrated by rotary evaporation (20 °C, 20 mmHg) to give 18 g of an orange oil. This material is purified by silica gel column chromatography (Note 7) to afford 14.8–15.4 g (86–89%) of bromoacetylene **1** as a light yellow oil (Note 8).

     B.   (5-(Benzyloxy)penta-1,3-diynyl)triisopropylsilane (2). A 500-mL, three-necked, round-bottomed flask equipped with a magnetic stir bar, glass stopper, rubber septum, and an argon inlet adapter is charged with bromoacetylene **1** (7.60 g, 33.8 mmol, 1.0 equiv), 230 mL of tetrahydrofuran (Note 3), and (triisopropylsilyl)acetylene (9.10 mL, 7.40 g, 40.5 mmol, 1.2 equiv) (Note 2). Dichlorobis(triphenylphosphine)palladium(II) (0.478 g, 0.68 mmol, 0.02 equiv) (Note 2) and copper(I) iodide (0.130 g, 0.68 mmol, 0.02 equiv) (Note 2) are added in one portion, and diisopropylamine (10.0 mL, 7.16 g, 70.8 mmol, 2.1 equiv) (Note 2) is added by syringe over 2 min. The resulting yellow slurry is stirred at 24 °C for 5 h, during which time the color of the slurry darkened progressively to brown (Note 9, 10). Saturated aq NH_4Cl solution (25 mL) is then added, and the resulting mixture is diluted with 250 mL of Et_2O and transferred to a 1-L separatory funnel. The organic layer is separated and washed with two 250 mL portions of brine, dried over anhydrous Na_2SO_4 (15 g, 10 min), filtered, and concentrated by rotary evaporation (25 °C, 20 mmHg) to give 13 g of viscous brown oil. Column chromatography on silica gel (Note 11) yields 8.13–8.41 g (74–76%) of TIPS-diyne **2** as a yellow oil (Note 12).

     C.   ((5-Bromopenta-2,4-diynyloxy)methyl)benzene (3). A 500-mL, three-necked, round-bottomed flask equipped with a rubber septum, glass stopper and an argon inlet adapter is charged with TIPS-diyne **2** (8.79 g, 26.9 mmol, 1.0 equiv), 135 mL of freshly distilled acetonitrile (Note 3), *N*-bromosuccinimide (5.75 g, 32.3 mmol, 1.2 equiv) and silver(I) fluoride (4.10 g, 32.3 mmol, 1.2 equiv) (Note 2). The reaction flask is fitted with a mechanical stirrer (Note 13) and wrapped in aluminum foil. The heterogeneous reaction mixture is stirred at 22-25 °C for 2 h (Note 14). The reaction mixture is filtered through a pad of 20 g of Celite with the aid of 100 mL of Et_2O (Note 6). The filtrate is diluted with 100 mL of Et_2O, transferred to a 1-L separatory funnel, washed with two 200 mL portions of brine, dried over Na_2SO_4 (15 g, 10 min), filtered, and concentrated by rotary evaporation (25 °C, 20 mmHg) to give 8.3 g of an orange oil. Column chromatography on silica gel (Note 15) gives 6.124 g (91%) of bromo diyne **3** (Note 16) as an orange oil (Note 17).

     D.   (7-(Benzyloxy)hepta-1,3,5-triynyl)triisopropylsilane (4). A 500-mL, two-necked, round-bottomed flask equipped with a stir bar, rubber septum, and argon inlet adapter is charged with alkynyl bromide **3** (5.90 g,

23.7 mmol, 1.0 equiv), 240 mL of tetrahydrofuran, and (triisopropylsilyl)acetylene (6.40 mL, 5.20 g, 28.5 mmol, 1.2 equiv). Dichlorobis(triphenylphosphine)palladium(II) (0.500 g, 0.71 mmol, 0.03 equiv) and copper(I) iodide (0.136 g, 0.71 mmol, 0.03 equiv) (Note 18) are added in one portion, and diisopropylamine (7.00 mL, 5.01 g, 49.5 mmol, 2.0 equiv) is added by syringe over 1 min. The resulting yellow slurry is stirred at 24 °C for 5 h, during which time the slurry darkens progressively to a deep brown color (Note 19). Saturated aq NH_4Cl solution (20 mL) is then added, and the resulting mixture is transferred to a 1-L separatory funnel and diluted with 250 mL of Et_2O. The organic layer is washed with two 250 mL portions of brine, dried over Na_2SO_4 (20 g, 10 min), filtered, and concentrated by rotary evaporation (rt, 20 mmHg) to give 5 g of a viscous black oil. Column chromatography on silica gel (Note 20) affords 3.19 g (38%) (Note 21) of triyne **4** as a brown oil (Note 22).

2. Notes

1. The checkers used flame-dried glassware and carried out the reaction under an atmosphere of argon. The submitters dried their apparatus in an oven at 80 °C for 8 h and performed the reaction under an atmosphere of nitrogen.

2. Benzyl propagyl ether (98.0%) was purchased from Acros Organics. *N*-Bromosuccinimide (99%) and silver(I) nitrate (99+%, ACS reagent) were purchased from Aldrich Chemical Co., Ltd. (Triisopropylsilyl)acetylene (98.0+%) was purchased from Fluka. Copper(I) iodide (99.999%), diisopropylamine (99.5%), and silver(I) fluoride (99%) were purchased from Aldrich Chemical Co., Ltd. The checkers purchased dichlorobis(triphenylphosphine)palladium(II) (99.99%) from Aldrich Chemical Co., Ltd. while the submitters obtained this catalyst (>98.0%) from Tokyo Chemical Industry Co., Ltd.

3. The checkers used acetone (HPLC grade, J. T. Baker) that was distilled under argon after drying over anhydrous K_2CO_3. The checkers purchased tetrahydrofuran (HPLC grade) from J.T. Baker and purified it by pressure filtration under argon through activated alumina. The checkers purchased acetonitrile (ChromAr grade) from Mallinckrodt and distilled it from CaH_2. The submitters employed acetone, tetrahydrofuran, and acetonitrile that was purchased from Burdick & Jackson and dried by distillation from anhydrous K_2CO_3, sodium/benzophenone ketyl, and CaH_2,

respectively. The submitters used tetrahydrofuran after bubbling nitrogen through it for 15 min.

4.   Because of the photosensitivity of NBS, AgF, and the brominated product, the reaction was protected from light by wrapping the flask with aluminum foil.

5.   The progress of the reaction was monitored by TLC analysis on EMD (Merck) pre-coated glass-backed silica gel 60 F-254 250 μm plates. The plates were eluted twice with 20:1 hexanes:EtOAc, and visualized by UV absorbance at 254 nm or with $KMnO_4$, PMA, or *p*-anisaldehyde stain. Benzyl propagyl ether has $R_f = 0.4$, and bromoacetylene **1** has $R_f = 0.5$.

6.   Diethyl ether was obtained by the checkers from Mallinckrodt (anhydrous, stabilized, AR®, ACS grade) and by the submitters from SK chemicals.   The checkers purchased $Na_2S_2O_3 \cdot 5H_2O$ (ACS grade, ≥99.5%) from Mallinckrodt.

7.   Flash column chromatography was carried out using Sorbent Technologies Standard Grade 60Å 230–400 mesh silica gel.   A glass column (6 x 40 cm) was slurry-packed with 300 g of silica gel.   The compound was loaded in a solution of 20:1 hexanes/EtOAc.   The column was eluted with 20:1 hexanes:EtOAc, collecting an initial 300 mL fraction, and then 30 mL fractions.   The fractions containing the desired product were combined and concentrated by rotary evaporation at room temperature (20 mmHg).

8.   The submitters report obtaining 16.19 g (94%) of bromoacetylene **1** as a colorless oil.   Characterization data for ((3-bromoprop-2-ynyloxy)methyl)benzene **(1)**: IR (film): 3065, 3031, 2853, 2214, 1497, 1454, 1353, 1091, 1029, 738, 698 cm^{-1}; ^{1}H NMR (400 MHz, CDCl$_3$) δ: 4.22 (s, 2H), 4.61 (s, 2H), 7.28–7.38 (m, 5H); ^{13}C NMR (100 MHz, CDCl$_3$) δ: 46.8, 58.2, 71.9, 76.4, 128.2, 128.3, 128.7, 137.3;   Anal. calcd for C$_{10}$H$_9$BrO: C, 53.36; H, 4.03. Found: C, 53.27; H, 3.91.

9.   When the reaction temperature was maintained at 30 °C, the reaction was completed within 1 h.   However, the formation of homocoupling product **6** increased significantly from 1 to 8%.

6

10.   The progress of the reaction was monitored by TLC analysis (two elutions with 50:1 hexanes/EtOAc).   Under these conditions (triisopropylsilyl)acetylene $R_f = 0.88$, TIPS-diyne **2** $R_f = 0.44$,

bromoacetylene **1** R_f = 0.35, and the homocoupling product **6** R_f = 0.26.

11. A glass column (6 x 40 cm) was slurry-packed with 300 g of silica gel. The product was loaded as a solution in 50:1 hexanes/EtOAc. The column was eluted with 50:1 hexanes:EtOAc, collecting an initial 500 mL fraction, and then 30-mL fractions. The fractions containing the desired product were combined and concentrated by rotary evaporation at room temperature (20 mmHg).

12. The submitters reported carrying out the reaction at 20 °C for 2 h, and obtained 8.50 - 9.10 g (77-83%) of TIPS-diyne **2** as a yellow oil. Characterization data for (5-(benzyloxy)penta-1,3-diynyl)triisopropylsilane (**2**): IR (film): 3066, 3032, 2944, 2891, 2866, 2105, 1497, 1462, 1384, 1352, 1241, 1073, 997, 883, 796, 736, 697, 665 cm^{-1}; ^{1}H NMR (400 MHz, CDCl$_3$) δ: 1.10 (s, 21H), 4.26 (s, 2H), 4.63 (s, 2H), 7.28-7.38 (m, 5H); ^{13}C NMR (100 MHz, CDCl$_3$) δ: 11.4, 18.7, 57.8, 71.9, 72.0, 73.1, 84.6, 89.2, 128.2, 128.3, 128.6, 137.3; UV (CH$_3$OH) λ_{max}, nm (ε): 243 (760), 255 (810), 269 (480); MS (EI) m/z (rel int) 326 (M+, 1); HRMS (EI) calcd for C$_{21}$H$_{30}$OSi(M+) 326.2066, found 326.2066; Anal. calcd for C$_{21}$H$_{30}$OSi: C, 77.24; H, 9.26. Found: C, 77.09; H, 9.37. The purity of TIPS-diyne **2** was determined by the submitters to be 97% according to HPLC analysis. HPLC analysis was performed on a Agilent 1200 series with YMC-Pack SIL column (250 x 4.6 mm), elution with EtOAc/hexane (gradient, 20% EtOAc/hexane, 25 min) at 1 mL/min while monitoring at 300 nm. The retention time of the TIPS-diyne **2** was 13.9 min.

13. The checkers found that vigorous stirring is critical to obtain consistently high yields. With magnetic stirring the checkers observed that longer reaction times (6–7 h) were required and the reaction proceeded in lower yield (30–40%). Silver(I) fluoride does not completely dissolve in the reaction mixture and without efficient stirring forms a sticky mass at the bottom of the reaction flask.

14. The progress of the reaction was monitored by TLC analysis (one elution with 20:1 hexanes/EtOAc; TIPS-diyne **2** R_f = 0.44, bromo diyne **3** R_f = 0.36).

15. A glass column (6 x 40 cm) was slurry-packed with 260 g of silica gel. The product was loaded as a solution in 20:1 hexanes/EtOAc. The column was eluted with 20:1 hexanes:EtOAc, collecting an initial 400 mL fraction, and then 225 mL fractions. The fractions containing the desired product were combined and concentrated by rotary evaporation (during concentration the colorless product slowly became yellow-orange).

16. When the reaction was carried out at half this scale (13.5 mmol) with magnetic stirring, the checkers obtained the bromo diyne in somewhat lower yield (2.83 g, 84%). The submitters report obtaining 5.91-6.24 g (88–93%) of bromo diyne **3** as a yellow oil. Bromo diyne **3** is stable to storage in a freezer (-20 °C) for at least a week. However, prolonged exposure to air and light at room temperature caused slow decomposition.

17. Characterization data for ((5-bromopenta-2,4-diynyloxy)methyl)benzene (**3**): IR (film): 3031, 2862, 2236, 2142, 1497, 1454, 1351, 1258, 1192, 1073, 1028, 905, 738, 697, 665 cm^{-1}; ^{1}H NMR (400 MHz, CDCl$_3$) δ: 4.24 (s, 2H), 4.62 (s, 2H), 7.28-7.39 (m, 5H); ^{13}C NMR (100 MHz, CDCl$_3$) δ: 41.6, 57.6, 65.1, 71.61, 71.63, 72.0, 128.3, 128.4, 128.7, 137.1; UV (CH$_3$OH) λ_{max}, nm (ε): 212 (1300), 265 (3000), 304 (800); HRMS (EI) calcd for C$_{12}$H$_9$BrO (M+) 247.9837, found 247.9839. The purity of bromo diyne **3** was determined by the submitters to be 98% by HPLC analysis. HPLC analysis was performed on a Agilent 1200 series with YMC-Pack SIL column (250 x 4.6 mm), elution with EtOAc/hexane (gradient, 15% EtOAc/hexane, 25 min) at 1 mL/min while monitoring at 300 nm. The retention time of the bromo diyne **3** was 9.0 min.

18. The checkers found that when grey CuI (99.999%, from Aldrich Chemical Co., Ltd) was used as received, the product was obtained in lower yield (30-33%) and the reaction was not complete after 6 h. Improved results were obtained (38-45% yield, reaction complete within 5 h) by using white CuI obtained by purification by Soxhlet extraction with THF followed by drying at 0.05 mmHg for 24 h.

19. The progress of the reaction was monitored by TLC analysis (one elution with 20:1 hexanes/EtOAc; (triisopropylsilyl)acetylene R_f = 0.86, triyne **4** R_f = 0.41, bromo diyne **3** R_f = 0.36, and homocoupling product **7** R_f = 0.17).

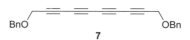

7

20. A glass column (6 x 40 cm) was slurry-packed with 270 g of silica gel. The product was loaded as a solution in 5 mL of 50:1 hexanes/EtOAc. The column was eluted with 50:1 hexanes/EtOAc, collecting an initial 600 mL fraction, and then 30 mL fractions. The fractions containing the desired product were combined and concentrated by rotary evaporation at room temperature (20 mmHg).

21. When the reaction was carried out at half this scale (11.8 mmol), the checkers obtained diyne **3** in improved yield (1.88 g 45%). The submitters obtained 3.65–4.15 g (44-50%) of triyne **4** as a deep brown oil and report that the yield of the reaction decreases considerably as the scale increases (0.3 g scale : 75%; 1.5 g scale = 67%; 5.0 g scale = 50%).

22. Characterization data for (7-(benzyloxy)hepta-1,3,5-triynyl) triisopropylsilane: IR (film): 3032, 2944, 2891, 2866, 2164, 2078, 1497, 1462, 1384, 1350, 1282, 1075, 1018, 997, 883, 736, 697, 678, 665 cm^{-1}; ^{1}H NMR (400 MHz, CDCl$_3$) δ: 1.108–1.113 (m, 21H), 4.26 (s, 2H), 4.62 (s, 2H), 7.28 7.38 (m, 5H); ^{13}C NMR (100 MHz, CDCl$_3$) δ: 11.4, 18.7, 57.7, 60.3, 63.9, 71.7, 72.0, 74.5, 85.5, 89.7, 128.25, 128.34, 128.7, 137.0; UV (CH$_3$OH) λ_{max}, nm (ε): 222 (75000), 285 (1000), 303 (1200), 354 (600); MS (EI) *m/z* (rel int) 350 (M+, 2); HRMS (EI) calcd for C$_{23}$H$_{30}$OSi (M+) 350.2066, found 350.2063; Anal. calcd for C$_{23}$H$_{30}$OSi: C, 78.80; H, 8.63. Found: C, 78.67; H, 8.43. The purity of triyne **4** was determined to be 96% by HPLC analysis under the conditions described in Note 17 (retention time of **4** was 13.9 min).

Safety and Waste Disposal Information

All hazardous materials should be handled and disposed of in accordance with "Prudent Practices in the Laboratory"; National Academy Press; Washington, DC, 1995.

3. Discussion

Unsymmetrically substituted conjugated diyne and polyyne units continue to attract widespread interest because of their unusual electrical, optical, and structural properties.[2] Consequently, the development of efficient synthetic approaches toward these rigid units remains an important challenge to synthetic organic chemists.[2-4]

The method most commonly used for the preparation of unsymmetrical diyne and polyyne compounds is the Cadiot-Chodkiewicz coupling reaction,[5] the metal-catalyzed cross-coupling of a 1-haloalkyne with a terminal alkyne. The major limitation of this coupling reaction, however, is that terminal diynes and higher polyynes required as coupling partners or precursors of 1-haloalkynes are often unstable.[2,3,6] To overcome this challenge, many elegant alternative methods have been explored.

We have developed an iterative strategy for the synthesis of unsymmetrically substituted polyynes as shown in Scheme 1. The key to the success of this iterative method lies in the *in situ* one-pot desilylative bromination, which avoids the complication encountered with the isolation of sensitive terminal alkynes.

In this process, the starting terminal alkyne was first converted to bromoalkyne under standard NBS/AgNO$_3$ conditions.[7] At this point, we employed TIPS-acetylene as a cross-coupling partner for the homologation reactions because it had previously been reported that TMS-acetylene decomposes under the basic conditions of the coupling reaction such that no desired cross-coupling product can usually be isolated.[8] When we used TIPS-acetylene, the desired cross-coupling product was obtained in good yield under the modified Sonogashira conditions. Our initial attempts to effect the in situ one-pot desilylative bromination of TIPS-diyne used the standard NBS/AgNO$_3$ conditions developed by Isobe and co-workers for the conversion of TMS-protected acetylenes to bromoacetylenes.[9] Unfortunately, these conditions led only to the recovery of the starting material. On the other hand, when we employed AgF instead of AgNO$_3$, we could obtain the desired bromo diyne in high yield. Repeating the two-step acetylene homologation sequence on this bromo diyne then generated the expected bromo triyne readily in good overall yield.

Scheme 1. General Iterative Protocol for Synthesis of Unsymmetrical Polyynes.

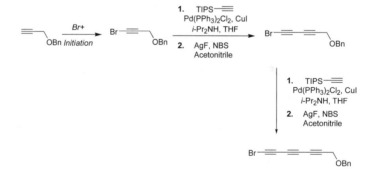

Scheme 2. Total synthesis of (S)-(E)-15,16-dihydrominquartynoic acid.

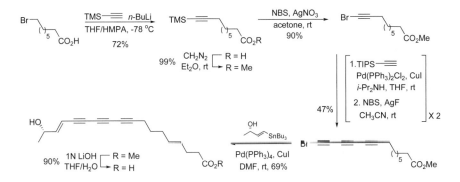

By employing our new iterative strategy, we accomplished the total synthesis of (S)-(E)-15,16-dihydrominquartynoic acid from a simple stating material in a high overall yield (Scheme 2).[10] In addition, we have developed a facile solid-phase synthetic pathway to generate a library of natural product-like polyynes.[11] These results demonstrate that our approach can be applied efficiently to the synthesis of various unsymmetrical polyynes.

1. College of Pharmacy, Seoul National University, San 56-1, Shilim, Kwanak, Seoul 151-742, Korea. *pennkim@snu.ac.kr*

2. (a) Ginsburg, E. J.; Gorman, C. B.; Grubbs, R. H. In *Modern Acetylene Chemistry*; Stang, P. J., Diederich, F., Eds.; VCH: Weinheim, 1995. (b) Brandsma, L. *Preparative Acetylenic Chemistry*; Elsevier: Amsterdam, 1988.

3. For recent papers that include a discussion of this topic, see: (a) Heuft, M. A.; Collins, S. K.; Yap, G. P. A.; Fallis, A. G. *Org. Lett.* **2001**, *3*, 2883–2886. (b) Shi Shun, A. L. K.; Chernick, E. T.; Eisler, S.; Tykwinski, R. R. *J. Org. Chem.* **2003**, *68*, 1339–1347.

4. For reviews, see: (a) Hartung, R. E.; Paquette, L. A. *Chemtracts* **2002**, *15*, 106–116. (b) Siemsen, P.; Livingston, R. C.; Diederich, F. *Angew. Chem., Int. Ed.* **2000**, *39*, 2632-2657.

5. Cadiot, P.; Chodkiewicz, W. In *Chemistry of Acetylenes*; Viehe, H. G., Ed., Marcel Dekker: New York, 1969; pp. 597–647.

6. (a) Haley, M. M.; Bell, M. L.; English, J. J.; Johnson, C. A.; Weakley,

T. J. R. *J. Am. Chem. Soc.* **1997**, *119*, 2956–2957. (b) Patel, G. N.; Chance, R. R.; Turi, E. A.; Khanna, Y. P. *J. Am. Chem. Soc.* **1978**, *100*, 6644–6649. (c) Haley, M. M.; Bell, M. L.; Brand, S. C.; Kimball, D. B.; Pak, J. J.; Wan, W. B. *Tetrahedron Lett.* **1997**, *38*, 7483–7486.

7. Hofmeister, H.; Annen, K.; Laurent, H.; Wiechert, R. *Angew. Chem. Int. Ed. Engl.* **1984**, *23*, 727-729.

8. (a) Marino, J. P.; Nguyen, H. N. *J. Org. Chem.* **2002**, *67*, 6841-6844. (b) Eastmond, R.; Walton, D. R. M. *Tetrahedron* **1972**, *28*, 4591-4599. (c) Eaborn, C.; Walton, D. R. M. *J. Organomet. Chem.* **1965**, *4*, 217-228.

9. Nishikawa, T.; Shibuya, S.; Hosokawa, S.; Isobe, M. *Synlett* **1994**, 485-486.

10. Kim, S.; Kim, S.; Lee, T.; Ko, H.; Kim, D. *Org. Lett.* **2004**, *6*, 3601-3604.

11. Lee, S.; Lee, T.; Lee, Y. M.; Kim, D.; Kim, S. *Angew. Chem. Int. Ed.* **2007**, *46*, 8422-8425.

Appendix
Chemical Abstracts Nomenclature; (Registry Number)

Benzyl propagyl ether; (4039-82-1)

N-bromosuccimide: NBS; (128-08-5)

Silver nitrate; (7761-88-8)

(Triisopropylsilyl)acetylene: Ethynyltriisopropylsilane; (89343-06-6)

Dichlorobis(triphenylphosphine)palladium(II):
    Bis(triphenylphosphine)palladium(II) Dichloride; (13965-03-2)

Copper(I) iodide: Cuprous iodide; (7681-65-4)

Diisopropylamine: DIPA; (108-18-9)

Silver(I) fluoride; (7775-41-9)

Sanghee Kim received his B.S. degree in pharmacy from Seoul National University in 1988, where he also obtained his M.S. degree in medicinal chemistry in 1990. After finishing military service as a lieutenant, he worked for a year as an assistant researcher at the Korea Institute of Science and Technology. In 1992, he joined Prof. Jeffery Winkler's group at the University of Pennsylvania and earned his Ph.D. in organic chemistry in 1997. After two years as a postdoctoral fellow with Prof. K.C. Nicolaou, at The Scripps Research Institute, he worked at Abbott Laboratories as a research scientist. He moved to Seoul National University in 1999, where he is currently a professor of college of pharmacy teaching various advanced courses in the field of medicinal and synthetic chemistry.

Soonho Hwang was born in 1983 in Daejeon, Korea and received his B.S. degree in pharmacy from Wonkwang University in 2007. In 2007, he began his graduate studies at the Seoul National University, under the guidance of Prof. Sanghee Kim. His research focuses on the total synthesis of natural products, mainly base on the palladium-catalyzed cross-coupling and [3,3]-sigmatropic Claisen rearrangement.

Hee Ryong was born in 1979 in Yesan, Korea. After finishing military service, he received his B.S. degree in chemistry from Kongju University in 2005. He completed his M.S. degree in 2007 at Seoul National University, where he worked on "Signal Regulator Synthesis Laboratory" under the supervision of Prof. Sanghee Kim. His thesis included discussions of the utilities of trialkylsilyl acetylene and AgF system. Currently, he works at the pharmaceutical company Chong Kun Dang as a medicinal chemist.

Olesya Haze graduated from the University of Rochester in 2006 with a B.S. degree in Chemistry and a B.A. degree in Mathematics. As an undergraduate she studied silane cation radical fragmentations and bonded exciplexes under the guidance of Professor Joseph Dinnocenzo. Currently Olesya is a graduate student at the Massachusetts Institute of Technology where she is developing photochemical benzannulation reactions for the synthesis of highly substituted heteroaromatic compounds in the laboratory of Professor Rick Danheiser.

ENANTIOSELECTIVE PREPARATION OF
DIHYDROPYRIMIDONES
[(S)-1-Benzyl-6-methyl-2-oxo-4-phenyl-1, 2, 3, 4-tetrahydropyrimidine-5-carboxylic methyl ester]

A.

B.

C.

D.

E.

Submitted by Jennifer M. Goss, Peng Dai, Sha Lou, and Scott E. Schaus.[1]
Checked by Tanja Brkovic and David Hughes.

1. Procedure

A. Allyl carbamate. An empty 3-necked, 1-L round-bottom flask is fitted with a stopper on the middle neck and septa on each of the outer necks. Through one septum of the empty flask is inserted a 6-mm glass tube which is connected to an ammonia gas cylinder via vacuum tubing. Through the other septum is inserted a 6-mm glass tube which is connected to the reaction flask via vacuum tubing. This empty flask serves as a trap to prevent backflow of the flask contents into the lecture bottle.

A trap to neutralize discharged ammonia gas is also constructed as follows. To a 3-necked, 1-L flask is added 400 mL of 10% HCl, which is gently stirred using a magnetic stir bar (3 cm). The middle neck of the flask is stoppered, one neck is left open to the atmosphere, and the third neck is fitted with a septum through which is inserted a 6 mm ID x 5 cm glass tube. The glass tube is situated well above the contents of the acid solution to prevent any back flow. The glass tube is connected via vacuum tubing to an empty 1-L 3-necked flask equipped with a stopper on the middle neck and septa on each of the outer necks, with each septa pieced with a 6 mm ID x 5 cm glass tube. The tube in one outer neck is connected via vacuum tubing to the HCl quench flask, while the tube in the other outer neck is connected via vacuum tubing to the reaction flask. The empty flask between the reaction flask and the flask containing aq. HCl ensures no back flow of the aq. solution or water vapor into the reaction flask.

An oven-dried, 1-L, three-necked round-bottom flask is equipped with an overhead mechanical stirrer. One neck of the flask is fitted with a septum through which has been inserted a 6-mm ID x 20 cm length glass tube that will extend into the contents of the flask (Note 1). A thermocouple probe is also inserted through this septum to monitor temperature. The outlet of this tube is connected to the empty 3-necked 1-liter flask and ammonia cylinder via vacuum tubing. The third neck of the reaction flask is similarly fitted with a septum pierced with a short 6 mm ID x 5 cm length glass tube, situated well above the contents of the flask, which is connected via the empty flask to the flask containing 10% aq. HCl.

To the reactor flask is charged by weight allyl chloroformate (84.0 g, 1.00 equiv, 697 mmol) and toluene (400 mL) (Notes 2, 3). The flask is immersed in a room temperature water bath. The solution is mechanically stirred at 300 rpm while bubbling ammonia gas through the solution at a rate to maintain the internal temperature below 45 °C. Ammonia addition is continued for 5.5 hours. Ammonium chloride (white solid) precipitates throughout the course of the reaction and the reaction mixture becomes quite thick. When complete consumption of the starting material is verified by NMR assay (Note 4), the reaction mixture is vacuum filtered through a 150-mL sintered glass funnel. The solids are washed with 3 x 70 mL portions of toluene and the resulting clear solution is concentrated under reduced pressure by rotary evaporation (30 mmHg, 35 °C water bath) to 75.4 g (Note 5). The remaining crude oil is distilled under vacuum at 65–67 °C (1–2 mmHg) to provide 63.2 g (89.7%) of allyl carbamate as a clear oil (Notes 6, 7).

B. *(Benzenesulfonyl-phenyl-methyl)-carbamic acid allyl ester.* An oven-dried, 1-L round-bottomed flask, equipped with a rubber septum and an oval 3-cm stir bar, is charged with benzenesulfinic acid sodium salt (25.6 g, 1.43 equiv, 154 mmol), allyl carbamate (15.5 g, 1.43 equiv, 154 mmol), methanol (100 mL) and water (200 mL). The reaction mixture is stirred at ambient temperature until homogeneous (about 5 min). Benzaldehyde (11.1 g, 1.00 equiv, 107 mmol) is added by syringe, weighing the syringe before and after addition, followed by addition of formic acid (39.5 g of 91 wt% solution, 7.3 equiv, 781 mmol). The reaction mixture is stirred for 4 days at room temperature (21–22 °C), during which time the formation of a heavy white precipitate occurs (Note 8). The mixture is vacuum filtered via a 150-mL medium porosity sintered glass funnel. The white solid is washed with two 40 mL portions of 3:1 water:methanol. The solid is air-dried to provide 21.8 g of (benzenesulfonyl-phenylmethyl)-carbamic acid allyl ester. The supernatant from the filtration is placed back in the 1-L flask and is stirred for an additional 4 days at ambient temperature. The mixture is then vacuum filtered via a 60-mL medium porosity sintered glass funnel, washed twice with 20 mL of 3:1 water:methanol, and air dried to afford 4.9 g as a second crop having comparable purity by NMR to the first crop. The total yield of the reaction is 26.7 g (75%) (Notes 9, 10).

C. *2-[(R)-Allyloxycarbonylamino-phenyl-methyl)-3-oxo-butyric acid methyl ester.* A 1-L, three-necked, round-bottom flask is equipped with an internal thermometer, an overhead mechanical stirrer and a 500-mL pressure-equalizing addition funnel. The three-necked flask is charged with (+)-cinchonine (1.77 g, 0.20 equiv, 6.0 mmol) and (benzenesulfonyl-phenyl-methyl)-carbamic acid allyl ester (9.97 g, 1.00 equiv, 30.1 mmol). Anhydrous dichloromethane (300 mL) is added, and the solution is stirred for about 5 min, during which time the majority of solids are dissolved. The

238

flask is submerged in an isopropyl alcohol bath cooled to −25 °C (Note 11) and the solution is mechanically stirred at 500 rpm. Once the reaction mixture has cooled to −15 °C (monitored by the internal thermometer), methyl acetoacetate (10.7 g, 3.0 equiv, 92.3 mmol) is added *via* syringe over 2 minutes. After 10 min, 300 mL of an aqueous Na₂CO₃/NaCl solution (15.0 g of sodium carbonate is dissolved in 300 mL of water, and then saturated with 150 g of sodium chloride) is added to the 500-mL pressure-equalizing addition funnel. The aqueous solution is added dropwise to the reaction mixture over 1 h, while maintaining an internal temperature of −15 °C (Note 12). The heterogeneous solution is mechanically stirred at 500 rpm for 27 h while maintaining an internal temperature of −15 °C. At the end of the reaction, the cold solution is transferred to a 1-L separatory funnel and the bottom organic layer is separated (Note 13). The aqueous layer is washed with two 250 mL portions of dichloromethane. The organic layers are combined, dried over sodium sulfate, filtered and concentrated by rotary evaporation (30 mmHg, 30 °C bath temperature). The remaining residue (17.8 g) is purified by column chromatography over silica gel (Note 14) to provide 8.69–9.00 g (95–98 %) of 2-[(*R*)-allyloxycarbonylamino-phenyl-methyl]-3-oxo-butyric acid methyl ester as a white solid (Notes 15, 16).

D. 2-[(R)-(3-Benzyl-ureido)-phenyl-methyl]-3-oxo-butyric acid methyl ester. An oven-dried, 500-mL round-bottom flask, equipped with a 3-cm oval magnetic stir bar, is charged with anhydrous BHT-free tetrahydrofuran (160 mL) and benzyl isocyanate (4.74 g, 1.46 equiv, 35.6 mmol). The flask is fitted with a vacuum adapter and is degassed via three vacuum/nitrogen purge cycles. Tetrakis(triphenylphosphine)palladium(0) (1.24 g, 0.045 equiv, 1.07 mmol) is then added and dissolved by stirring the mixture. The vacuum adapter is replaced with a rubber septum pierced with a needle connected to a nitrogen inlet. A second oven-dried, 250-mL round-bottom flask equipped with a 2-cm oval magnetic stir bar is charged with 1,3-dimethylbarbituric acid (2.09 g, 0.53 equiv, 12.9 mmol) and 2-[(*R*)-allyloxycarbonylamino-phenyl-methyl)-3-oxo-butyric acid methyl ester (7.44 g, 1.00 equiv, 24.40 mmol) (Note 17). Anhydrous BHT-free tetrahydrofuran (80 mL) is added, yielding a homogeneous solution upon stirring. The flask is fitted with a vacuum adapter and degassed with three vacuum/nitrogen purge cycles. The vacuum adapter is then removed and replaced with a septum. The solution in the second flask is transferred dropwise *via* cannula to the stirring solution of tetrakis(triphenylphosphine)palladium(0), benzyl isocyanate and anhydrous tetrahydrofuran over 5 min followed by a 5 mL tetrahydrofuran rinse of the flask. An exotherm of about 5 °C occurs over 10 min. The reaction is stirred for 5 h at 21–22 °C, during which time the solution changes from yellow to a pale orange (Note 18). The solution is concentrated by rotary evaporation (30 mmHg, 30 °C bath), and the resulting residue is purified by column chromatography over silica gel (Note 19) to afford 6.07 g (70%) of 2-[(S)-(3-benzyl-ureido)-phenyl-methyl]-3-oxo-butyric acid methyl ester (Note 20) as a pale yellow oil.

E.   *(S)-1-Benzyl-6-methyl-2-oxo-4-phenyl-1,2,3,4-tetrahydropyrimidine-5-carboxylic methyl ester.* An oven-dried, 250-mL round-bottom flask, equipped with 2-cm oval magnetic stir bar and a water condenser, is charged with 2-[(R)-(3-benzyl-ureido)-phenyl-methyl]-3-oxo-butyric acid methyl ester (6.62 g having 85% purity (Note 20), 5.63 g corrected for purity, 1.00 equiv, 15.9 mmol), anhydrous 200-proof ethanol (15 mL) and anhydrous acetic acid (45 mL). The reaction solution is stirred at room temperature until homogeneous. The 250-mL round-bottomed flask is submerged in an oil bath and the solution is heated to reflux (107 °C) for 5 min. The solution is cooled to room temperature and transferred to a 250-mL separatory funnel. Dichloromethane (100 mL) and 5% aqueous brine (50 mL) are added, the two phases are mixed well by shaking, then the lower organic phase is separated. The organic layer is extracted with 2 x 50 mL portions of 5% aqueous brine (Note 21). The remaining organic layer is dried over sodium sulfate, filtered and concentrated via rotary evaporation (30 mmHg, 30 °C bath temperature) to an oil (7.2 g). The resulting residue is purified by column chromatography over silica gel (Note 22) to afford 4.13 g (77%) of (S)-1-benzyl-6-methyl-2-oxo-4-phenyl-1,2,3,4-tetrahydropyrimidine-5-carboxylic methyl ester as a white solid (Notes 23–25).

2. Notes

1. The inlet tube has to be sufficiently wide to prevent plugging as ammonium chloride buildup results in a thick mixture during the reaction. The tube is positioned to supply ammonia gas subsurface but not interfere with the stir blade.

2. The submitters used benzene in their procedure.

3. Allyl chloroformate (97%), formic acid (88%), cinchonine (85%), methyl acetoacetate (99%), sodium carbonate (>99.5%), 1,3-

dimethylbarbituric acid (99%), benzyl isocyanate (99%), benzaldehyde (99+% redistilled), toluene (99%), methanol (99%), ethanol (99.5%), ethyl acetate (99.5%), hexanes (99.5%) and glacial acetic acid (99+%) were obtained from Aldrich Chemical Co., Inc. and were used as received. Sodium chloride (99%) was obtained from VWR and used as received. Tetrakis(triphenylphosphine)palladium(0) (99%) was obtained from Strem Chemicals Inc. and used as received. Benzenesulfonic acid sodium salt (97%) was obtained from Acros. Ammonia gas was obtained from Linde Gas LLC and was used as received. The submitters obtained all anhydrous solvents from Thermo Fisher Scientific Inc. and purified through use of a dry solvent system (pressure filtration under argon through activated alumina). The checkers used anhydrous solvents as received from Aldrich Chemical Co.

4. The reaction was followed by ^{1}H NMR by diluting a sample into CDCl$_3$ and integrating the methylene protons of the allyl group of the starting material and product. The reaction progressed as follows: 50% conversion at 1.5 h, 85% conversion at 3 h, and >99% conversion at 5 h.

5. NMR analysis indicated the mixture contained 89 % allyl carbamate by weight, along with 11 wt% toluene. The distillate from the concentration was analyzed by NMR and contained no product.

6. Two fractions were collected, a small forecut at 65–66 °C (3.87 g) and the main cut, 66–67 °C, (59.36 g). The fractions were combined after NMR analysis indicated similar purity. Approximately 3 mL remained as pot residue, and about 10 mL of toluene was collected in the dry-ice vacuum trap.

7. Allyl carbamate has the following physical properties: clear oil; bp 207-208 °C. ^{1}H NMR (CDCl$_3$, 400 MHz) δ: 4.55 (dd, J = 5.6, 1.5 Hz, 2 H), 5.2 (br, 2 H, NH$_2$), 5.21 (dd, J = 10.5, 1.4 Hz, 1 H), 5.31 (dd, J = 17.1, 1.4 Hz, 1 H), 5.90 (m, 1 H); ^{13}C NMR (CDCl$_3$, 100 MHz) δ: 65.8, 117.9, 132.8, 157.2; IR (thin film, cm^{-1}): 2400, 1728, 1216; HRMS m/z 124.0368 [(M + Na$^+$) calcd for C$_4$H$_7$NO$_2$Na$^+$: 124.0374]. The purity (>99%) was determined by GC with a Agilent J&W HP-5 column (0.32 mm × 30 m)(oven temperature: 110 °C; head pressure: 60 kPa; retention time: 3.8 min).

8. The reaction is monitored by ^{1}H NMR by taking an aliquot of the reaction, diluting in DMSO-d$_6$, and integrating the benzaldehyde resonance at 10 ppm relative to the total aromatic protons. The reaction is about 80% complete in 4 days.

9. The submitters dried the material by addition of 100 mL toluene and concentrating to a solid by rotary evaporation.

10. (Benzenesulfonyl-phenyl-methyl)-carbamic acid allyl ester has the following physical properties: white solid; mp 148–150 °C. (Submitters report mp 162-165 °C). ^{1}H NMR (CDCl$_3$, 400 MHz) δ: 4.42 (d, J = 5.2 Hz, 2 H), 5.22 (m, 2 H), 5.76 (m, 1 H), 6.01 (dd, J = 10.8, 10.5 Hz, 2 H), 7.39–7.46 (m, 5 H), 7.53 (m, 2 H), 7.66 (m, 1 H), 7.86 (m, 2 H); ^{13}C NMR (CDCl$_3$, 75.0 MHz) δ: 66.7, 74.7, 118.7, 129.0 (degenerate), 129.0, 129.4, 129.7, 130.1, 132.1, 134.3, 136.8, 154.7; IR (thin film, cm^{-1}): 3334, 3063, 1730, 1527, 1496, 1448, 1308, 1235, 1141, 1081, 691; HRMS m/z 354.0796 [(M + Na$^+$) calcd for C$_{17}$H$_{17}$NO$_4$NaS$^+$: 354.0776]. The purity (>95%) was determined by ^{1}H NMR.

11. Chilling system Thermo NESLAB CB-60 with cryotrol probe was used by the submitters; a Julabo FT 901 chiller was used by the checkers The isopropyl alcohol bath temperature is monitored using a thermometer.

12. Maintaining the temperature near –15 °C during the addition of base is critical. In one run by the checkers, the temperature rose to –6 °C during the addition, which resulted in a decrease in the ee (85% vs 91%).

13. The initial extraction is carried out at 0 °C to prevent racemization of the product. The 2-phase mixture contains solids (which were determined to be cinchonine and related by-products by NMR). The solids are kept with the upper aqueous phase in all the separations.

14. The residue is dissolved in 40 mL of anhydrous dichloromethane and is loaded onto a 3-in.×12-in. column, wet-packed (10% ethyl acetate in hexanes) with 300 g of silica gel (submitters used Sorbent Technologies, 60 Å; checkers used EM Sciences, EM60, 230-400 mesh), and eluted with a gradient of ethyl acetate in hexanes (1 L of 20%, 2.5 L of 30%). The desired product is collected in fractions of 100-mL volume. TLC analysis is performed on silica gel with 30% ethyl acetate in hexanes as eluent, visualization with ultraviolet light and by staining with ceric ammonium molybdate. R$_f$ = 0.3. Methyl acetoacetate elutes just before the product. If less silica gel is used, methyl acetoacetate is not fully separated from the product. ^{1}H NMR analysis of the combined fractions indicates the product is a ~1:1 mixture of diastereomers.

15. A single diastereomer can be isolated by crystallization of the diastereomeric mixture, as follows. 400 mg of the diastereomeric mixture is added to 20 mL hexanes and heated to reflux. Ethyl acetate (1.5 mL) is added to fully dissolve all solids, then the solution is allowed to cool to

ambient temperature with stirring. After stirring overnight (15 hours) the mixture is filtered and washed with 5 mL hexanes to provide 320 mg of white needles after drying in ambient air. ^{1}H NMR analysis indicated an 88:12 mixture of diastereomers. The high recovery (80%) and the fact that the supernatant contained a 1:1 mixture of diastereomers suggested a racemization/crystallization process was occurring, funneling the mixture to the less soluble diastereomer. One additional recrystallization under the same procedure provided one diastereomer (97.5:2.5 diastereomeric ratio and >99% ee by the chiral HPLC method outlined below). 2-[(R)-Allyloxycarbonylamino-phenyl-methyl)-3-oxo-butyric acid methyl ester has the following physical properties: white solid, single diastereomer; mp 96–98 °C. ^{1}H NMR (CDCl$_3$, 400 MHz) less soluble diastereomer δ: 2.15 (s, 3 H), 3.70 (s, 3 H), 4.07 (d, J = 5.5 Hz, 1 H), 4.55 (dd, J = 4.3, 1.0 Hz, 2 H), 5.20 (d, J = 10.3 Hz, 1 H), 5.27 (d, J = 17.0 Hz, 1 H), 5.48 (dd, J = 6.2, 9.1 Hz, 1 H), 5.88 (m, 1 H), 6.12 (br d, J = 9 Hz, 1 H), 7.24–7.35 (m, 5 H). More soluble diastereomer δ: 2.33 (s, 3 H), 3.65 (s, 3 H), 4.02 (br s, 1 H), 4.55 (m, 2 H), 5.20 (d, J = 10.0 Hz, 1 H), 5.29 (d, J = 16.5 Hz, 1 H), 5.58 (m, 1 H), 5.90 (m, 1 H), 6.40 (s, 1 H), 7.22–7.37 (m, 5 H); ^{13}C NMR (100 MHz, CDCl$_3$) less soluble diastereomer, δ: 31.0, 53.0, 54.6, 63.2, 66.0, 117.9, 126.3, 126.6, 128.0, 128.7, 128.9, 132.9, 139.6, 155.7, 167.9, 203.3; more soluble diastereomer, δ: 29.1, 52.6, 53.5, 64.3, 66.0, 118.3, 126.3, 126.6, 128.0, 128.5, 128.9, 132.8, 139.4, 155.9, 169.2, 201.0; IR (thin film, cm^{-1}), both diastereomers reported: 3374, 2955, 1718, 1527, 1434, 1360, 1248, 1048, 993, 904, 730. HRMS m/z 328.1167 [(M + Na$^+$) calcd for C$_{16}$H$_{19}$NO$_5$Na$^+$: 328.1161]. The purity (>98%) was determined by HPLC-ELSD (210 nm).

16. The four diastereomers were separated by a normal phase HPLC method using a Chiralpak AD-H column (250 x 4.6 mm, 5 micron) with isocratic elution consisting of 15% (1:1 MeOH:EtOH) and 85% heptane, a flow rate of 1.0 mL/min and detection at 210 nm. Elution times: major enantiomeric pair (13.5 and 16 min), minor enantiomeric pair (10.5 and 11.5 min). The enantiomeric ratio was determined to be 95:5 on the diastereomeric mixture isolated from the silica gel chromatography. The material that was recrystallized twice showed none of the minor enantiomer (detection limit 0.5%). This diastereomer reverts back to a 1:1 mixture in a solution of methanol over a 24 hour period. The opposite enantiomeric pair was prepared by carrying out the reaction using cinchonidine instead of

cinchonine. The reaction with cinchonidine carried out at 0–5 °C provided an 89:11 enantiomeric ratio.

Elution times: 13.5 and 16 min

Elution times: 10.5 and 11.5 min

17. The weight is corrected for 4% residual ethyl acetate present from the previous step.

18. The reaction was followed by TLC on silica gel with 1:1 ethyl acetate/hexanes as eluent (R_f = 0.3). The reaction was >90% complete by TLC analysis in 2 hours.

19. The residue is combined with 6 mL of dichloromethane to make the oil mobile and is loaded onto a 3-in.×12-in. column, wet-packed (10% ethyl acetate in hexanes) with 380 g of silica gel (submitters used Sorbent Technologies, 60 Å; checkers used EM Sciences EM60, 230-400 mesh), and eluted with a gradient of ethyl acetate in hexanes (2 L of 25%, 1 L of 33%, 2 L of 40%, 1.5 L of 50%). The desired product is collected in fractions of 100-mL volume. TLC analysis is performed on silica gel with 50% ethyl acetate in hexanes as eluent, visualization with ultraviolet light and by staining with ceric ammonium molybdate. R_f = 0.30.

20. The weight of the isolated oil is 7.14 g. The yield of 6.07g is corrected based on a purity of 85%, which includes 10% ethyl acetate by weight and an estimated 5% impurity. 2-[(R)-(3-Benzyl-ureido)-phenyl-methyl]-3-oxo-butyric acid methyl ester has the following physical properties: white solid; mp 100-102 °C. ^{1}H NMR (CDCl$_3$, 400 MHz, both diastereomers reported) δ: 2.22 (s, 3 H), 2.30 (s, 3 H), 3.53 (s, 3 H), 3.60 (s, 3 H), 4.01 (d, J = 8.4 Hz, 1 H), 4.06 (d, J = 4.5 Hz, 1 H), 4.28 (m, 4 H), 5.38 (t, J = 5.8 Hz, 1 H), 5.47 (t, J = 5.8 Hz, 1 H), 5.62 (dd, J = 9.4, 7.7 Hz, 1 H), 5.79 (dd, J = 9.7, 4.8 Hz, 1 H), 6.13 (d, J = 9.4 Hz, 1 H), 6.40 (d, J = 9.8 Hz,

244

1 H), 7.17–7.30 (m, 20 H). ^{13}C NMR (CDCl$_3$, 100 MHz, both diastereomers reported) δ: 29.1, 30.1, 44.5, 44.6, 52.3, 52.4, 52.7, 53.8, 64.2, 64.6, 126.3, 126.7, 127.5, 127.6, 128.7, 128.8, 139.3, 139.4, 140.1, 140.3, 157.5, 157.6, 168.1, 169.8, 202.2, 204.0. IR (thin film, cm^{-1}): 3408, 3019, 1740, 1709, 1687, 1527, 1453, 1364, 1216, 929, 909, 700. HRMS m/z 355.1690 [(M + H$^+$) calcd for C$_{20}$H$_{22}$N$_2$O$_4^+$: 355.1613]. The purity (>90%) was determined by HPLC-ELSD (210 nm).

21. When just water is employed for the extractions, the separation is very slow and incomplete.

22. The residue is loaded onto a 3-in.×12-in. column, wet-packed (10% ethyl acetate in hexanes) with 220 g of silica gel (submitters used Sorbent Technologies, 60 Å; checkers used EM Sciences EM60, 230-400 mesh), and eluted with a gradient of ethyl acetate in hexanes (1 L of 20%, 1L of 30%, 500 mL of 40%, 1 L 50%). The desired product is collected in fractions of 100-mL volume. TLC analysis on silica gel with 50% ethyl acetate in hexanes as eluent and visualization with ultraviolet light and stained with ceric ammonium molybdate. R$_f$ = 0.6.

23. (S)-1-Benzyl-6-methyl-2-oxo-4-phenyl-1,2,3,4-tetrahydro-pyrimidine-5-carboxylic methyl ester has the following physical properties: white solid; mp 136–137 °C. ^{1}H NMR (CDCl$_3$, 400 MHz) δ: 2.45 (s, 3 H), 3.64 (s, 3 H) 4.88 (d, J = 16.4 Hz, 1 H), 5.21 (d, J = 16.1 Hz, 1 H), 5.45 (s, 1 H), 5.99 (s, 1 H), 7.11 (d, J = 7.7 Hz, 2 H), 7.22–7.29 (m, 10 H). ^{13}C NMR (CDCl$_3$, 100 MHz) δ: 16.7, 46.2, 51.6, 54.0, 104.9, 126.5, 126.6, 127.4, 128.0, 128.90, 128.93, 138.1, 143.2, 149.5, 154.2, 166.7. IR (thin film, cm^{-1}): 3234, 2948, 1685, 1623, 1456, 1387, 1257, 1203, 1164, 1106, 696. HRMS m/z 359.1375 [(M + Na$^+$) calcd for C$_{20}$H$_{20}$N$_2$O$_3^+$: 359.1372]. [α]$^{23}_D$ – 29.8 (c 1.00, CHCl$_3$). The purity (>95%) was determined by HPLC-ELSD (210 nm).

24. The enantiomeric ratio of the product was determined to be 95:5 using the following reverse phase HPLC method: Chiralcel OD-RH, (150 x 4.6 mm), 5 micron, isocratic elution, A: 0.1% H$_3$PO$_4$, B: MeCN, A: 45: B: 55; flow of 1.0 mL/min., ambient temp., detection at 210 nm. Major enantiomer elutes at 6.5 min, minor at 8 min. A normal phase HPLC method can also be employed: Chiralcel OD, (250 x 4.6 mm), 10 micron, isocratic elution, A: 2-propanol, B: heptane, A: 5: B: 95, flow of 1.0 mL/min., ambient temp., detection at 210 nm; Major enantiomer elutes at 31 min, minor at 39 min.

25. The checkers used the following recrystallization procedure to upgrade the final product to optical purity: In a 100-mL round-bottomed flask equipped with a 1.5 cm oval magnetic stir bar is added (*S*)-1-benzyl-6-methyl-2-oxo-4-phenyl-1,2,3,4-tetrahydropyrimidine-5-carboxylic methyl ester (2.00 g, 95:5 er) and ethyl acetate (20 mL). The mixture is warmed to 50 °C in a water bath with stirring to dissolve the solids. While warm, *n*-heptane (20 mL) is added dropwise with stirring over 5 min, resulting in crystallization of a white solid. The mixture is cooled over 20 min to ambient temperature and is stirred for 1 h to afford a thick slurry. The mixture is vacuum filtered through a 30 mL sintered glass funnel and washed with 10 mL of 1:1 heptane/EtOAc to provide 1.59 g (80%) of white needles after air drying. The ee was determined to be >99.5% based on the limit of detection of the reverse phase HPLC method in Note 24.

The submitters used the following recrystallization procedure to upgrade the final product to optical purity: In a 150-mL Erlenmeyer flask, (*S*)-1-benzyl-6-methyl-2-oxo-4-phenyl-1,2,3,4-tetrahydropyrimidine-5-carboxylic methyl ester (3.5 g, 95:5 er) is added to boiling diethyl ether (60 mL). The solution is boiled until most of the solid has dissolved. The solution is cooled in an ice-water bath until complete crystallization is observed (approximately 20 minutes). The pure dihydropyrimidone crystals are vacuum filtered, and are rinsed with two 20 mL portions of cold diethyl ether. The remaining mother liquor is concentrated under reduced pressure, and the resulting solid is transferred to a 150-mL Erlenmeyer flask containing boiling diethyl ether (30 mL). The solution is boiled until most solid has dissolved, and subsequently cooled in an ice-water bath until complete crystallization is observed (approximately 20 minutes). The crystals are filtered, and rinsed with 2 x 10 mL portions of cold diethyl ether. The solids are combined and dried under reduced pressure to yield 2.70 g (77 %) of (*S*)-1-benzyl-6-methyl-2-oxo-4-phenyl-1,2,3,4-tetra-hydropyrimidine-5-carboxylic methyl ester. The enantiomeric ratio of the product was determined to be >99:1 using the normal phase HPLC method in Note 24.

Safety and Waste Disposal Information

All hazardous materials should be handled and disposed of in accordance with "Prudent Practices in the Laboratory"; National Academy Press: Washington, DC, 1995.

3. Discussion

Chiral dihydropyrimidones are an important class of heterocycles that range in biological and pharmacological behavior.[2] While racemic dihydropyrimidones are easily prepared through use of the Biginelli reaction,[3] few methods provide dihydropyrimidinones in enantioenriched form.[4] The Mannich addition of β-ketoesters to acyl imines catalyzed by the cinchona alkaloids provides a chiral amine precursor to the enantioenriched dihydropyrimidinone core.[5]

The Mannich reaction proceeds well for aryl imines, however, in the case of aliphatic imines, tautomerization to the enamine restricts reactivity for nucleophilic addition. Recent methodology overcomes this challenge through the development of a biphasic cinchona alkaloid-catalyzed Mannich reaction utilizing α-amido sulfones.[6] The bench-stable and easily prepared α-amido sulfones serve as precursors to acyl imines; the acyl imine is formed *in situ* in the presence of sodium carbonate and the cinchona alkaloid catalyst.[7] Utilization of the α-amido sulfones in the cinchona alkaloid-catalyzed asymmetric Mannich reaction provides good control of enantioselectivity and scalability.

Conversion of the chiral amine precursor to the asymmetric dihydropyrimidone proceeds through two high yielding synthetic steps. Formation of the chiral primary amine and addition of benzyl isocyanate provides the benzyl ureido intermediate in high yield. The heterocycle is formed using reflux or microwave conditions in the presence of acetic acid and ethanol. Both methods provide the dihydropyrimidone in high yield with retention of stereochemistry.

With this methodology, a library of dihydropyrimidones was synthesized with three points of diversity (Table 1). The α-amido sulfone, β-ketoester and isocyanate were each altered to provide a diverse set of heterocycles in high yields and enantioselectivities. The stereochemical configuration of the dihydropyrimidone is dictated by the choice of cinchona alkaloid catalyst used in the Mannich reaction.

Table 1. Diverse Library of Asymmetric Dihydropyrimidones

product	yield (%) er	product	yield (%) er
	71 95.5: 4.5		68 95.5 ; 4.5*
	72 96 : 4*		67 95.5 ; 4.5
	71 95.5 : 4.5		68 96 : 4*
	67 95 : 5		66 95 : 5
	70 95 : 5		63 95 : 5

*cinchonidine was used in place of cinchonine

1. Department of Chemistry, Boston University, Boston, MA 02215.
2. For reviews on dihydropyrimidones, see: (a) Kappe, C. O. *Acc. Chem. Res.* **2000**, *33*, 879-888. (b) Kappe, C. O. *Eur. J. Med. Chem.* **2000**, *35*, 1043-1058.
3. (a) Biginelli, P. *Gazz. Chim. Ital.* **1893**, *23*, 360-416. (b) Kappe, C. O. Tetrahedron **1993**, *49*, 6937-6963.
4. (a) Chen, X.; Yu, X.; Liu, H.; Chun, L.; Gong, L. *J. Am. Chem. Soc.* **2006**, *128*, 14802-14803. (b) Huang, Y.; Yang, F.; Zhu, C. *J. Am. Chem. Soc.* **2005**, *127*, 16386-16387. (c) Schnell, B.; Strauss, U.T.; Verdino, P.; Faber, K ; Kappe, C. O.; *Tetrahedron: Asymmetry* **2000**, *11*, 1449-1453. (d) Kleidernigg, O.P.; Kappe, C.O. *Tetrahedron: Asymmetry* **1997**, *8*, 2057-2067. (e) Lewandowske, K.; Murer, P.; Svec, F.; Frechet, J.M. J. *J. Comb. Chem.* **1999**, *1*, 105-112. (d) Kappe, C. O.; Uray, G.; Roschger, P.; Lindner, W.; Kratky, C.; Keller, W. *Tetrahedron* **1992**, *48*, 5473-5480. (e) Kontrec, D.; Vinkovic, V.; Sunjic, V.; Schuiki, B.; Fabian, W. M.; Kappe, C. O. *Chirality* **2003**, *15*, 550. (f) Muñoz-Muñiz, O.; Juaristi, E. *ARKIVOC* **2003**, *11*, 16-26. (g) Guillena, G.; Ramon, D.; Yus, M. *Tetrahedron: Asymmetry* **2007**, *18*, 693-700.
5. (a) Lou, S.; Taoka, B. M.; Ting, A.; Schaus, S. E. *J. Am. Chem. Soc.* **2005**, *127*, 11256-11257. (b) Ting, A.; Lou, S.; Schaus, S. E. *Org. Lett.* **2006**, *8*, 2003-2006. (c) Lou, S.; Dai, P.; Schaus, S. E. *J. Org. Chem.* **2007**, *ASAP*, DOI: 10.1021/jo701777g.
6. Review: Petrini, M. *Chem. Rev.* **2005**, *105*, 3949-3977.
7. (a) Chemla, F.; Hebbe, V.; Normant, J.-F. *Synthesis* **2000**, 75. (b) Kanazawa, S. M.; Denis, J. N.; Greene, S. E. *J. Org. Chem.* **1994**, *59*, 1238-1240. (c) Zawadzki, S.; Zwierzak, A. *Tetrahedron Lett.* **2004**, *45*, 8505-8506. (d) Morton, J.; Rahim, A.; Walker, E. R. H. *Tetrahedron Lett.* **1982**, *23*, 4123-4126.

Chemical Abstracts Nomenclature (Registry Number)

Carbamic acid, 2-propen-1-yl ester; (2114-11-6)
Carbamic acid, *N*-[phenyl(phenylsulfonyl)methyl]-, 2-propen-1-yl ester; (921767-12-6)
(β *R*)-Benzenepropanoic acid, α-acetyl-β-[[(2-propen-1-yloxy)carbonyl]amino]-, methyl ester; (921766-57-6)

(β R)-Benzenepropanoic acid, α-acetyl-β-
[[[(phenylmethyl)amino]carbonyl]amino]-, methyl ester; (865086-76-6)
(4S)-5-Pyrimidinecarboxylic acid, 1,2,3,4-tetrahydro-6-methyl-2-oxo-4-phenyl-1-(phenylmethyl)-, methyl ester; (865086-56-2)

Scott E. Schaus studied chemistry at Boston University, where he completed his undergraduate degree in 1995. He received his Ph.D. in organic chemistry from Harvard University in 1999 under the direction of Professor Eric N. Jacobsen. His graduate work focused on the development of chiral salen transition metal catalysts and reactions for use in synthesis. He carried out his postdoctoral research as an NIH Postdoctoral Fellow in Professor Andrew G. Myers's laboratories studying the use of genomic technologies to facilitate drug target identification. In 2001 he joined the Department of Chemistry at Boston University as an Assistant Professor and, in 2002, he became one of the co-principal investigators of the Center for Chemical Methodology and Library Development at Boston University. His research interests include the development of asymmetric catalytic reactions for synthesis, new methodologies for library synthesis, and drug target identification and validation.

Jennifer Goss was born in Houston, Texas, in 1983. She received a B.S. in Chemistry in at Texas A&M University in College Station, Texas, where she performed undergraduate research in polymer chemistry with Professor Steven Miller. She began pursuing a doctorate degree in Organic chemistry in 2005 in the lab of Professor Scott E. Schaus. Her research focuses on catalytic asymmetric syntheses of dihydropyrimidones.

Peng Dai obtained his B. S. degree from the University of Science and Technology of China in 1997. He completed his Ph. D. studies in December 2004 under the guidance of Prof. Patrick H. Dussault at the University of Nebraska-Lincoln, where he studied asymmetric synthesis of peroxide natural products. He joined the CMLD-BU in November 2005 and works on the synthesis of dihydropyrimidones and related chemical libraries.

Sha Lou was born in He Bei, China, in 1979. He received a B.S. in Chemistry at Beijing University of Chemical Technology in 2002, where he conducted undergraduate research in fullerene functionalizations with Professor Shen-yi, Yu. He is currently pursuing a Ph.D. degree under the direction of Professor Scott E. Schaus. His research has focused on transition metal- and organic molecule-catalyzed asymmetric carbon-carbon bond forming reactions and synthesis.

ONE-POT MULTICOMPONENT PREPARATION OF TETRAHYDROPYRAZOLOQUINOLINONES AND TETRAHYDROPYRAZOLOQUINAZOLINONES

Submitted by Toma N. Glasnov and C. Oliver Kappe.[1]

Checked by MaryAnn T. Robak and Jonathan A. Ellman.

Caution! *During microwave heating using sealed vessel technology the reaction mixtures are heated well above their boiling points generating an internal pressure of 7-15 bar. Only special microwave process vials supplied by the vendor that are designed to withhold elevated temperatures and pressures must be used. After completion of the experiment, the vessel must be allowed to cool down to a temperature well below the boiling point of the solvent (ca. 50 °C) before removal from the microwave cavity and opening to the atmosphere. A dedicated microwave reactor for organic synthesis with appropriate online temperature and pressure monitoring must be employed.*

Org. Synth. **2009**, *86*, 252-261
Published on the Web 4/14/2009

1. Procedure

A.   7,7-Dimethyl-3-phenyl-4-p-tolyl-6,7,8,9-tetrahydro-1H-pyrazolo [3,4-b]quinolin-5(4H)-one (**1**). Into a dedicated 20-mL one-necked Pyrex microwave process vial equipped with a magnetic stirring bar (Note 1), 10 mL of dry ethanol (Note 2), triethylamine (981 µL, 712 mg, 7.04 mmol, 1.6 equiv) (Note 3), 5-phenyl-1*H*-pyrazol-3-amine (700 mg, 4.40 mmol, 1.0 equiv) (Note 4) and 5,5-dimethyl-1,3-cyclohexanedione (617 mg, 4.40 mmol, 1.0 equiv) (Note 5) are added and stirred vigorously for 2 min at room temperature to form a slightly brownish solution. *p*-Tolualdehyde (519 µL, 529 mg, 4.40 mmol, 1.0 equiv) (Note 6) is then added. The reaction vial is tightly sealed with a Teflon septum inserted into an aluminum crimp (Note 7) and transferred to a Biotage single-mode microwave reactor for microwave processing at 150 °C for 30 min (Notes 8 and 9). After the reaction mixture has been processed, the vial is cooled down to 50 °C using the gas-jet cooling feature of the instrument (5 min) and subsequently removed from the microwave cavity by the robotic arm. After decrimping of the sealed process vial, the dark yellow reaction mixture is quickly transferred into a 250-mL Erlenmeyer flask filled with 200 mL of water at room temperature under vigorous stirring, resulting in a bright yellow turbid suspension. The pH of the reaction mixture is brought to ca. 2 by careful addition of 7 mL of 6M HCl within ca. 1 min, which leads to the formation of a yellow precipitate. After stirring for an additional 60 min at room temperature the precipitate is collected by suction filtration on a Büchner funnel with a coarse glass frit and then is triturated with water (3 × 20 mL) by turning off the vacuum, adding the solvent, crushing the solid and mixing thoroughly with a spatula, and then turning on the vacuum to remove the solvent. The product is subsequently dried overnight in a drying oven at 50 °C under vacuum. For purification the crude product is transferred to a Büchner funnel with a medium glass frit and triturated with dichloromethane (3 × 20 mL), applying gentle suction to remove the filtrate after each trituration. The solid is then dried for 1 h in a drying oven at 50 °C under vacuum to yield 1.19–1.22 g of pale yellow powdery solid. The solid is then transferred to a 50-mL Erlenmeyer flask equipped with a magnetic stirbar and the flask is heated to 90 °C, adding 25–26 mL of EtOH slowly until the solid is dissolved. The hot solution is then quickly filtered (hot filtration) by suction filtration through filter paper on a Büchner funnel (Note 10). The clear yellow solution is reheated in the 90 °C oil bath, and then the heat and magnetic stirring are turned off and the solution is allowed to slowly cool to room temperature and crystallize overnight. The resulting crystals are collected by suction filtration on a Büchner funnel, washed with 2 × 1 mL of EtOH, and dried overnight in a drying oven at 50 °C under vacuum to provide 0.78–0.85 g (46-50%) of 7,7-dimethyl-3-phenyl-4-*p*-tolyl-6,7,8,9-tetrahydro-1*H*-pyrazolo[3,4-*b*]quinolin-5(4*H*)-one (**1**) as yellow crystals in 99% purity (Notes 11 and 12).

    B. 6,6-Dimethyl-2-phenyl-9-p-tolyl-5,6,7,9-tetrahydropyrazolo[5,1-b]quinazolin-8(4H)-one (**2**). Into a dedicated 20-mL one-necked Pyrex microwave process vial equipped with a magnetic stirring bar (Note 1), 10 mL of acetonitrile (Note 13), 5-phenyl-1*H*-pyrazol-3-amine (1.39 g, 8.74 mmol, 2.0 equiv) (Notes 4 and 14) and 5,5-dimethyl-1,3-cyclohexanedione (1.23 g, 8.74 mmol, 2.0 equiv) (Note 5) are added and stirred vigorously for 1 min at room temperature to form a yellow suspension. Chlorotrimethylsilane (222 µL, 190 mg, 1.75 mmol, 0.40 equiv) (Note 15) is added, sometimes resulting in the formation of a white precipitate that dissolves upon stirring. After an additional stirring period of 1 min at room temperature, *p*-tolualdehyde (515 µL, 525 mg, 4.37 mmol, 1.0 equiv) (Note 6) is added, whereupon the suspension turns into a yellow solution. The checkers noted that a small amount of solid sometimes remains undissolved. The formed reaction mixture is stirred for an additional 2 min and then the reaction vial is tightly sealed with a Teflon septum inserted into an aluminum crimp (Note 7) and transferred to a Biotage single-mode microwave reactor for microwave processing at 170 °C for 30 min (Notes 16 and 17). After the reaction mixture has been processed, the vial is cooled down to 50 °C using the gas-jet cooling feature of the instrument (ca. 5 min) and is subsequently removed from the microwave cavity by the robotic arm. After decrimping of the sealed process vial, the dark orange reaction mixture is quickly transferred into a 500-mL Erlenmeyer flask containing 300 mL of a vigorously stirred ethanol-water-NaOH mixture (Note 18) to adjust the pH to 8-9. The resulting suspension is stirred for an additional 2 h and is then collected by suction filtration on a Büchner funnel with a coarse glass frit and triturated with a H_2O/EtOH mixture (2:1, 3 × 20 mL) by turning off the vacuum, adding the solvent, crushing the solid and mixing thoroughly with a spatula, and then turning on the vacuum to remove the solvent. The precipitate is dried overnight in a drying oven at 50 °C under vacuum. The crude product is then transferred to a 250-mL Erlenmeyer flask containing 100 mL of acetonitrile and stirred vigorously for 10 min to break up the

254   *Org. Synth.* **2009**, *86*, 252-261

solid into a fine suspension. The stirring mixture is then heated to boiling for 5 min, allowing the desired product to dissolve while leaving a side product suspended as a white solid. The hot mixture is then filtered through a medium glass-fritted Büchner funnel, applying gentle suction to assist in the filtration. The clear yellow filtrate is reheated to boiling, and 100 mL of boiling water is added. The mixture is allowed to cool to room temperature with stirring for 1 h, then the flask is placed in an ice bath and stirring is continued for an additional 1 h. During the recrystallization of product **2**, the solution becomes cloudy within 1-2 min after adding the boiling water, and crystallization proceeds as the solution begins to cool. The resulting white solid is collected by suction filtration on a Büchner funnel with a coarse glass frit, washed with 2 × 10 mL of a cold (0 °C) acetonitrile/H_2O mixture (1:1) and dried overnight in a vacuum drying oven at 50 °C to provide 1.22–1.28 g (73–76%) of 6,6-dimethyl-2-phenyl-9-*p*-tolyl-5,6,7,9-tetrahydropyrazolo[5,1-*b*]quinazolin-8(4*H*)-one (**2**) as a white fibrous solid (Notes 19 and 20).

2. Notes

1.   A 20-mL microwave process vial containing the appropriate magnetic stirring bar (length 17 mm) for the Optimizer EXP microwave reactor from Biotage AB (Sweden) was used (see Figure 1).

2.   Anhydrous ethanol (water < 0.2 % (K.F.)) was obtained from Acros Organics and used as received.

3. Triethylamine (99% purity) was obtained from Acros Organics and used as received.

4.   5-Phenyl-1*H*-pyrazol-3-amine (97% purity) was obtained from Maybridge and used as received.

5.   5,5-Dimethyl-1,3-cyclohexanedione (>99% purity) was obtained from Fluka and used as received.

6.   *p*-Tolualdehyde (97% purity) was obtained from Aldrich and used as received.

7.   The aluminum crimp tops/Teflon septa are commercially available from Biotage AB (Sweden). An appropriate crimper/decapper was used for sealing and opening of the process vials.

8.   The submitters employed a 300 W Biotage Optimizer Sixty EXP single-mode microwave reactor (Biotage AB, Sweden) (Figure 1). The checkers employed a 400 W Biotage Initiator EXP microwave reactor

(Biotage AB, Sweden). The instrument is programmed as follows: reaction temperature – 150 °C, reaction time – 1800 sec (30 min), hold time – on, pre-stirring – 10 sec, absorption level – high. Alternative microwave instruments may include a Discover reactor (CEM, Matthews, NC, USA).

9.   An internal reaction pressure of 10–12 bars was observed.

10.   Filters No 42 from Whatman were used.

11.   The physical properties of 7,7-dimethyl-3-phenyl-4-p-tolyl-6,7,8,9-tetrahydro-1H-pyrazolo[3,4-b]-quinolin-5(4H)-one (**1**) were as follows: yellow crystalline solid, mp 194–195 °C, R_f = 0.66 (9:1 dichloromethane/methanol, TLC on silica gel 60 F_{254} plates, UV detection); IR (film): 3631, 3139, 3056, 3016, 2955, 2904, 1585, 1569, 1539, 1504, 1486, 1419, 1376, 1257, 1214, 1141 cm^{-1}; ^{1}H NMR (600 MHz, DMSO-d_6) δ: 0.84 (s, 3 H), 1.00 (s, 3 H), 1.95 (d, J = 16.1, 1 H), 2.10–2.19 (m, 4 H), 2.36 (d, J = 16.6, 1 H), 2.43–2.53 (m, 1 H), 5.30 (s, 1 H), 6.89 (d, J = 8.0, 2 H), 7.00 (d, J = 8.0, 2 H) 7.28 (t, J = 7.4, 1 H), 7.38 (apparent t, J = 7.5 Hz, 2 H), 7.53 (d, J = 7.6, 2 H), 9.90 (s, 1 H), 12.56 (s, 1 H); ^{13}C NMR (151 MHz, DMSO-d_6) δ: 20.5, 26.5, 29.0, 31.9, 34.7, 40.8, 50.4, 103.4, 107.9, 125.9, 127.3, 127.8, 128.2, 128.7, 129.5, 134.2, 137.1, 144.6, 148.2, 152.0, 192.7; MS (ESI) m/z (relative intensity): 384 (M+H, 100). Calcd for $C_{25}H_{25}N_3O$:   C, 78.30; H, 6.57; N, 10.96.   Found: C, 78.03; H, 6.47; N, 10.97.

12.   The submitters reported purification of the crude precipitate by crystallization from hot EtOH:H$_2$O (2:1) to give 0.92–1.05 g (55–63%) of product. However, elemental analysis was not provided by the submitters, and in the checkers' hands, this procedure failed to give product of satisfactory purity based on ^{1}H NMR and elemental analysis. The checkers found that trituration of the crude product with CH$_2$Cl$_2$ could remove most of the impurities, providing material with a single side product in approximately 10:1 ratio of desired product to side product by ^{1}H NMR analysis. Including this trituration step prior to recrystallization resulted in higher isolated yields under both the EtOH:H$_2$O recrystallization conditions provided by the submitters and the EtOH recrystallization conditions investigated by the checkers. Recrystallization from EtOH:H$_2$O (2:1) provides material with a 97:3 ratio of desired product to side product, while recrystallization from EtOH provides material with a 99:1 ratio of desired product to side product by ^{1}H NMR analysis, sufficiently pure for elemental analysis.

13. Acetonitrile (HPLC-grade, Acros Organics) was used by the submitters. The checkers used HPLC-grade acetonitrile from Fisher Scientific.

14. If equimolar ratios of the starting materials are used a different product distribution is obtained, with typically more of the Hantzsch-type product **1** being formed. In addition more side products are visible by HPLC analysis and the benzaldehyde starting material is not completely consumed. Other acidic catalysts may be applied (HCl or Lewis acids such as LaCl$_3$), however the best result is obtained when using chlorotrimethylsilane as a reaction mediator.

15. Chlorotrimethylsilane (98% purity) was obtained from Aldrich and used as received.

16. The submitters employed a 300 W Biotage Optimizer Sixty EXP single-mode microwave reactor (Biotage AB, Sweden) (Figure 1). The checkers employed a 400 W Biotage Initiator EXP microwave reactor (Biotage AB, Sweden). The instrument is programmed as follows: reaction temperature – 170 °C, reaction time – 1800 sec (30 min), hold time – on, pre-stirring – 10 sec, absorption level – high. Alternative microwave instruments may include a Discover reactor (CEM, Matthews, NC, USA).

17. An internal reaction pressure of 13–14 bars was observed.

18. The solvent mixture was prepared by mixing 200 mL of distilled water, 100 mL of ethanol and 160 mg of NaOH. Sodium hydroxide pearls (99% purity) were obtained from Acros Organics (submitters) or EMD Chemicals (checkers) and used as received.

19. The physical properties of 6,6-dimethyl-2-phenyl-9-*p*-tolyl-5,6,7,9-tetrahydropyrazolo[5,1-*b*]-quinazolin-8(4*H*)-one (**2**) were as follows: white solid, mp 235–238 °C; R$_f$ = 0.74 (9:1 dichloromethane/methanol, TLC on silica gel 60 F$_{254}$ plates, UV detection); IR (film): 2960, 2921, 2873, 1642, 1582, 1569, 1430, 1383, 1361, 1344,1250 cm^{-1}; ^{1}H NMR (500 MHz, DMSO-d$_6$) δ: 0.93 (s, 3 H), 1.05 (s, 3 H), 2.04 (d, J = 16.1 Hz, 1 H), 2.16–2.26 (m, 4 H), 2.43–2.51 (m, 1 H), 2.56 (d, J = 16.7 Hz, 1 H), 6.16 (s, 1 H), 6.18 (s, 1 H), 7.04 (d, J = 8.2 Hz, 2 H), 7.07 (d, J = 8.2 Hz, 2 H), 7.26 (t, J = 7.3 Hz, 1 H), 7.34 (apparent t, 2 H), 7.71 (d, J = 7.7 Hz, 2 H), 10.50 (br s, 1 H); ^{13}C NMR (126 MHz, DMSO-d$_6$) δ: 20.6, 26.7, 28.7, 32.1, 49.8, 57.3, 85.4, 105.0, 125.0, 126.6, 127.6, 128.5, 128.6, 133.1, 136.3, 138.2, 140.2, 149.1, 150.1, 192.3; MS (ESI) *m/z* (relative intensity): 384 (M+H, 100). Calcd for C$_{25}$H$_{25}$N$_3$O: C, 78.30; H, 6.57; N, 10.96. Found: C, 78.39; H, 6.31; N, 11.00.

20. The submitters reported purification of the crude precipitate by crystallization from hot 2-propanol:H_2O (2:1). However, elemental analysis was not provided by the submitters, and in the checkers hands, this procedure failed to give product of satisfactory purity based on ^1H NMR and elemental analysis. The checkers found that the main side product which remained after crystallization from 2-propanol:H_2O exhibited low solubility in hot acetonitrile. Therefore, hot filtration of an acetonitrile solution of the crude mixture followed by crystallization from acetonitrile:H_2O (1:1) was found to provide analytically pure product (^1H NMR and elemental analysis).

Figure 1. Optimizer Sixty EXP single-mode microwave reactor with 20 mL reaction vessel (Biotage AB, Sweden).

Waste Disposal Information

All hazardous materials should be handled and disposed of in accordance with "Prudent Practices in the Laboratory"; National Academy Press; Washington, DC, 1995.

3. Discussion

Multicomponent condensation reactions (MCRs) of 5-aminopyrazoles with cyclic 1,3-diketones and aromatic aldehydes can lead to the formation of several different tricyclic reaction products due to the presence of at least three non-equivalent nucleophilic reaction centers in the aminopyrazole building block (N1, C4 and NH_2).[2,3] The resulting partially hydrogenated azoloazines belong to a class of interesting target structures owing to their specific role in several biological processes and their diverse physiological activities.[4] Therefore, these MCRs have recently attracted the interest of the synthetic community and several reports have already discussed the formation of isomeric reaction products of type **1** or **2** from these three-component condensations.[3] In general these MCRs lead to the formation of product mixtures of pyrazoloquinolinones (Hantzsch-type product **1**) or pyrazoloquinazolinones (Biginelli-type product **2**) with difficult to control selectivities.[2,3] We have recently found that by employing high-temperature microwave processing, the reaction can be effectively tuned toward the formation of either the Hantzsch or Biginelli-type condensation products using a basic (Hantzsch) or acidic (Biginelli) reaction medium.[2]

In recent years, microwave-assisted organic synthesis (MAOS) has attracted considerable attention, and has become a very popular and convenient tool for performing organic reactions at high temperatures in sealed vessels.[5,6] In particular, the use of dedicated single-mode microwave reactors that enable the rapid and safe thermal processing of reaction mixtures in sealed vessels under controlled conditions with concurrent temperature and pressure monitoring has greatly increased the general acceptance of this method in the scientific community. Microwave dielectric heating not only often enables a dramatic reduction in reaction time, but also enhances reactions in terms of yield, purity and reproducibility in comparison to conventional thermal heating.[5,6]

1. Christian Doppler Laboratory for Microwave Chemistry (CDLMC) and Institute of Chemistry, Karl-Franzens-University Graz, Heinrichstrasse 28, A-8010 Graz, Austria. oliver.kappe@uni-graz.at
2. (a) Chebanov, V. A.; Saraev, V. E.; Desenko, S. M.; Chernenko, V. N.; Shishkina, S. V.; Shishkin, O. V.; Kobzar, K. M.; Kappe, C. O. *Org. Lett.* **2007**, *9*, 1691. (b) Chebanov, V. A.; Saraev, V. E.; Desenko, S.

M.; Chernenko, V. N.; Knyazeva, I. V.; Groth, U.; Glasnov, T. N.; Kappe, C. O. *J. Org. Chem.* **2008**, *73*, 5110.

3. (a) Drizin, I.; Altenbach, R. J.; Buckner, S. A.; Whiteaker, K. L.; Scott, V. E.; Darbyshire, J. F.; Jayanti, V.; Henry, R. F., Coghlan, M. J.; Gopalakrishnan, M.; Carroll, W. A. *Bioorg. Med. Chem.* **2004**, *12*, 1895. (b) Quiroga, J.; Insuasty, B.; Hormaza, A.; Saitz, C.; Jullian, C. *J. Heterocycl. Chem.* **1998**, *35*, 575. (c) Quiroga, J; Mejia, D.; Insuasty, B.; Abonia, R.; Nogueras, M.; Sanchez, A.; Cobo, J.; Low, J. N. *Tetrahedron* **2001**, *57*, 6947. (d) Drizin, I; Holladay, M. W.; Yi, L.; Zhang, G. Q.; Gopalakrishnan, S.; Gopalakrishnan, M.; Whiteaker, K. L.; Buckner, S. A.; Sullivan, J. P.; Carroll, W. A. *Bioorg. Med. Chem. Lett.*, **2002**, *12*, 1481.

4. (a) Ma, Z.; Huang, Q.; Bobbitt, J. M. *J. Org. Chem.* **1993**, *58*, 4837. (b) Vicente, J.; Chicote, M. T.; Guerrero, R.; de Arellano, M. C. R. *Chem. Commun.* **1999**, 1541. (c) Alajarin, R.; Avarez-Builla, J.; Vaquero, J. J.; Sunkel, C.; Fau de Casa-Juana, M.; Statkow, P.; Sanz-Aparicio, J. *Tetrahedron Asymm.* **1993**, *4*, 617. (d) Bossert, F.; Vater, W. *Med. Res. Rev.*, **1989**, *9*, 291. (e) Triggle, D. J.; Langs, D. A.; Janis, R. A. *Med. Res. Rev.* **1989**, *9*, 123. (f) Atwal, K. S.; Moreland, S. *Bioorg. Med. Chem. Lett.* **1991**, *1*, 291.

5. (a) Kappe, C. O. *Angew. Chem. Int. Ed.* **2004**, 43, 6250. (b) Hayes, B. L. *Aldrichim. Acta* **2004**, *37(2)*, 66. (c) Roberts B. A.; Strauss C. R. *Acc. Chem. Res.* **2005**, *38*, 653. (d) de la Hoz, A.; Diaz-Ortiz, A.; Moreno, A. *Chem. Soc. Rev.* **2005**, *34, 164*.

6. (a) Kappe, C. O.; Dallinger, D.; Murphree, S. S. *Practical Microwave Synthesis for Organic Chemists - Stategies, Instruments, and Protocols*, Wiley-VCH: Weinheim, Germany, 2009. (b) *Microwaves in Organic Synthesis, 2nd ed.;* Loupy, A., Ed.; Wiley-VCH: Weinheim, Germany, 2006. (c) Kappe, C. O.; Stadler, A. *Microwaves in Organic and Medicinal Chemistry,* Wiley-VCH: Weinheim, Germany, 2005.

Appendix
Chemical Abstracts Nomenclature; (Registry Number)

Triethylamine: *N,N*-diethylethanamine; (121-44-8)
5-Phenyl-1H-pyrazol-3-amine: 1*H*-pyrazol-3-amine, 5-phenyl-; (1571-10-7)
Dimedone: 1,3-cyclohexanedione, 5,5-dimethyl-; (126-81-8)

p-Tolualdehyde: benzaldehyde, 4-Methyl-; (104-87-0)

7,7-Dimethyl-3-phenyl-4-*p*-tolyl-6,7,8,9-tetrahydro-1*H*-pyrazolo[3,4-*b*]-
quinolin-5(4*H*)-one: 5*H*-Pyrazolo[3,4-*b*]quinolin-5-one, 1,4,6,7,8,9-
hexahydro-7,7-dimethyl-4-(4-methylphenyl)-3-phenyl-; (904812-68-6)

Chlorotrimethylsilane: Silane, Cholorotrimethyl-; (75-77-4)

C. Oliver Kappe received his undergraduate and graduate education at the Universiy of Graz, Austria under Professor Gert Kollenz (1992). After periods of postdoctoral research work with Professor Curt Wentrup at the University of Queensland (1993-1994) and with Professor Albert Padwa at Emory University (1994-1996), he moved back to the University of Graz in 1996 to start his independent academic career. In 1999 he became Associate Professor and in 2006 was appointed Director of the Christian Doppler Laboratory for Microwave Chemistry at the University of Graz. His research interests include the development of new synthetic methods, combinatorial and high-throughput organic synthesis and heterocyclic chemistry.

Toma Glasnov was born in 1977 in Bjala Slatina, Bulgaria, and studied pharmacy at the Medical University of Sofia, Bulgaria. After obtaining his master degree in pharmacy in 2002 and work experience at Sopharma AD, he started his doctoral studies under the supervision of Professor Ivo C. Ivanov at the Pharmaceutical Faculty, Medical University of Sofia. In 2003 he moved to the University of Graz, Austria, with an Ernst-Mach grant from the Austrian Academic Exchange Service to perform research in the field of microwave-assisted organic synthesis. In 2007 he received his Ph.D. under the supervision of Professor C. Oliver Kappe.

MaryAnn Robak was born in 1982 in Binghamton, NY. She studied as an undergraduate at the University at Buffalo, State University of New York, where she completed B.S. degrees in Chemistry and Medicinal Chemistry, working on undergraduate research in the labs of Dr. Joseph A. Gardella Jr. and Dr. Michael R. Detty. Currently, she is a fourth year graduate student at University of California, Berkeley, working under the direction of Dr. Jonathan A. Ellman. Her research is on the development of sulfinyl-based hydrogen-bonding organocatalysts.

(3R,7aS)-3-(TRICHLOROMETHYL)TETRAHYDROPYRROLO[1,2-C]OXAZOL-1(3H)-ONE: AN AIR AND MOISTURE STABLE REAGENT FOR THE SYNTHESIS OF OPTICALLY ACTIVE α-BRANCHED PROLINES

Submitted by Gerald D. Artman III, Ryan J. Rafferty, and Robert M. Williams.[1]
Checked by Gregory L. Aaron, Matthew M. Davis, and Kay M. Brummond.

1. Procedure

A. *(3R,7aS)-3-(Trichloromethyl)tetrahydropyrrolo[1,2-c]oxazol-1(3H)-one.* To a suspension of L(-)-proline (11.55 g, 100.3 mmol) (Note 1) in chloroform (500 mL) (Note 2) in a 1000-mL, single-necked, round-bottomed flask equipped with a magnetic stirring bar is added 2,2,2-trichloro-1-ethoxyethanol (23.27 g, 120.3 mmol) (Note 3 and 4). A 25-mL Dean-Stark trap topped with a reflux condenser, fitted with an argon adapter, is attached to the reaction vessel and the reaction mixture is heated at reflux using a heating mantle until L(-)-proline is no longer visibly suspended and consumption is observed by reverse phase TLC (Note 5). Heating is discontinued and the volatile organics are removed under reduced pressure on a rotary evaporator (40 °C, 20–25 mmHg). The resulting brown, crystalline solid is recrystallized from ethanol. Boiling ethanol (30 mL) is added to the crude residue in the reaction flask warmed to 50 °C (bath temperature). The resultant mixture is stirred magnetically with heating on a hot plate until the mixture becomes homogenous. The solution is quickly poured into a 125-mL Erlenmeyer flask. The flask is fitted loosely with a septa and cooled slowly to room temperature then in an ice/water bath for 1 h. The resulting crystals are collected by suction filtration on a Büchner funnel and washed with 15 mL of ice-cold ethanol. The crystals are then transferred to a round-bottomed flask and dried overnight at 0.06 mmHg to provide *(3R,7aS)-3-(trichloromethyl)tetrahydropyrrolo[1,2-c]oxazol-1(3H)-one* (15.19–15.96 g, 62–65%) as colorless to light brown crystals (Note 6, 7, 8).

B.   *(3R,7aR)-7a-Allyl-3-(trichloromethyl)tetrahydropyrrolo[1,2-c]oxazol-1(3H)-one.* A flame-dried, 500-mL, single-necked, round-bottomed flask equipped with a magnetic stirring bar and an adaptor with an argon inlet, is charged with *N,N*-diisopropylamine (10.0 mL, 71.4 mmol) (Note 9) and tetrahydrofuran (THF, 140 mL) (Note 10). The reaction vessel is cooled to –78 °C before *n*-butyllithium in hexane (1.6M, 46.0 mL, 73.6 mmol) (Note 11) is added via syringe. The reaction mixture is stirred for an additional 30 min at –78 °C. In a separate 250-mL single-necked, round-bottomed flask equipped with a magnetic stirbar under argon, (3R,7aS)-3-(trichloromethyl)tetrahydropyrrolo[1,2-c]oxazol-1(3H)-one (12.2 g, 49.9 mmol) is dissolved in THF (100 mL). This solution is cooled to 0 °C and stirred for 10 min. A cannula is used to rapidly deliver this THF solution to the LDA solution at –78 °C under argon over 5 min (Note 12). The resulting solution is stirred for an additional 30 min at –78 °C before the addition of allyl bromide (7.8 mL, 90 mmol) (Note 13) via syringe in a single portion. The reaction mixture is placed in a CO_2/CH_3CN bath to warm to –40 °C, where it is maintained for an additional 30 min (Note 14). The reaction mixture is then poured into a 1-L separatory funnel containing 300 mL of water. The aqueous solution is extracted with chloroform (3 x 300 mL). The combined organic extracts are dried over Na_2SO_4 and concentrated using a rotary evaporator (40 °C, 20–25 mm Hg) to afford *(3R,7aR)-7a*-allyl-3-(trichloromethyl)tetrahydropyrrolo[1,2-c]oxazol-1(3H)-one (11.34–11.92 g, 80–82%) as a brown oil (Note 15).

C. (R)-Methyl 2-allylpyrrolidine-2-carboxylate hydrochloride. A 500-mL, three-necked, round-bottomed flask is equipped with a magnetic stirring bar, a reflux condenser fitted with an argon inlet, a 300-mL pressure-equalizing additional funnel fitted with a rubber septum, and a glass stopper. The glass stopper is removed and the flask is charged with (*3R,7aR*)-7a-allyl-3-(trichloromethyl)tetrahydropyrrolo[1,2-c]oxazol-1(3*H*)-one (8.20 g, 29.0 mmol) and methanol (100 mL) (Note 16). Sodium metal (420 mg, 18.3 mmol) (Note 17) is added slowly (~1 piece every 2 min) over 30 min by removal of the glass stopper. The reaction mixture is stirred for an additional 30 min until sodium pieces are no longer visible (Note 18). The reaction vessel is cooled in an ice/water bath and the pressure-equalizing addition funnel is charged with acetyl chloride (40 mL, 563 mmol) (Note 19), which is added dropwise into the reaction mixture over 1 h (Note 20). The funnel is removed and replaced with a glass stopper and both stoppers are secured using Keck® clips. The resulting milky brown solution is heated to reflux until only baseline material is evident by thin layer chromatography (Note 21). The volatile organics are then removed using a rotary evaporator (40 °C, 20–25 mm Hg). The resulting oily solid is diluted with methylene chloride (50 mL). The precipitated sodium chloride is removed via filtration through a Büchner funnel washing with additional methylene chloride (10 mL). The filtrate is concentrated under reduced pressure by rotary evaporation (40 °C, 20–25 mm Hg) This process is repeated two additional times to afford (*R*)-methyl 2-allylpyrrolidine-2-carboxylate hydrochloride as an oil. Purification of the crude hydrochloride salt is achieved using flash silica gel chromatography eluting with a gradient of 95:5→90:10 CH₂Cl₂:MeOH (Note 22) to afford *(R)*-methyl 2-allylpyrrolidine-2-carboxylate hydrochloride (4.16–4.44 g, 71–74%) as an oil, which solidifies under reduced pressure (Note 23). An enantiomeric excess of >99% for the desired product was determined through synthesis of the Mosher amide under Schotten-Baumann conditions followed by NMR spectroscopy and HPLC analysis (Note 24).

2. Notes

1.   L(-)-Proline (99+%) was used as received from Acros Organics.

2.   Chloroform (ACS grade) was used without further purification from Fisher Scientific.

3. The original procedure reported by Germanas employed trichloroacetaldehyde or chloral. However, this reagent is regulated and difficult to obtain. The submitters have found that commercially available 2,2,2-trichloro-1-ethoxyethanol can be used as a masked form of chloral.

4. 2,2,2-Trichloro-1-ethoxyethanol (98%) was used as commercially available and was obtained from Alfa Aesar.

5. Disappearance of L(-)-Proline (R_f = 0.89) and formation of (3R,7aS)-3-(trichloromethyl)tetrahydropyrrolo[1,2-c]oxazol-1(3H)-one (R_f = 0.26) was observed via reverse phase thin layer chromatography performed on Partisil® KC18 Silica Gel 60Å (200 µm thickness) on glass backed plates (1:1 H_2O/CH_3CN) visualizing with $KMnO_4$ TLC Stain (yellow spots). The reaction requires 15–19 h to reach completion, during which time a color change from a milky opaque to an orange solution is observed.

6. (3R,7aS)-3-(Trichloromethyl)tetrahydropyrrolo[1,2-c]oxazol-1(3H)-one[2,3] displays the following physical and spectral characteristics: mp 108–109 °C (lit.[3] 107–109 °C); optical rotation: [α]D = +34.0 (c 2, C_6H_6), lit.[3] [α]D = +33 (c 2, C_6H_6); ¹H NMR (500 MHz, $CDCl_3$) δ: 1.70–1.79 (m, 1 H), 1.90–1.97 (m, 1 H), 2.08–2.14 (m, 1 H), 2.19–2.27 (m, 1 H), 3.11–3.15 (m, 1 H), 3.42 (ddd, J = 11, 7.5, 6 Hz, 1 H), 4.12 (dd, J = 9, 4.5 Hz, 1 H), 5.17 (s, 1 H); ¹³C NMR (100 MHz, $CDCl_3$) δ: 25.3, 29.9, 57.9, 62.4, 100.6, 103.6, 175.5; IR (thin film) 2978, 2962, 2899, 2871, 1782, 1327, 1178, 1009, 959, 815, 791, 744 cm⁻¹; Anal. Calcd for $C_7H_8Cl_3NO_2$: C, 34.39; H, 3.30; N, 5.73. Found: C, 34.47; H, 3.28; N, 5.65.

7. Unlike the Seebach pivaldehyde/proline condensate, this product is air- and moisture-stable and can be stored upon the bench top with no decomposition by NMR spectroscopy after more than 30 days.

8. Following the submission of this procedure, (3R,7aS)-3-(trichloromethyl)tetrahydropyrrolo[1,2-c]oxazol-1(3H)-one is now commercially produced by AK Scientific, California, USA.

9. N,N-Diisopropylamine (99%) was purchased from Fisher Scientific and was freshly distilled from $CaCl_2$ prior to use.

10. Tetrahydrofuran (THF, 99.5%) was purchased from Sigma-Aldrich and was purified via a Sol-Tek ST-002 solvent purification system.

11. 1.6 M n-Butyllithium in hexanes was purchased from Sigma-Aldrich and freshly titrated using the method developed by Love and Jones.[4]

12. A color change is apparent as the enolate is formed. The LDA solution changes from light yellow, to dark red, to dark brown upon the addition of the oxazolinone.

13. Allyl bromide (98%) was used as received from Alfa Aesar.

14. Thin layer chromatography (TLC) on silica gel F_{254} (200 μm thickness) glass backed plates was used to monitor the alkylation. Developing the plate in 1:7 EtOAc:hexanes separates the product ($R_f = 0.44$) from the starting material ($R_f = 0.27$). Both the product and starting material can be visualized with $KMnO_4$ TLC stain (yellow spots).

15. The allylated product is of sufficient purity to be used in the next step. However, an analytical sample was obtained by purifying 145 mg of the crude material via flash silica gel chromatography (Column inner diameter 1 cm; packed length 12.5 cm) eluting 1:7 EtOAc/hexanes to afford 111 mg of alkylated product. (3R,7aR)-7a-Allyl-3-(trichloromethyl)-tetrahydropyrrolo[1,2-c]oxazol-1(3H)-one displays the following physical and spectral characteristics: pale yellow oil, which slowly crystallizes upon standing at -10 °C; optical rotation: $[\alpha]_D = +44$ (c 2, $CHCl_3$), $[\alpha]_D = +60$ (c 2, C_6H_6), lit.[3] $[\alpha]_D = +44.6$ (c 2, $CHCl_3$); ^1H NMR (500 MHz, $CDCl_3$) δ: 1.60–1.70 (m, 1 H), 1.86–1.93 (m, 1 H), 2.02 (ddd, J = 13.5, 10.0, 7.0 Hz, 1 H), 2.13 (ddd, J = 13.0, 8.0, 3.0 Hz, 1 H), 2.55 (dd, J = 14.0, 8.5 Hz, 1 H), 2.62 (dd, J = 14.0, 6.5 Hz, 1 H), 3.14–3.24 (m, 2 H), 4.98 (s, 1H), 5.17 (d, J = 6.0 Hz, 1 H), 5.20 (s, 1 H), 5.85–5.93 (m, 1 H); ^{13}C NMR (100 MHz, $CDCl_3$) δ: 25.2, 35.2, 41.5, 58.3, 71.3, 100.4, 102.3, 119.9, 131.9, 176.2; IR (thin film) 3079, 2976, 2897, 1801, 1640, 1457, 1438, 1355, 1324, 1192, 1129, 1104, 1020, 922, 838, 803, 746 cm^{-1}; Anal. Calcd for $C_{10}H_{12}Cl_3NO_2$: C, 42.21; H, 4.25; N, 4.92. Found: C, 42.51; H, 4.31; N, 4.84.

16. Methanol (MeOH, ACS grade) was used as received and obtained from Fischer Scientific. Anhydrous MeOH can be employed, but the yield was unchanged as observed by the submitters.

17. Sodium metal cubes (99.95%) in mineral oil were supplied by Sigma-Aldrich. The sodium metal was cut into small pieces using a razor blade and weighed into a tared beaker containing hexanes to remove residual mineral oil prior to addition.

18. The deprotection was monitored by TLC using silica gel F_{254} (200 μm thickness) glass backed plates, 1:5 EtOAc:hexanes, $KMnO_4$ TLC stain (yellow spots) observing consumption of the starting material ($R_f = 0.57$) and formation of the N-formyl ester intermediate ($R_f = 0.11$). It was observed that a full equivalent of sodium methoxide is not necessary for the opening of the lactone to the N-formyl ester intermediate.

19. Acetyl chloride (98%) was used as received from Sigma-Aldrich.

20. The addition of acetyl chloride must be conducted at a slow rate to avoid an exothermic reaction and loss of HCl gas. The submitters observed that if the addition is too fast, an additional quantity of acetyl chloride (~20 mL) generally has to be added to the reaction mixture once the solution is brought to reflux.

21. The reaction was monitored by TLC using silica gel F_{254} (200 μm thickness) glass backed plates, 1:1 EtOAc:hexanes, $KMnO_4$ TLC stain, (yellow spots) for the disappearance of the intermediate N-formyl ester (R_f = 0.25), and other intermediate compounds until only the hydrochloride salt (R_f = 0.00) remains.

22. Flash silica gel chromatography of the final product employed a column with specifications of: inner diameter: 2.5 inches; packed length: 6 inches. Fractions of ~27 mL were collected in 16 x 150 mm test tubes. Fractions containing the desired product (R_f = 0.47) were determined by TLC (90:10 CH_2Cl_2:MeOH) with $KMnO_4$ TLC staining. These fractions were combined and concentrated under reduced pressure (40 °C, 20–25 mm Hg).

23. (R)-Methyl 2-allylpyrrolidine-2-carboxylate hydrochloride displays the following physical and spectral characteristics: brown oil that slowly solidifies to a brown solid (99 % ee); optical rotation: $[\alpha]_D$ = -83 (c 2, CH_2Cl_2); ^1H NMR (700 MHz, $CDCl_3$) δ: 1.93 (bs, 1 H), 2.14 (bs, 2 H), 2.45 (bs, 1 H), 2.83–2.90 (m, 1 H), 3.03–3.10 (m, 1 H), 3.54 (bs, 1 H), 3.62 (bs, 1 H), 5.22 (d, J = 9.8 Hz, 1 H), 5.32 (d, J = 16.8 Hz, 1 H), 5.83-5.92 (m, 1 H), 9.55 (bs, 1 H), 10.64 (bs, 1 H); ^{13}C NMR (176 MHz, $CDCl_3$) δ: 22.4, 34.5, 39.2, 45.6, 53.7, 72.5, 121.4, 130.2, 170.0; IR (thin film) 3404, 2956, 2719, 2491, 1745, 1642, 1452, 1236 cm^{-1}; Anal. Calcd for $C_9H_{16}ClNO_2$: C, 52.56; H, 7.84; N, 6.81; Found: C: 52.24; H: 7.69; N: 6.66.

24. The ee of the final product was determined via conversion of the final product to the Mosher amide using commercially available (S)-(+)-α-methoxy-α-trifluoromethylphenylacetyl chloride (Note 25) under Schotten-Baumann conditions: In a 5-mL round-bottomed flask with a magnetic stir bar, (R)-methyl 2-allylpyrrolidine-2-carboxylate hydrochloride (26 mg, 0.15 mmol) was partitioned between CH_2Cl_2 (0.75 mL) and water (0.75 mL). NaOH (30 mg, 0.75 mmol) was added followed by commercially available (S)-(+)-α-methoxy-α-trifluoromethylphenylacetyl chloride (0.03 mL, 0.16 mmol). The reaction mixture was stirred open to the air for 1 h before being transferred to a 30-mL separatory funnel using CH_2Cl_2 (20 mL) and diluted with H_2O. The aqueous layer was separated and the resulting organic layer

was washed with saturated NaHCO$_3$ (10 mL), 2 M HCl (10 mL), and brine (10 mL). The organic phase was dried over Na$_2$SO$_4$, filtered and concentrated *in vacuo*. The resulting crude product (47 mg) was analyzed by NMR spectroscopy and HPLC. ^{1}H NMR spectroscopy of the crude material observed a single amide rotamer at room temperature. An analytical sample was obtained by purifying the crude material via flash silica gel chromatography (Inner diameter 1 cm; Packed Length 11.5 cm) eluting 2.5:97.5 to 10:90 EtOAc/hexanes to afford 25 mg of the Mosher amide. ^{1}H NMR (700 MHz, CDCl$_3$) δ: 1.59–1.71 (m, 2 H), 1.88 (ddd, *J* = 13.3, 7.0, 4.9 Hz, 1 H), 2.03 (ddd, *J* = 13.3, 9.8, 7.0 Hz, 1 H), 2.83 (dd, *J* = 14.0, 7.0 Hz, 1 H), 3.07 (ddd, *J* = 11.2, 7.0, 4.2 Hz, 1 H), 3.18 (dd, *J* = 14.0, 7.7 Hz, 1 H), 3.32 (ddd, *J* = 11.2, 8.4, 6.3 Hz, 1 H), 3.71 (s, 3 H), 3.79 (s, 3 H), 5.15 (dd, *J* = 9.8, 0.7 Hz, 1 H), 5.20 (d, *J* = 16.8 Hz, 1 H), 5.81 (ddt, *J* 17.5, 9.8, 7.0 Hz, 1 H), 7.40–7.41 (m, 3 H), 7.58–7.60 (m, 2 H); ^{13}C NMR (176 MHz, CDCl$_3$) δ: 24.0, 34.1, 37.8, 48.7, 52.2, 55.4, 69.7, 84.5 (q, *J* = 25.0 Hz), 119.7, 123.5 (q, *J* = 290.4 Hz), 127.3, 127.9, 129.3, 132.8, 132.9, 164.4, 173.5; ^{19}F NMR (376 MHz, CDCl$_3$) δ: –69.8; HPLC purity of the amide was determined by dissolving a sample in CH$_3$CN and passing it through a Phenomenex Luna 3 micron particle size C18 column (Length 100 mm; Diameter 4.6 mm) using a 60:40 solution of 0.1 % TFA in H$_2$O and 0.01% TFA in CH$_3$CN at 1 mL/min over 70 min. The desired product was observed as a single peak at 42.48 min (>99% pure) that was compared to a mixture of both Mosher amide diastereomers (Note 26). See attached chromatograph below.

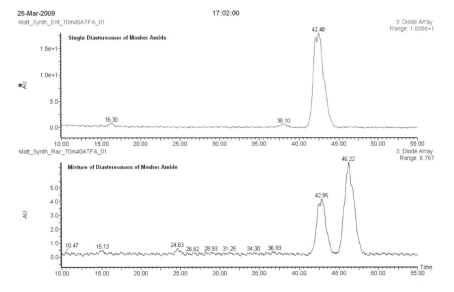

25. (*S*)-(+)-α-Methoxy-α-trifloromethylphenylacetyl chloride (>99.5% *ee*) was purchased from Aldrich and used without further purification.

26. The amine L-proline methyl ester hydrochloride was protected as the *tert*-butyloxycarbamate using di-*tert*-butyl dicarbonate and triethylamine. Subsequent alkylation and epimerization was accomplished by deprotonating with lithium diisopropylamide followed by addition of allyl bromide. The amine was liberated by treatment with trifluoroacetic acid. The amine was converted to the Mosher amide under Schotten-Baumann conditions.

Waste Disposal Information

All toxic materials were disposed of in accordance with "Prudent Practices in the Laboratory"; National Academy Press; Washington, DC, 1995.

3. Discussion

The synthesis of optically active amino acids and derivatives continues to be an important area of research for academic laboratories and the pharmaceutical industry. In 1995, Seebach reported an *Organic Syntheses* procedure for the synthesis of a proline/pivaldehyde condensate

(3), which could be employed for the synthesis of optically active α-branched proline amino acids (*cf.* **4**).[5] However, difficulties are often encountered during the preparation of **3**. The condensation of proline (**1**) and pivaldehyde (**2**) requires long reaction times in a low boiling solvent, which systematically needs to be replaced over 3-7 days. Since product **3** is extremely sensitive to air and moisture, rigorously anhydrous conditions are required to ensure that no moisture enters the system. In addition, pivaldehyde (**2**) is required in large excess (~6-7 equivalents) making this procedure economically prohibitive due to its high cost.

Scheme 1

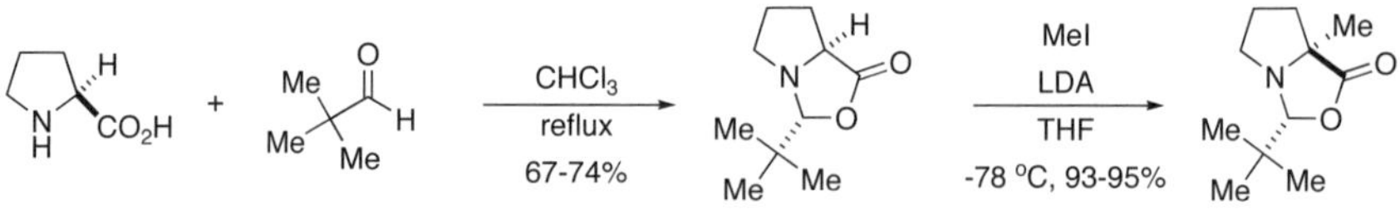

Interestingly, Germanas and Wang reported an alternative to the Seebach oxazolinone, (3*R*,7*aS*)-3-(trichloromethyl)tetrahydropyrrolo[1,2-*c*]oxazol-1(3*H*)-one (**6**), to generate optically active α-branched proline derivatives in good yields (*cf.* **7**).[2b] Unlike the Seebach compound **3**, the trichloro oxazolinone **6** is an air- and moisture-stable crystalline solid, which can be stored on the bench top for greater than 1 month with no decomposition or observed loss of optical purity. Furthermore, the preparation of **6** requires only a small excess of a chloral (**5**) or chloral hydrate.

Scheme 2

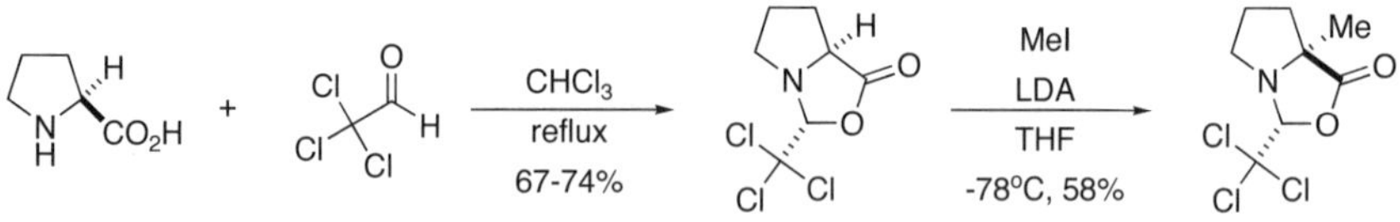

Despite the advantages of the trichloro oxazolinone **6** over the Seebach compound **3**, it has seen little use in synthesis.[3,6] Chloral (**5**) is a regulated substance greatly limiting its commercially availability even for small (10–20 g) quantities. Secondly, the cleavage of the trichloro auxiliary from **7**, though reported by Germanas to proceed in high yield is generally

270

reported by other groups to require >24h and proceeds in moderate to low yields.

As such, we have found that 2,2,2-trichloro-1-ethoxyethanol, which is commercially available, can be used as a chloral synthon resulting in a scalable procedure for the synthesis of the oxazolinone **6**. In addition, we have discovered that the initial opening of the lactone to the *N*-formyl methyl ester intermediate is slow when using refluxing HCl in methanol. By employing the one-pot procedure described, exposure of the alkylated product (*cf.* **7**) to sodium methoxide results in rapid conversion to the *N*-formyl methyl ester at room temperature. This compound is much more amenable to cleavage of the *N*-formyl group under refluxing HCl in methanol to reproducibly afford the desired *R*-allyl prolinate hydrochloride salt on a multigram scale.

1. Department of Chemistry, Colorado State University, Fort Collins, CO 80523-1872.
2. (a) Orsini, F.; Pelizzoni, F.; Forte, M.; Sisti, M.; Bombieri, G.; Benetollo, F. *J. Heterocyclic Chem.* **1989**, *26*, 837-841. (b) Wang, H.; Germanas, J. P. *Synlett* **1999**, 33-36.
3. Harris, P. W. R.; Brimble, M. A.; Muir, V. J.; Lai, M. Y. H.; Trotter, N. S.; Callis, D. J. *Tetrahedron* **2005**, *61*, 10018-10035.
4. Love, B. E.; Jones, E. G. *J. Org. Chem.* **1999**, *64*, 3755-3756.
5. (a) Seebach, D.; Boes, M.; Naef, R.; Schweizer, W. B. *J. Am. Chem. Soc.* **1983**, *105*, 5390-5398. (b) Beck, A. K.; Blank, S.; Job, K.; Seebach, D.; Sommerfield, Th. *Org. Syn.* **1995**, *72*, 62.
6. (a) Hoffman, T.; Lanig, H.; Waibel, R.; Gmeiner, P. *Angew Chem., Int. Ed.* **2001**, *40*, 3361-3364. (b) Bittermann, H.; Einsiedel, J.; Hübner, H.; Gmeiner, P. *J. Med. Chem.* **2004**, *47*, 5587-5590. (c) Bittermann, H.; Gmeiner, P. *J. Org. Chem.* **2006**, *71*, 97-102.

Appendix
Chemical Abstracts Nomenclature (Registry Number)

(*S*)-Proline; (147-85-3)

2,2,2-Trichloro-1-ethoxyethanol; (515-83-3)

(3*R*,7*aS*)-3-(Trichloromethyl)tetrahydropyrrolo[1,2-*c*]oxazol-1(3*H*)-one; (97538-67-5)

n-Butyllithium; (109-72-8)

N,N-Diisopropylamine; (108-18-9)

Allyl bromide; (106-95-6)

(3R,7aR)-7a-Allyl-3-(trichloromethyl)tetrahydropyrrolo[1,2-c]oxazol-
    1(3H)-one (220200-87-3)

Sodium; (7440-23-5)

Acetyl chloride; (75-36-5)

(R)-Methyl 2-allylpyrrolidine-2-carboxylate hydrochloride (112348-46-6)

Robert M. Williams was born in New York in 1953 and attended Syracuse University where he received the B.A. degree in Chemistry in 1975. He obtained the Ph.D. degree in 1979 at MIT (W.H. Rastetter) and was a post-doctoral fellow at Harvard (1979-1980; R.B. Woodward/Yoshito Kishi). He joined Colorado State University in 1980 and was named a University Distinguished Professor in 2002. His interdisciplinary research program (over 250 publications) at the chemistry-biology interface is focused on the total synthesis of biomedically significant natural products, biosynthesis of secondary metabolites, studies on antitumor drug-DNA interactions, HDAC inhibitors, amino acids and peptides.

Gerald Artman III was born in Michigan in 1978. He received his B.Sc. in Chemistry from Eastern Michigan University in 1999. Gerald moved to the Pennsylvania State University at University Park for his graduate studies. Under the guidance of Professor Steven Weinreb, he explored new methodology development and alkaloid total synthesis. As a NIH Postdoctoral Fellow in the lab of Professor Robert M. Williams, Gerald completed the total synthesis of the stephacidin alkaloids. Since 2007, he has been employed at the Novartis Institutes for BioMedical Research in Cambridge, MA.

Ryan J. Rafferty was born in Denver, CO in 1976. He received his B.Sc. in Chemistry (Biochemistry Emphasis) from the University of Northern Colorado in 2000, where he remained to receive this B.Sc. in Biology and M.Sc. in Biochemistry under the supervision of Prof. Richard Hyslop. His master's degree focused on the toxicology and kinetic studies and development of metabolite assays of 6-thiopurine and its analogs. He is currently pursuing his Ph.D. at Colorado State University under the supervision of Prof. Robert M. Williams. His research is focused on the total synthesis of the antifungal alkaloid ambiguine family.

Gregory Aaron was born in 1985 in Franklin, Pennsylvania. In 2008 he received his B.S. degree in chemistry from the University of Pittsburgh. While pursuing his undergraduate degree he carried out research on a number of projects under the supervision of Prof. Kay Brummond.

Matthew Davis was born in 1981 in Park Forest, Illinois. In 2004 he received his B.S. degree in chemistry from Hope College in Holland, Michigan. He is currently pursuing graduate studies at the University of Pittsburgh, under the guidance of Prof. Kay Brummond. His research currently focuses on expanding the scope of the Rh(I)-catalyzed cyclocarbonylation reaction of allene-ynes

VINYLATION WITH INEXPENSIVE SILICON-BASED REAGENTS: PREPARATION OF 3-VINYLQUINOLINE AND 4-VINYLBENZOPHENONE

Submitted by Scott E. Denmark and Christopher R. Butler.[1]
Checked by Andreas Pfaltz and David H. Woodmansee.

1. Procedure

A. Preparation of 3-Vinylquinoline. A 250-mL Schlenk flask (Note 1) equipped with a magnetic stir bar and a glass stopper fitted with a PTFE O-ring (Note 2) is flame-dried under vacuum (0.3 mmHg). The flask is then back-filled with an inert atmosphere of argon and allowed to cool to room temperature at which time the flask is charged with palladium bromide (398 mg, 1.49 mmol, 0.05 equiv) (Note 3) and 2-(di-*tert*-butylphosphino)biphenyl (901 mg, 3.02 mmol, 0.10 equiv) (Note 4) under argon purge. The flask is sealed with a septum and the contents of the flask placed under full vacuum (0.3 mmHg) followed by back-filling with argon, and this evacuation/backfill process is repeated three times. 1,3,5,7-Tetramethyl-1,3,5,7-tetravinylcyclotetrasiloxane (D_4^V) (6.36 mL, 6.34 g, 18.3 mmol, 0.61 equiv) (Note 5) is added neat *via* syringe. A solution of tetrabutylammonium fluoride trihydrate in tetrahydrofuran (60 mL, 1.0 M, 60 mmol, 2 equiv) (Notes 6 and 7) is transferred *via* cannula to the Schlenk flask. The resulting red orange solution is stirred for 15 minutes at room temperature during which time a color change from red to yellow occurs. 3-Bromoquinoline (4.03 mL, 6.24 g, 30.0 mmol, 1 equiv) (Note 8) is added neat *via* syringe, the septum is replaced with a depth adjustable glass

Org. Synth. **2009**, *86*, 274-286
Published on the Web 4/23/2009

thermometer adapter equipped with a standard laboratory alcohol thermometer and a PTFE faced silicone washer (Note 9) and fitted with a PTFE O-ring (Note 2) ensuring a good seal with the Schlenk flask. The adapted thermometer assembly is secured in place with a metal clamp and the flask is placed in a preheated 50 °C oil bath and is stirred at that temperature. During the first 10 minutes of heating the internal temperature rises past the bath temperature mildly to 62 °C for approximately 10 minutes and gradually returns to 50 °C. The starting material is consumed after 2.5 h (Note 10) and the mixture is allowed to cool to room temperature. Diethyl ether (100 mL) is added and the resulting suspension is then filtered through a short column of silica (Note 11) which is further eluted with an additional 400 mL of diethyl ether. The fractions are combined and concentrated on a rotary evaporator (35 °C, distillation is carried out from 375 mmHg to distill bulk solvent to 10 mmHg to remove final traces of higher boiling volatiles) resulting in 10.4 g of a crude mixture of 3-vinylquinoline contaminated with a siloxane byproduct (Note 12). The crude material is purified by silica gel chromatography (Note 13) eluting with pentane (1000 mL), followed by pentane/ethyl acetate, 9:1 (2000 mL), and final elution with 4:1 pentane/ ethyl acetate (2000 mL). Concentration of the eluate by rotary evaporation (500 mmHg and 40 °C to remove pentane, 10 mmHg and 40 °C to remove ethyl acetate) affords 4.31 g of 3-vinylquinoline which is further purified by Kugelrohr distillation (bp 85-100 °C at 0.6 mmHg) to provide 4.23 g (91%) of 3-vinylquinoline as a colorless liquid (Notes 14, 15, 16).

B. *Preparation of 4-Vinylbenzophenone.* A 250-mL Schlenk flask (Note 1) equipped with a magnetic stir bar and a glass stopper fitted with a PTFE O-ring (Note 2) is flame-dried under vacuum (0.3 mmHg). The flask is then back-filled with an inert atmosphere of argon and allowed to cool to room temperature at which time the flask is charged with potassium trimethylsilanolate (13.5 g, 105 mmol, 3.5 equiv) (Note 17), palladium tris(dibenzylideneacetone) (685 mg, 0.75 mmol, 0.025 equiv) (Note 18), and triphenylphosphine oxide (418 mg, 1.5 mmol, 0.05 equiv) (Note 19) under a very gentle argon purge. The flask is sealed with a septum and the contents of the flask are placed under full vacuum (0.3 mmHg) followed by back-filling with argon, and this evacuation/backfill process is repeated three times. Dry, degassed tetrahydrofuran (60 mL) (Note 7) is added *via* syringe. The stirring is turned off and the septum removed under a gentle argon purge and 4-bromobenzophenone (7.83 g, 30.0 mmol, 1.0 equiv) (Note 20) is added as a solid. The septum is replaced, stirring is restarted and 1,3-

divinyltetramethyldisiloxane (7.63 mL, 6.17 g, 33.1 mmol, 1.1 equiv) (Note 21) is added neat *via* syringe. The septum is replaced with a depth adjustable glass thermometer adapter equipped with a standard laboratory alcohol thermometer and a PTFE faced silicone washer (Note 9) and fitted with a PTFE O-ring (Note 2) ensuring a good seal with the Schlenk flask. The adapted thermometer assembly is secured in place with a metal clamp and the flask is placed in a preheated 70 °C oil bath and is stirred at that temperature. The internal temperature rises to approximately 66 °C and the resulting dark solution is stirred for 5 h (Note 22). The flask is removed from the oil bath and allowed to cool to room temperature at which time the flask is opened and 50 mL of diethyl ether is added and the mixture stirred for 5 min. The resulting suspension is filtered through a short column of silica gel which is eluted with an additional 200 mL of ether (Note 23). The fractions are combined and concentrated on a rotary evaporator (35 °C, distillation was carried out from 375 mmHg to distill bulk solvent to 10 mmHg to remove final traces of higher boiling volatiles) resulting in 8.5 g of crude material which is purified by column chromatography (Note 24) eluting with pentane (1000 mL), followed by pentane/ dichloromethane, 50:50 (2000 mL), and finishing with dichloromethane (1500 mL). The solvent is removed from clean fractions on a rotary evaporator (35 °C, distillation is carried out from 375 mmHg to distill bulk solvent to 10 mmHg to remove final traces of higher boiling volatiles), resulting in 5.15 g of 4-vinylbenzophenone. Mixed fractions of the product and aryl bromide (0.5 g) are combined and purified by chromatography under proportional conditions (Note 25), the purified fractions are combined with previously isolated material and the solvent removed on the rotavap (35 °C, distillation is carried out from 375 mmHg to distill bulk solvent to 10 mmHg to remove final traces of higher boiling volatiles) to provide a total of 5.62 g (90%) of 4-vinylbenzophenone (Note 26) as a white solid.

2. Notes

1. Schlenk flask (250 mL with 29/32 female joint) can be purchased from Aldrich, part number Z515760-1ea. The submitters used a 250-mL single-necked, round-bottomed flask equipped with a T-shaped side-arm adapter.
2. The Checkers used GLINDEMANN®-sealing rings (PTFE) purchased from AMSI-Glas AG.

3. Checkers and submitters purchased palladium(II) bromide (99%) from Strem Chemicals and used the chemical as received.

4. Checkers purchased 2-(di-*tert*-butylphosphino)biphenyl (99%) from Strem Chemicals and used the chemical as received. Submitters purchased 2-(di-*tert*-butylphosphino)biphenyl (97%) from Aldrich Chemical Company and used the chemical as received.

5. Checkers purchased 1,3,5,7-Tetramethyl-1,3,5,7-tetravinyl-cyclotetrasiloxane (D_4^V) from Aldrich Chemical Company and used the chemical as received. Submitters purchased 1,3,5,7-tetramethyl-1,3,5,7-tetravinylcyclotetrasiloxane from Gelest and used the chemical as received.

6. Checkers and submitters purchased tetrabutylammonium fluoride trihydrate (97%) from Fluka Chemical Corporation as a solid. A 1M solution of tetrabutylammonium fluoride was prepared in a glove box with absolute THF (Note 7) and used immediately to avoid contamination with moisture and air.

7. The Checkers purchased HPLC-grade THF from VWR and dried the solvent using a Pure-Solv™ system in accordance with Pangborn, A. B.; Giardello, M. A.; Grubbs, R. H.; Rosen, R. K.; Timmers, F. J. *Organometallics* **1996**, *15*, 1518-1520. The submitters purchased HPLC grade THF from Fisher which was dried by percolation through a column packed with neutral alumina and a column packed with Q5 reactant, a supported copper catalyst for scavenging oxygen, under a positive pressure of argon.

8. Checkers purchased 3-bromoquinoline ($\geq$ 97%) from Fluka Chemical Corporation and used the chemical as received. Submitters purchased 3-bromoquinoline 98% from Alfa-Aesar and used the chemical as received.

9. A thermometer adapter complete with PTFE faced silicone washer can be purchased from Aldrich Chemical Company, part number Z551805-1EA for 29/42 ground glass joint. The submitters used a Teflon-coated digital thermometer inserted through a septum in place of the thermometer adapter, available from Omega Instruments, Digicator Model 400a with K-type thermocouple leads.

10. Checkers followed the progress of the reaction by quickly removing the thermometer adapter assembly under a very gentle purge of argon and removing a drop of reaction solution with the tip of a pipette. The reaction solution was transferred to a micropipette by simple capillary action and the solution was spotted on a TLC plate along with starting material and

a cospot. The TLC plate was run in pentane/ ethyl acetate 9/1 and the R_f for 3-bromoquinoline is 0.7, R_f for 3-vinylquinoline is 0.3. When no discernable starting material remained a small aliquot of 100 µL was removed and the sample was concentrated under a stream of nitrogen, the remaining amorphous material was taken up in 1 mL of ether and passed through a plug of silica (150 mg of silica loaded into a Pasteur pipette plugged with glass wool). The filtrate was analyzed by GC/MS on an HP6890 gas chromatograph with a HP5970A detector equipped with a Machery and Nagel Optima5 5% polyphenylmethylsiloxane column, 25 m x 0.2 mm id and 35 µM film thickness, flow set to 20 psi of hydrogen carrier gas, a 20/1 split ratio. The oven was programmed for a starting temperature of 100 °C, a 2 minute holding time at that temperature, a 10 °C/minute ramp with a final temperature of 270 °C and a holding time of 10 minutes at that temperature. Starting material elutes at 10.8 minutes and product elutes at 10.6 minutes. Submitters monitored the reaction as follows: A 50-µL aliquot was removed *via* syringe and quenched into 4 drops of an aqueous *N,N*-dimethyl-2-aminoethanethiol solution (10% w/w). The aliquot was extracted with 1 mL of ethyl acetate and the organic layer was passed through a pipette plug of silica gel. GC analysis: product, t_R 4.69 min; 3-bromoquinoline, 4.91 min (HP-1, 100 ° to 250 °C, 15 psi H_2, 20 °C /min).

11. Checkers purchased silica gel from Fluka with a 0.040 – 0.063 mm particle size and 0.1% Ca stabilizer. Checkers used for a rough purification 50 g of silica gel loaded as a pentane slurry into a 60 mm diameter column. Submitters silica gel was purchased from Aldrich Chemical Company, (Merck, grade 9385, 230-400 mesh). For the purification submitters used 50 g of silica gel loaded as a diethyl ether slurry into a 65 mm diameter glass column.

12. A byproduct was found to make up a significant amount of the initial isolated weight. The byproduct distilled with the product and was not detectable on TLC plates under UV light. The byproduct stained poorly with standard stains but a very faint iodine stain was observed to streak just before the product on TLC. Upon isolation the byproduct showed the following physical properties: ^{1}H-NMR (400.1 MHz, CDCl$_3$, 298 K) δ: 0.47-0.45 (broad m, 2H), 0.93-0.89 (m, 10 H), 1.01-0.98 (m, 3 H), 1.31-1.25 (m, 10 H), 1.42-1.37 (m, 6 H), 1.78-1.71 (m, 13 H), 2.4-2.36 (m, 6 H), 2.78 (t, *J* = 7.4 Hz, 2 H), 3.6-3.49 (m, 1 H); GC/MS elution time of 6.2 min (Note 10); MS (EI) *m/z* 143 (4), 142 (46), 101 (2), 100 (30), 84 (3), 58 (10), 44 (14).

Org. Synth. **2009**, *86*, 274-286

13. Checkers dissolved 10.4 g of crude material in 100 mL of dichloromethane in a 500-mL round-bottomed flask with 35 g of silica gel and removed the solvent at the rotovap (35 °C, distillation is carried out from 375 mmHg to distill bulk solvent to 10 mmHg to remove final traces of low boiling volatiles) resulting in a free flowing powder suitable for dry loading (if the resulting powder is not free flowing after sonication 5 g of silica gel and 100 mL of DCM are added and the process repeated). Silica gel (230 g) is loaded as a pentane slurry into a 60-mm glass column and packed under nitrogen pressure (150 mmHg) until the stationary phase forms a solid column and 10 cm of pentane is left on top of the packed silica gel to cushion the packing of the crude absorbed silica gel. The product on silica gel is gently added to the column with the help of a glass funnel and packed under pressure. Samples are collected in 50 mL test tubes, the first fractions of product are found to contain a byproduct (Note 12) which is not observable by TLC but clear in the ^{1}H NMR and GC/MS. Mixed fractions are combined and chromatographed under proportional conditions and the purified fractions are combined with the rest of the pure material. All attempts to purify the product from the byproduct by Kugelrohr resulted in codistillation.

14. Boiling points (bp) correspond to uncorrected air-bath temperatures in the Buchi GKR-50 Kugelrohr. Submitters used a slightly higher pressure (1.2 mmHg) and found a slightly higher boiling point range (100-120 °C). Checkers recommend using the best vacuum available to avoid excessive heat and the resulting polymerization.

15. The product displayed the following physical properties ^{1}H-NMR (400.1 MHz, CDCl$_3$, 298 K) δ: 5.47 (d, J = 11.0 Hz, 1 H), 6.00 (d, J = 17.7 Hz, 1 H), 6.87 (dd, J = 17.7, 11.0 Hz, 1 H), 7.54 (t, J = 7.9 Hz, 1 H), 7.68 (t, J = 8.3 Hz, 1 H), 7.80 (d, J = 8.1 Hz, 1 H), 8.04–8.13 (m, 2 H), 9.02 (d, J = 2.2 Hz, 1 H). ^{13}C-NMR (100.6 MHz, CDCl$_3$, 298 K) δ: 116.4, 127.0, 127.9, 128.0, 129.1, 129.3, 130.3, 132.6, 133.7, 147.5, 149.0. IR (NaCl) 3063, 3007, 1632, 1618, 1568, 1492, 1461, 1429, 1413, 1369, 1327, 1124, 987, 974, 908, 860, 786, 752, 699, 609 cm^{-1}. MS (EI) m/z: 156.00 (m+1)/z (12.29 %), 155.00 (m)/z (100.0 %), 154.00 (6.59 %), 127.95 (12.45 %), 126.95 (20.20 %), 102.00 (0.90 %), 77.00 (0.96 %), 75.00 (1.07 %), 64.00 (1.17 %), 63.05 (1.20 %), 50.95 (1.58 %), 49.95 (1.21 %). TLC: R$_f$ = 0.30 (SiO$_2$, pentane/EtOAc, 9:1); Anal. Calcd. for C$_{11}$H$_9$N: C, 85.13; H, 5.85; N, 9.03. Found C, 84.64; H, 6.10; N, 9.26. The submitters elemental analysis found C, 84.83; H, 5.93; N, 9.07.

16. A second run made on a half scale yielded 2.04 g (88% yield) of product matching the previously synthesized material perfectly. The product is fairly prone to polymerization with light and heat, so great care should be exercised if absolutely pure sample is required. The product's freezing point is below –20 °C and it remains a liquid at most regular freezer temperatures. A 50 mg sample left in a 4 °C refrigerator over a period of two days was found to leave 10 mg of residue after distillation. The samples also turn a light brown fairly quickly, presumably from CO_2 absorption.

17. The checkers purchased potassium trimethylsilanolate, tech. ~90% from Fluka and used the chemical as received. The submitters purchased potassium trimethylsilanolate, tech. ~90% from Aldrich Chemical Company and used the chemical as received. The source of the potassium trimethylsilanolate is critical to the successful outcome of this reaction. Different product distributions were obtained depending upon the supplier. In reactions using potassium trimethylsilanolate purchased from Gelest, Inc. (two distinct lots), the reduction of the aryl bromide to the corresponding arene was observed, whereas reactions using potassium trimethylsilanolate purchased from Aldrich (three distinct lots) provided the desired vinylation product. Therefore, it is highly recommended to purchase the $KOSiMe_3$ from Aldrich Chemical Company for the vinylation of aryl bromides.

18. The checkers purchased tris(benzylideneacetone)dipalladium (0) from Strem Chemicals and used the chemical as received. The submitters purchased tris(benzylideneacetone)dipalladium (0), minimum 21.5 % Pd, from Alfa-Aesar and used the chemical as received.

19. The checkers purchased triphenylphosphine oxide (≥ 98%) from Fluka and used the chemical as received. The submitters purchased triphenylphosphine oxide (98%) from Aldrich Chemical Company and used the chemical as received.

20. The checkers purchased 4-bromobenzophenone (98%) from Aldrich Chemical Company and used the chemical as received. The submitters purchased 4-bromobenzophenone (98%) from Alfa-Aesar and used the chemical as received.

21. The checkers purchased 1,3-divinyltetramethyldisiloxane (97%) from Aldrich Chemical company and used the chemical as received. Submitters purchased 1,3-divinyltetramethyldisiloxane from Gelest and used the chemical as received

22. The checkers monitored the progress of the reaction by the disappearance of starting material on TLC (Note 10). The checkers used a

280   *Org. Synth.* **2009**, *86*, 274-286

mobile phase of dichloromethane/pentane in 50:50 ratio and noted a very tight separation, with the 4-vinylbenzophenone exhibiting $R_f = 0.32$ and the 4-bromobenzophenone exhibiting $R_f = 0.44$. After 4 hours a faint spot of the 4-bromobenzophenone remained and an aliquot was removed and treated as in note 10 under identical GC/MS parameters. A trace of 4-bromobenzophenone (t_r 15.6 min) remained with a smaller amount of benzophenone (t_r 12.6 min) and mostly product so the reaction was allowed to proceed for an additional 30 min and a second aliquot was worked up. No visible change in the ratios of 4-bromobenzophenone, benzophenone, or product was observed for the second aliquot so the reaction was stopped to prevent loss from thermal polymerization. The submitters monitored the reaction as follows: A 50-μL aliquot was removed *via* syringe and quenched into 4 drops of an aqueous *N,N*-dimethyl-2-aminoethanethiol solution (10% w/w). The aliquot was extracted with 1 mL of ethyl acetate and the organic layer was passed through a pipette plug of silica gel. GC analysis: 4-vinylbenzophenone, t_R 7.16 min; 4-bromobenzophenone, 7.41 min.

23. The checkers used 100 g of silica gel loaded into a 60-mm diameter column as a pentane slurry. For the purification the submitters used 75 g of silica gel loaded as a diethyl ether slurry into a 65 mm diameter glass column.

24. The checkers dry loaded the 4-vinylbenzophenone crude as in Note 13 except using 40 g of silica gel for absorption which was loaded as in Note 13 onto a 240 gram silica column packed and prepared in a 50-mm diameter glass column. The submitters used for the purification 200 g of silica gel loaded as a hexane slurry into a 50-mm diameter glass column.

25. The 0.5 g of mixed fractions are absorbed onto 3 g of silica gel and loaded onto a 20 gram column packed as a pentane slurry on a 15 mm diameter column.

26. The product exhibited the following physical properties: mp: 47.8-48.5 °C. ^{1}H-NMR (400.1 MHz, CDCl$_3$, 298 K) δ: 5.41 (d, $J = 10.8$ Hz, 1 H), 5.90 (d, $J = 17.6$ Hz, 1 H), 6.78 (dd, $J = 17.6$, 10.8 Hz, 1 H), 7.47–7.52 (m, 4 H), 7.59 (tt, $J = 7.5$, 1.3 Hz, 1 H), 7.78-7.81 (m, 4 H). ^{13}C-NMR (100.6 MHz, CDCl$_3$, 298 K) δ: 116.6, 126.0, 128.3, 129.9, 130.5, 132.3, 136.0, 136.6, 137.7, 141.5, 196.2. IR (NaCl) 3082, 3060, 3007, 1650, 1603, 1554, 1446, 1402, 1316, 1277, 1176, 1148, 1115.7, 1073, 1027, 1000, 991, 937, 923, 857, 797, 755, 702, 597 cm^{-1}. MS (EI) *m/z*: 209.05 (m+1)/z (1.18%), 208.05 (m)/z (6.85%), 131.95 (10.22%), 130.95 (100.00%), 105.05 (3.38%), 103.00 (2.35%), 78.00 (0.65%), 77.00 (7.95%), 76.10 (0.59%), 50.95

(3.40%), 49.95 (1.21%). TLC: R_f 0.32 (SiO$_2$, pentane/DCM, 1:1). Anal. Calcd. for C$_{15}$H$_{12}$O: C, 86.51; H, 5.81; Found C, 85.93; H, 5.96. The submitters elemental analysis found C, 86.22; H, 5.85. A second run on half scale provided 2.72 g (87%) matching in excellent agreement to the material made previously.

Waste Disposal Information

All toxic materials were disposed of in accordance with " Prudent Practices in the Laboratory", National Academy Press; Washington, DC, 1995.

3. Discussion

A number of new synthetic methods to prepare styrenes have been developed that involve the installation of the vinyl group by palladium-catalyzed cross-coupling reactions. The most common organometallic donors for this process include, among others, vinyltri-*n*-butyltin,[2] a trivinylboroxane-pyridine complex,[3] substituted vinylboronic esters,[4] potassium vinyltrifluoroborate,[5] and vinyltrimethylsilane.[6]

Recent disclosures from these laboratories have described a range of cost-efficient, non-toxic, widely available polyvinylsiloxanes that can serve as highly efficient vinyl donors in the cross-coupling reaction with aryl iodides.[7] Tetrabutylammonium fluoride (TBAF) serves as an activator for these coupling reactions. Of these, 1,3,5,7-tetramethyl-1,3,5,7-tetravinylcyclotetrasiloxane (D$_4^V$) was chosen based upon efficiency of vinyl transfer and its low cost. The cross coupling reaction of D$_4^V$ provides high yields for a variety of aryl iodides.

Extension of this reaction to include aryl bromides as substrates has been accomplished using D$_4^V$ as the vinyl donor and TBAF as the most effective promoter.[8] The use of a bulky phosphine ligand, 2-(di-*tert*-butylphosphino)biphenyl,[9] was required, likely to facilitate oxidative addition of the aryl bromide onto the palladium center (which is known to be more challenging for a carbon-bromine bond, relative to a carbon-iodine bond). The reaction requires 0.5 equiv of the tetrameric D$_4^V$, implying that two of the four vinyl groups on the donor are effectively transferred. The reactions occur under mild conditions (50 °C) and generally take less than 12 hours. The reaction has high tolerance for a range of functional groups

282

and substitution patterns, as shown in Scheme 1. The presence of heteroatoms or sterically bulky substituents does not impact the yield significantly, although reaction times are often longer for more hindered substrates.

Scheme 1

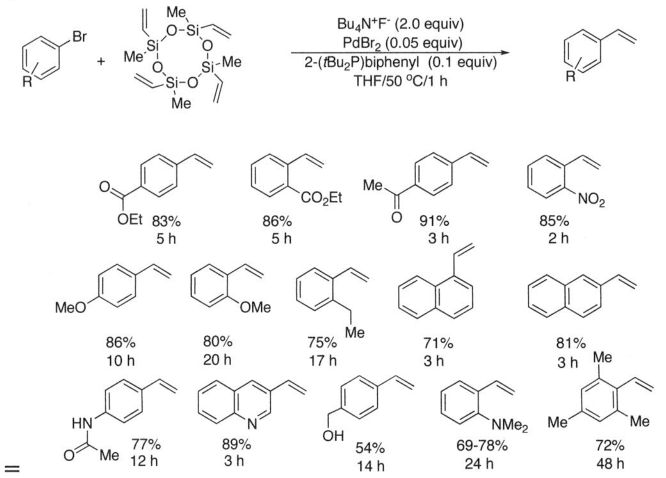

A complementary method for the vinylation of aryl halides using organosilane reagents has been developed which does not require fluoride activation.[10] In this case, two equivalents of potassium vinyldimethylsilanolate are generated in situ by a silanolate exchange between potassium trimethylsilanolate and divinyltetramethyldisiloxane (DVDS). This exchange has been shown to occur readily in both THF and DMF. The vinyldimethylsilanolate generated reacts with aryl iodides and bromides (using a palladium catalyst) to afford the corresponding styrenes in moderate to excellent yield. The reactions with aryl iodides occur under very mild conditions (room temperature, in 30 min to 12 h) and employ Pd(dba)$_2$ as catalyst and DMF as solvent (Scheme 2). These conditions often lead to exothermic reactions.

Scheme 2

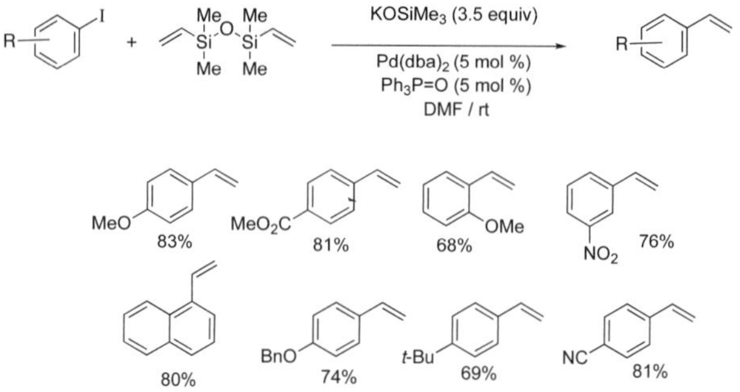

The vinylations using aryl bromides are performed using either of two protocols. The first employs potassium tri*ethyl*silanolate and DVDS, and requires a bulky phosphine ligand, 2-(di-*tert*-butylphosphino)biphenyl, and allylpalladium chloride dimer as the catalyst in DMF. These reactions occur at lower temperatures (ambient to 40 °C) and have a broad substrate scope (Scheme 3).

Scheme 3

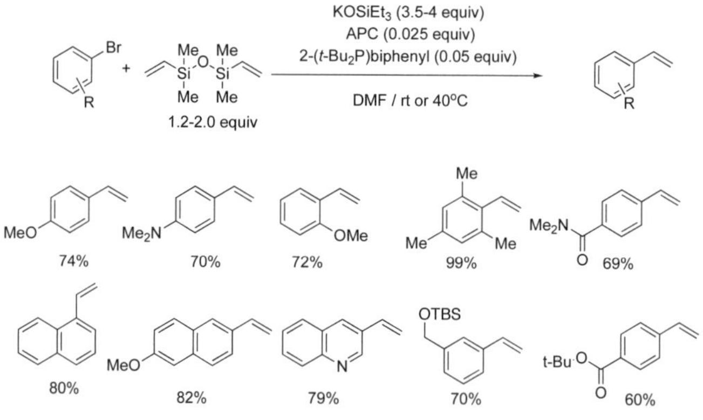

The second protocol employs the commercially available potassium tri*methyl*silanolate and DVDS as the potassium vinyldimethylsilanolate precursors. These reactions are performed using Pd(dba)$_2$ or Pd$_2$(dba)$_3$ as the palladium source, triphenylphosphine oxide as the ligand in THF and require elevated temperatures (67 °C). Successful couplings using this protocol are shown in Scheme 4.

284

Scheme 4

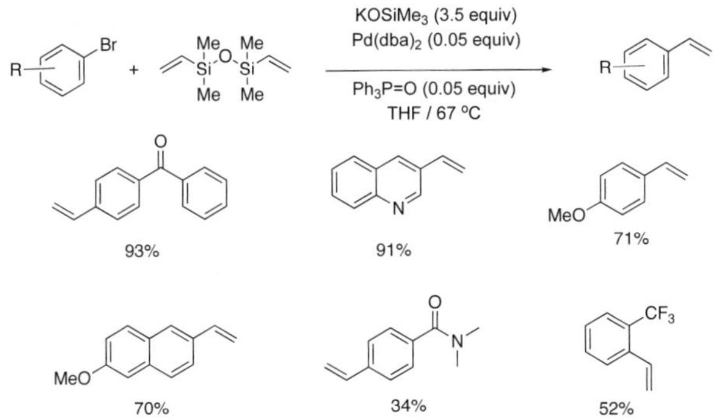

These complementary protocols (fluoride and non-fluoride activation) provide a set of practical, mild, cost-efficient, and high-yielding alternatives to current vinylation methods.

1. Department of Chemistry, Roger Adams Laboratory, University of Illinois at Urbana-Champaign. Urbana, IL, 61801, email: denmark@scs.uiuc.edu.
2. (a) Littke, A. F.; Schwarz, L.; Fu, G. C. *J. Am. Chem. Soc.* **2002**, *124*, 6343-6348. (b) McKean, D. R.; Parrinello, G.; Renaldo, A. F.; Stille, J. K. *J. Org. Chem.* **1987**, *52*, 422-424.
3. (a) Kerins, F.; O'Shea, D. F. *J. Org. Chem.* **2002**, *67*, 4968-4971. (b) Cottineau, B.; Kessler, A.; O'Shea, D. F *Org. Synth.* **2006**, *83*, 45-48.
4. Lightfoot, A. P.; Twiddle, S. J. R.; Whiting, A. *Synlett* **2005**, 529-531.
5. Molander, G. A.; Brown, A. R. *J. Org. Chem.* **2006**, *71*, 9681-9686.
6. Hatanaka, Y.; Hiyama, T. *J. Org. Chem.* **1988**, *53*, 918-920.
7. Denmark, S. E.; Wang, Z. *Synthesis* **2000**, 999-1002.
8. Denmark, S. E.; Butler, C. R. *Org. Lett.* **2006**, *8*, 63-66.
9. (a) Wolfe, J. P.; Singer, R. A.; Yang, B. H.; Buchwald, S. L. *J. Am. Chem. Soc.* **1999**, *121*, 9550-9561. (b) Barder, T. E.; Walker, S. D.; Martinelli, J. R.; Buchwald, S. L. *J. Am. Chem. Soc.* **2005** *127*, 4685-4696.
10. Denmark, S. E.; Butler, C. R. *J. Am. Chem. Soc.* **2008** *130*, 3690-3704.

4-Vinylbenzophenone: Methanone, (4-ethenylphenyl)phenyl-; (3139-85-3)

3-Vinylquinoline: Quinoline, 3-ethenyl-; (67752-31-2)

1,3,5,7-Tetramethyl-1,3,5,7-tetravinylcyclotetrasiloxane; (2554-06-5)

Palladium tris(dibenzylideneacetone); (48243-18-1)

2-(Di-*tert*-butylphosphino)biphenyl: Phosphine, [1,1'-biphenyl]-2-ylbis(1,1-dimethylethyl)-; (224311-51-7)

Scott E. Denmark was born in Lynbrook, New York in 1953. He obtained an S.B. degree from MIT in 1975 and his D.Sc.Tech. (with Albert Eschenmoser) from the ETH Zürich in 1980. That same year he began his career at the University of Illinois. He was promoted to associate professor in 1986, to full professor in 1987 and since 1991 he has been the Reynold C. Fuson Professor of Chemistry. He is currently on the Board of Editors of *Organic Syntheses* where he holds the modern record for most procedures checked. As of 2008 he became the Editor in Chief of *Organic Reactions*. He served for six years as an Associate Editor of *Organic Letters* and is co-Editor of *Topics in Stereochemistry*. He counts among his honors the ACS Award for Creative Work in Synthetic Organic Chemistry (2003), the Yamada-Koga Prize (2006), election as a Fellow of the Royal Society (2006), the Prelog Medal (2007) and a crushing victory from pole in the GT-3R class at Road America (2007).

Christopher Butler received a bachelor's degree in Chemistry from Illinois Wesleyan University in 2000. He then worked for three years in drug discovery at Johnson & Johnson Pharmaceutical Research & Development in San Diego CA, before beginning graduate school at the University of Illinois working with Professor Scott E. Denmark in 2003. His thesis work focuses on the development of palladium-catalyzed vinylation of aryl and heteroaryl bromides using inexpensive polyvinylsiloxanes.

David H. Woodmansee was born in San Diego, California in 1973. He received his bachelors in 1997 and masters in 2000 under the tutelage of Patrick J. Walsh. After working a few years in the San Diego biotech industry as a medicinal chemist including a three-year stay at the Genomics Institute for the Novartis Research Foundation David returned to graduate school and is currently working in the group of Professor Andreas Pfaltz. David's thesis topic is in the area of catalyst design and the elucidation of structure activity relationships in asymmetric catalysis.

LOW PRESSURE CARBONYLATION OF EPOXIDES TO β-LACTONES

Submitted by John W. Kramer, Daniel S. Treitler, and Geoffrey W. Coates.[1]
Checked by Scott E. Denmark and Andrew J. Hoover.

1. Procedure

Caution: Carbon monoxide is a highly toxic gas. All manipulations with carbon monoxide must be performed in a well-ventilated fume hood in the presence of a carbon monoxide detector.

In an argon-filled glove box, racemic 1,2-epoxy-3-phenoxypropane (5.00 g, 33.3 mmol) (Note 1) and dimethoxyethane (DME, 8.5 mL) (Note 2) are combined in an oven-dried, 85-mL Lab-Crest® Pressure Reaction Vessel (Note 3) equipped with a magnetic stir bar. The catalyst, [(salph)Cr(THF)$_2$][Co(CO)$_4$] (**1**, 0.150 g, 0.166 mmol, 0.497 mol %) (Note 4) is weighed into an oven-dried, 20-mL vial and dissolved in 8.0 mL of DME. This catalyst solution is drawn into an oven-dried, 10-mL gastight syringe, and the syringe needle is embedded into a septum to exclude air upon removal from the glove box.

In a well-ventilated fume hood, the Lab-Crest® Pressure Reaction Vessel is connected to a cylinder of carbon monoxide (Notes 3 and 5). The reactor is then submerged into a water bath at 20 °C and concurrently pressurized with CO to 20 psi (Note 6). After five minutes in the water bath, the catalyst solution is added to the reactor through a septum over the injection port (Note 7). The reactor is then pressurized to 100 psi CO and the dark red solution is stirred while the temperature of the water bath is held between 20–26 °C (Note 8). The CO pressure is maintained by repressurizing the reactor to 100 psi when the pressure drops to 85 psi. After

Org. Synth, **2009**, *86*, 287-297
Published on the Web 5/5/2009

6 h, the reactor is carefully vented in the fume hood and the reaction mixture is transferred to a 100-mL round-bottomed flask.

The crude reaction mixture is concentrated by rotary evaporation (20 °C, 5 mmHg) and is then dissolved in 13 mL of dichloromethane (Note 9). This solution is poured into a medium-pore frit (9 cm diameter) packed with dry silica gel (4 cm) (Note 10). The lactone product is eluted with dichloromethane (1.1 L) and the filtrate is concentrated by rotary evaporation (15 °C, 14 mmHg) and then under high vacuum (26 °C, 0.33 mmHg) to afford 5.61–5.77 g (95–97%) of 4-phenoxymethyl-2-propiolactone as a white solid (Note 11). The lactone is dissolved in boiling diethyl ether (250 mL) in a 500-mL round-bottomed flask and the solution is allowed to cool at 25 °C for 15 min and is then submerged in a 10 °C water bath. Crystals formed within 7 min, and after 1 h in the bath the flask is placed in a freezer at –20 °C for 12 h. The supernatant is carefully decanted and the crystals are washed with 10 mL of cold (–40 °C) diethyl ether, which is decanted. The residual solvent is removed under vacuum (25 °C, 0.5 mmHg) to afford 5.13 g (86%) of 4-phenoxymethyl-2-propiolactone. The mother liquor is concentrated under vacuum (25 °C, 0.2 mmHg) to afford 568 mg of an off white solid, which is recrystallized in the same manner as described above except that 21 mL of boiling ether is used. The solution is cooled at –20 °C for 2.5 h and the crystals are washed with 3 mL of –40 °C diethyl ether to afford 0.48 g (8%) of 4-phenoxymethyl-2-propiolactone. The two crops are combined and residual solvent is removed under vacuum (25 °C, 0.325 mmHg) to afford 5.53 g (93%) of analytically pure 4-phenoxymethyl-2-propiolactone as colorless needles (Note 12).

2. Notes

1. 1,2-Epoxy-3-phenoxypropane is purchased from Aldrich Chemical Co., Inc. (99%) and is dried by stirring over CaH_2 under a nitrogen atmosphere for at least one week, degassed with three freeze-pump-thaw cycles, vacuum distilled, and transferred into an argon glove box prior to use. The submitters performed the reaction in a nitrogen-filled glove box.

2. DME is vacuum-distilled from a sodium/benzophenone ketyl and transferred into an argon glove box prior to use.

3. See Figures 1 and 2 for the assembly of the pressure apparatus. The Lab-Crest® pressure reaction vessel (**A**) and shield (**B**) are purchased from Andrews Glass Co. (part # 110207 0003, includes **A**, **B**, and fittings **C-F**)

288

and fitted with a needle valve adaptor (SS 1/8" NPT, Andrews Glass Co. part #110957 0001, **G**), which is fitted with a connector (male, SS, 1/8" NPT, 1/4" Swagelok Tube fitting, **H**). This connector is coupled to a union cross (SS, 1/4" Swagelok Tube fittings, **I**). Two of the union cross joints are coupled to two Swagelok needle valves (SS, 1/4" Swagelok Tube fittings, **J-K**). Valve **K** is fitted with a septum (**L**), and valve **J** is connected to a tube from a carbon monoxide tank. The last union cross joint is coupled to an elbow (SS, the two ends are 1/4" Swagelok tube fitting and 1/4" NPT male, **M**), which is connected to a tee-joint (SS, all female 1/4" joints, **N**). The tee-joint is connected to a pressure release valve (SS, 1/4" NPT, Swagelok part #SS-RL3M4-S4, **O**) and a pressure gauge (brass, 0-200 psi, Wika part #111.11.53, **P**). 1" SS tube sections (**Q**) were used to couple all Swagelok tube fittings. The threads of **H**, **M**, **O** and **P** were wrapped with Teflon tape. Always shield glass reactors when under pressure.

4. The submitters prepared catalyst **1** according to reference 5f. The checkers purchased catalyst **1** from Aldrich Chemical Co., Inc. (catalog # 674680), opened it only in an argon glove box, and used it as received.

5. The submitters purchased research-grade carbon monoxide (99.99% min) from Matheson and used it as received. The checkers purchased CP grade carbon monoxide (99.5%) from Matheson-Trigas and used it as received.

6. Pressurization of the cylinder is done by overpressurizing the apparatus, closing valve **J** and adjusting the internal pressure with the release valve **O**, Figure 3.

7. The catalyst solution is added via syringe through the septum covered port in valve **K**, Figure 4.

8. Pressurization of the cylinder is done by overpressurizing the apparatus, closing valve **J** and adjusting the internal pressure with the release valve **O**, Figure 5.

9. The submitters used dichloromethane without purification. The checkers purchased dichloromethane (ACS reagent grade, ≥99.5%) from Aldrich Chemical Co., Inc. and used it as received.

10. The submitters used MP SiliTech silica gel (neutral, 45-55 μm particle size, pore diameter of 60 Å) purchased from MP Biomedicals. The checkers used Merck silica gel (grade 9385, 60 mesh) purchased from Aldrich.

11. In one experiment, the checkers found that the product is contaminated with 1% of a ketone side product.

12. The product exhibits the following physicochemical properties: mp 75–76 °C; ^{1}H NMR (500 MHz, CDCl$_3$) δ: 3.57 (dd, J = 16.3, 4.8 Hz, 1 H), 3.61 (dd, J = 16.3, 5.8 Hz, 1 H), 4.23 (dd, J = 11.0, 4.5 Hz, 1 H), 4.34 (dd, J = 11.0, 3.5 Hz, 1 H), 4.86 (dddd, J = 5.0 Hz, 1 H), 6.93 (d, J = 8.0 Hz, 2 H), 7.01 (t, J = 7.3 Hz, 1 H), 7.31 (t, J = 8.0 Hz, 2 H); ^{13}C NMR (126 MHz, CDCl$_3$) δ: 40.4, 67.6, 68.6, 115.0, 122.1, 130.0, 158.3, 167.5; IR (CH$_2$Cl$_2$) cm^{-1}: 3063 (w), 3044 (w), 2925 (w), 2870 (w), 1833 (s), 1600 (m), 1591 (m), 1497 (m), 1453 (w), 1409 (w), 1366 (w), 1334 (w), 1303 (m), 1245 (s), 1175 (w), 1114 (s), 1052 (w), 972 (m), 926 (w), 887 (w), 838 (m), 739 (m), 692 (m); MS (EI), m/z (relative intensity): 179 (11), 178 (100), 176 (11), 136 (31), 108 (11), 107 (70), 94 (62), 85 (11), 79 (14), 77 (63), 65 (16); exact mass calcd for C$_{10}$H$_{10}$O$_3$: 178.0630. Found: 178.0632; Anal. Calcd for C$_{10}$H$_{10}$O$_3$: C, 67.41; H, 5.66. Found: C, 67.10; H, 5.62.

Safety and Waste Disposal Information

All hazardous materials should be handled and disposed of in accordance with "Prudent Practices in the Laboratory"; National Academy Press; Washington, DC, 1995.

3. Discussion

β-Lactones find a wide variety of uses in organic synthesis as both attractive intermediates[2] and natural product targets.[3] Since the initial report of epoxide carbonylation to β-lactones over a decade ago,[4] the method has emerged as a versatile synthetic route and has been elaborated to include a variety of catalyst systems applicable under a range of conditions.[5]

However, until recently[5f] all of these systems required high pressures of CO (>100 psi) for efficient carbonylation, necessitating the use of stainless steel reactors and other high-pressure equipment. The procedure presented herein allows for the carbonylation of epoxides to β-lactones with comparable efficiency to other systems but at significantly lower CO pressures.

Our initial work focused on epoxide carbonylation at 100 psi CO which allows the use of a glass-sealed reactor, a significantly cheaper and more accessible alternative to a stainless steel reactor. This method worked well under optimized conditions on a 2-mmol scale for a variety of functionally diverse epoxides (Table 1, entries 1-11). In each case, β-lactone is formed as the exclusive product and is isolated from the catalyst in high yield using the method described in the Procedure. Further, the reaction proceeds with retention of stereochemistry (entry 3),[6] so enantiopure β-lactones can be prepared from enantiopure epoxides.

Table 1. Synthesis of functionally diverse β-lactones at 100 psi CO[a]

Entry	Epoxide	β-Lactone	Yield[b] (%)
1[c]			63[e]
2[d]			93
3[d]	(R)	(R)	92[f]
4[d]			92
5[d]			98
6[d]			94
7			94
8			84
9			82[g]
10[d]			99
11			86

[a] All reactions performed on a 2 mmol scale with 1 mol % **1**. [b] Isolated yields of β-lactones. [c] Diethyl ether is reaction solvent and reaction time is 2 h. [d] 10 mL Hexanes added to crude reaction mixture before filtering through silica. [e] The volatility of BBL caused the anomalously low isolated yield upon purification. [f] Both epoxide and β-lactone >99% ee; determined by chiral GC. [g] Contains 3% γ-lactone.[7]

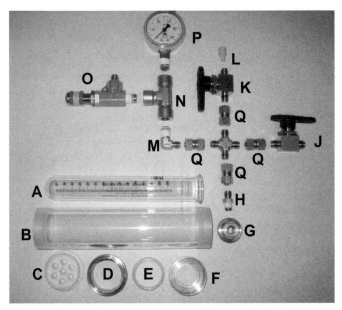

Figure 1. Components of pressure reactor.

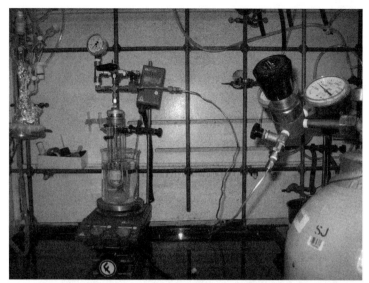

Figure 2. Assembled pressure reactor with CO cylinder.

Figure 3. Adjusting pressure to 20 psi.

Figure 4. Valve orientation for injecting the catalyst solution.

Figure 5. Pressure maintained at 100 psi throughout the procedure.

1. Department of Chemistry and Chemical Biology, Baker Laboratory, Cornell University, Ithaca, New York 14853-1301. Email: gc39@cornell.edu

2. For a review of β-lactones in organic chemistry, see: Wang, Y.; Tennyson, R. L.; Romo. D. *Heterocycles* **2004**, *64*, 605-658.

3. Pommier, A.; Pons, J.-M. *Synthesis* **1995**, 729-744.

4. Drent, E.; Kragtwijk, E. European Patent Application EP 577206; *Chem. Abstr.* **1994**, *120*, 191517c.

5. (a) Lee, J. T.; Thomas, P. J.; Alper, H. *J. Org. Chem.* **2001**, *66*, 5424-5426. (b) Getzler, Y. D. Y. L.; Mahadevan, V.; Lobkovsky, E. B.; Coates, G. W. *J. Am. Chem. Soc.* **2002**, *124*, 1174-1175. (c) Mahadevan, V.; Getzler, Y. D. Y. L.; Coates, G. W. *Angew. Chem., Int. Ed.* **2002**, *41*, 2781-2784. (d) Allmendinger, M.; Zintl, M.; Eberhardt, R.; Luinstra, G. A.; Molnar, F.; Rieger, B. *J. Organomet. Chem.* **2004**, *689*, 971-979. (e) Schmidt, J. A. R.; Mahadevan, V.; Getzler, Y. D. Y. L.; Coates, G. W. *Org. Lett.* **2004**, *6*, 373-376. (f) Kramer, J. W.; Lobkovsky, E. B.; Coates, G. W. *Org. Lett.* **2006**, *8*, 3709-3712. (g) Rowley, J. M.; Lobkovsky, E. B.; Coates, G. W. *J. Am. Chem. Soc.* **2007**, *129*, 4948-4960.

6. A thorough mechanistic study of a closely related catalyst system hasbeen completed. See: Church, T. L.; Getzler, Y. D. Y. L.; Coates, G. W. *J. Am. Chem. Soc.* **2006**, *128*, 10125-10133.

7. γ-Lactones are observed as rearranged side products during the carbonylation of glycidyl esters. For mechanistic insight, see: Schmidt, J. A. R.; Lobkovsky, E. B.; Coates, G. W. *J. Am. Chem. Soc.* **2005**, *127*, 11426-11435.

Appendix
Chemical Abstracts Nomenclature; (Registry Number)

1,2-Epoxy-3-phenoxypropane: oxirane, 2-(phenoxymethyl)-; (122-60-1)
[*N,N'*-Bis(3,5-di-*tert*-butylsalicylidene)-1,2-phenylenediamino-chromium-di-tetrahydrofuran]tetracarbonylcobaltate (**1**); (909553-60-2)
Carbon monoxide; (630-08-0)

Geoffrey W. Coates obtained a B.A. degree in chemistry from Wabash College in 1989 and a Ph.D. in organic chemistry from Stanford University in 1994. He was an NSF Postdoctoral Fellow with Robert H. Grubbs at the California Institute of Technology. In 1997 he joined the faculty of Cornell University as an Assistant Professor of Chemistry. He was promoted to Associate Professor in 2001, and to Professor in 2002. He was Associate Chair of Chemistry from 2004 to 2008 and was appointed as the first Tisch University Professor in 2008. His research focuses on the development of catalysts for organic and polymer synthesis.

John W. Kramer was born in 1980 in East Troy, Wisconsin. In 2002 he earned his B.A. in Chemistry and Environmental Studies from Grinnell College, working under the guidance of Professor T. Andrew Mobley. In 2003 he began his graduate studies at Cornell University and joined the laboratory of Geoffrey W. Coates. His work in the Coates group has focused on the ring-expanding carbonylation of heterocycles for the synthesis of biologically relevant molecules. He is currently a Research Scientist at Dow Chemical in Midland, Michigan.

Daniel S. Treitler was born in Denville, New Jersey, in 1985. He joined the Coates group in 2005 as a sophomore at Cornell University. His research focused on the synthesis of substituted β-lactones and their polymerization to form new polyesters. From 2006 to 2007, he worked for Novomer, LLC where he optimized the large-scale synthesis of catalyst **1**. Dan is currently pursuing his Ph.D. in organic chemistry at Columbia University.

Andrew J. Hoover was born in New London, Connecticut in 1987. He enrolled at the University of Illinois in 2005, and beginning in 2006 he conducted research on deoxyribozyme catalysis with Professor Scott Silverman. As an Amgen scholar during the summer of 2007 he worked with Professor Richmond Sarpong at UC Berkeley on efficient syntheses of cyclopentenones, and in the fall of 2007 he joined the research group of Scott Denmark, where he currently is investigating the use of Lewis bases as catalysts for enantioselective functionalization of alkenes. He will graduate in December 2008 with a chemistry B.S. degree, and will enroll in graduate school in 2009.

TETRAKIS(DIMETHYLAMINO)ALLENE

Submitted by Alois Fürstner, Manuel Alcarazo, and Helga Krause.[1]
Checked by Peter Wipf and Carl Deutsch.[2]

Caution! This procedure should be carried out in a well ventilated hood because of the evolution of HCl gas (step A), dimethylamine (step B) and NH₃ (step C), which are corrosive, noxious and irritant agents.

1. Procedure

A. 1,3-Dichloro-1,3-bis(dimethylamino)propenium chloride. A 250-mL, two-necked, round-bottomed, flame-dried flask equipped with a Teflon-coated stir bar (4 cm), a glass stopper, and a reflux condenser (20 cm) fitted with an argon inlet is charged with dichloromethylene-dimethyliminium chloride (15.50 g, 95.41 mmol, 2 equiv) (Note 1), dimethylacetamide (4.15 g, 47.71 mmol, 1 equiv) (Note 1) and dichloromethane (100 mL) (Note 2). The suspension is heated at reflux for 3 h in an oil bath at 47 °C. During this time, a constant evolution of HCl gas is observed while all solid materials dissolve and the solution develops an intense yellow to orange-red color. After reaching ambient temperature, the reflux condenser is replaced by a

298

short path distillation head (10 cm) and all volatile materials are distilled off under reduced pressure (10 mmHg, bath temperature ca. 40 °C) to give 1,3-dichloro-1,3-bis(dimethylamino)-propenium chloride as bright yellow, air-sensitive crystals that are dried in vacuo (10^{-3} mmHg). The crude material is used in the next step without further purification (10.68 g, 97%) (Note 3).

   B. 1,1,3,3-Tetrakis(dimethylamino)propenium tetrafluoroborate. A 250-mL, two-necked, round-bottomed, flame-dried flask equipped with a Teflon-coated stir bar (4 cm), an argon inlet, and a glass stopper is charged under argon with crude 1,3-dichloro-1,3-bis(dimethylamino)propenium chloride (10.4 g, 45 mmol, 1 equiv) (Note 3). The flask is immersed into a dry ice/acetone cooling bath at −78 °C.

   The volume of 23 mL is clearly marked on the outside of a one-necked Schlenk tube (50 mL, 15 cm length) (Note 4) equipped with a glass stopper and connected to the argon/vacuum line via its side arm. The flask is flame dried in vacuo (0.3 mmHg), back-filled with argon, and allowed to reach ambient temperature. The glass stopper is replaced by a gas inlet and the flask is immersed into a dry ice/acetone bath. Dimethylamine (ca. 23–26 mL, 350–390 mmol, 8 equiv) (Note 1) is then gently condensed from a commercial cylinder into the flask (Note 5). By applying slight argon overpressure, the condensed dimethylamine is rapidly (< 1 min) transferred via a teflon tube (2 mm diameter) into the round-bottomed flask containing the chilled 1,3-dichloro-1,3-bis(dimethylamino)propenium chloride. The resulting mixture is vigorously stirred for 1 h at −78 °C and for 1 h at −20 °C using an external thermometer in a dry ice/acetone bath. During this time, the yellow solid disappears and a white precipitate is formed. The suspension is then warmed to room temperature, allowing the excess dimethylamine to evaporate through the argon bubbler (Note 6). The remaining white solid is dissolved in dichloromethane (150 mL), and the resulting solution transferred into a separation funnel (250 mL) and washed with a saturated solution of $NaBF_4$ (4 x 40 mL) (Note 1). The aqueous phase is extracted with dichloromethane (50 mL), and the combined organic layers are dried with $MgSO_4$ (6 g). The solid drying agent is filtered off and washed with an additional 50 mL of dichloromethane, and the combined filtrates are concentrated on a rotary evaporator (10 mmHg, bath temperature ca. 40 °C) to afford 1,1,3,3-tetrakis(dimethyl-amino)propenium tetrafluoroborate as a beige solid (10.5 g, 77%) (Note 7).

   C. Tetrakis(dimethylamino)allene. A 100-mL, three-necked, round-bottomed, flame-dried flask equipped with a teflon-coated stir bar (3 cm), a glass stopper, a tubing adapter connected to an ammonia tank, and a cold finger connected to a gas outlet (Note 8) is charged under argon with 1,1,3,3-tetrakis(dimethylamino)propenium tetrafluoroborate (8.20 g, 27.3 mol, 1.0 equiv) and sodium amide (1.36 g, 35 mmol, 1.2 equiv) (Note 1). The cold finger is filled with a mixture of dry ice/acetone and the flask is immersed into a dry ice/acetone cooling bath (−78 °C) before dry ammonia (ca. 40 mL) (Note 9) is condensed into the flask. The cooling bath is removed, causing the ammonia to reflux back from the cold finger kept at −78 °C into the reaction mixture. After 1 hour, the cooling of the condenser is discontinued such that the ammonia gently evaporates while the reaction mixture reaches room temperature (Note 10). Once the excess ammonia has evaporated, the reflux condenser is quickly replaced by a small distillation head (Note 11) and the residue is distilled in vacuum (bp 41–46 °C/10^{-3} mmHg; bp 64–68 °C, 0.4 mmHg) (Note 12) to afford tetrakis(dimethylamino)allene as a colorless oil that can be stored at −20 °C under argon for months without noticeable decomposition (4.98 g, 85%) (Notes 13 and 14).

2. Notes

1.   Dichloromethylene-dimethyliminium chloride (technical grade, ≥ 95% Acros), $NaBF_4$ (98%) (Aldrich Chemical Company, Inc.), and dimethylamine (≥ 99%, Fluka) were used as received. Dimethylacetamide (≥ 99%, Aldrich Chemical Company, Inc.) was dried over BaO prior to use and stored over 4 Å molecular sieve; KFC-Titration showed 14 ppm of water. $NaNH_2$ (technical grade, Aldrich Chemical Company, Inc., ≥ 95% Acros) was used as received and transferred under Ar.

2.   The CH_2Cl_2 was distilled over CaH_2 and transferred under argon.

3.   The product salt is ca. 90% pure and contains impurities mainly derived from the starting material that is technical grade. It is prone to hydrolysis when kept in air and must be stored under argon. Under argon, the compound stays stable for more than a week (if exposed to air, the solid decomposes during ca. 4 h). Characteristic physicochemical properties: ^{1}H NMR (300 MHz, CDCl₃) δ: 3.55 (s, 12 H), 5.76 (s, 1 H); ^{13}C NMR (75 MHz, CDCl₃) δ: 44.8, 89.6, 158.8; IR (neat) 2944, 1626, 1559, 1429, 1304, 1204, 1127, 764 cm^{-1}; HRMS m/z calcd for $C_7H_{13}N_2Cl_2$ 195.0456, found 195.0449. An additional broad singlet at 12.07 ppm in the ^{1}H NMR might be due to the technical grade starting material or compounds derived from

partial hydrolysis of the product; it did not appear to have an effect on the next reaction.

4. The Checkers used Teflon tape to mark the line up to which the Schlenk tube was filled. To enhance the condensation of dimethylamine, 80% of the length of the tube was placed inside the cooling bath.

5. The Checkers found it useful to condense 26 mL of dimethylamine to facilitate stirring at the beginning of the reaction. On small scale (half-scale) this was not necessary.

6. Excess dimethylamine was destroyed by bubbling the gas stream through dilute sulfuric acid (250 mL, ca. 0.5 M,).

7. The product is $\geq$ 95% pure (NMR) and has the following physicochemical properties: ^{1}H NMR (400 MHz, CDCl$_3$) δ: 2.94 (s, 24 H), 3.56 (s, 1 H); ^{13}C NMR (100 MHz, CDCl$_3$) δ: 40.9, 72.4, 169.2; IR (neat) 2950, 1526, 1395, 1028, 764, 688 cm^{-1}; HRMS m/z calcd for C$_{11}$H$_{25}$N$_4$ 213.2079, found 213.2078.

8. The following set-up was used:

9. Ammonia (anhydrous) was used without further purification.

10. The evaporation of the ammonia took approximately 2 h after cooling was discontinued

11. A distillation head with a thermometer and a water-jacketed condenser arm (ca. 5 cm length) was used.

12. No fractions were taken during the distillation, which was continued until a very thick salt slurry had formed. If distillation was continued beyond that point and the slurry was distilled to dryness, yellow byproducts were also collected in the distillate, and the resulting thick yellow oily distillate was subjected to a second distillation without decreasing the yield. Because the residual slurry may contain traces of unreacted $NaNH_2$, it was decomposed upon suspension in hexanes (100 mL) and careful addition of isopropanol (30 mL), followed by the addition of water (50 mL).

13. The product is $\geq$ 95% pure (NMR) and has the following physicochemical properties: ^{1}H NMR (300 MHz, C_6D_6) δ: 2.62 (s, 24 H); ^{13}C NMR (75 MHz, C_6D_6) δ: 42.1, 142.8, 157.0; HRMS m/z calcd for $C_{11}H_{25}N_4$ $(M+H)^+$ 213.2079, found 213.2077. The product fumes when exposed to air and must be kept under argon. Because of this sensitivity, an accurate elemental analysis could not be obtained, and the checkers did not measure an IR.

14. The submitters obtained 3.77–4.52 g (65–78%). IR (neat) 2864, 1894, 1613, 1544, 1507, 1424, 1370, 1326, 1149, 1114, 1021, 924, 762, 681 cm^{-1}.

Waste Disposal Information

All hazardous materials were disposed in accordance with "Prudent Practices in the Laboratories"; National Academy Press; Washington, DC, 1965.

3. Discussion

Attachment of electron releasing substituents to the termini of an allene induces a significant polarization of the π-bonds towards the central carbon atom.[3] The substantial contribution of the resulting dipolar mesomeric extremes (for example, 1' and 1'') is clearly expressed in the peculiar physical and chemical properties of such compounds.

Specifically, their central C-atoms resonate at unusually high field (compare $\delta_C = 141.8$ ppm in **1** compared with $\delta_C = 211.4$ ppm in the parent allene $H_2C=C=CH_2$). Even more striking is the fact that tetra-amino(alkoxy)allenes either (strongly) deviate from linearity or can be bent to a very significant degree at almost no energetic cost.[3,4] This propensity allowed such units to be incorporated even into a five-membered (!) ring system by virtue of the fact that the aromatic resonance form **3'** holds an important share of the ground state structure.[5]

Figure 1 Comparison of Allene Ground States

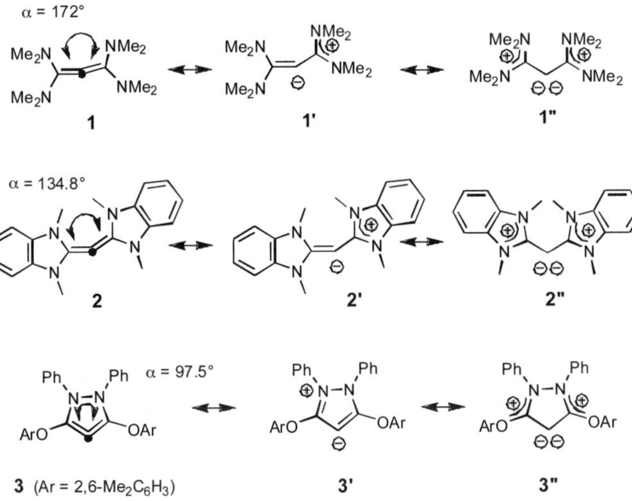

The particular bonding situation of tetraaminoallenes also dominates their chemical behavior. Specifically, compound **1** forms stable dipolar adducts with CO_2 (**4**), CS_2 or SO_2,[6] and reacts with metal templates such as $[(Ph_3P)Au]^+$ to give the structurally unusual η^1-adduct **5** rather than the η^2-bound π-complex to be expected for a normal allene ligand.[7] Furthermore, it has been shown that related tetraaminoallenes can be protonated twice at the central C-atom.[8] These experimental findings are in good accord with computational studies which suggest that lateral donor substituents impart considerable "carbodicarbene" character onto an allene motif. The resonance extreme of a "carbodicarbene" can be described as consisting of a formally zerovalent carbon atom – featuring two orthogonal electron pairs – which is stabilized by coordination to two flanking diaminocarbene units.[3,9]

Scheme 1

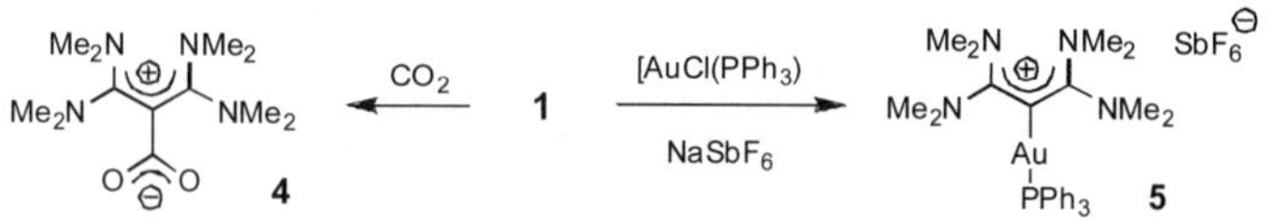

Figure 2 Stabilization of "Carbodicarbene"

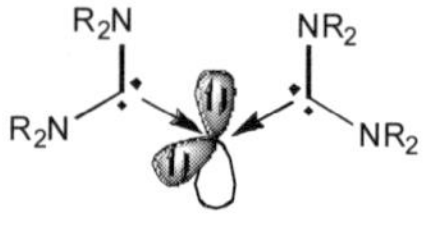

The preparation of tetrakis(dimethylamino)allene as the parent compound of this series described herein is a safe and convenient modification of the original method described by Viehe and coworkers.[6,10] The essential differences are the use of (i) tetrafluoroborate salts instead of perchlorates,[11] and (ii) $NaNH_2$ in liquid ammonia[12] rather than n-BuLi in THF as the base for the final deprotonation reaction. The BF_4^- counterion ensures that the tetraamino-substituted allyl cation produced in step B is sparingly water soluble and therefore can be easily separated from the dimethyl ammonium chloride byproduct by an aqueous work up. Moreover, the purification of the final product is more convenient when $NaNH_2$ in liquid ammonia is used,[12] as this combination leaves no organic solvent behind and hence avoids a cumbersome fractionation. Since the generated $NaBF_4$ salt also does not need to be separated, no further manipulation of the air sensitive tetraaminoallene **1** is necessary; simple distillation of the crude mixture provides the desired compound in analytically pure form in good yield. In contrast, a distillation in the presence of perchlorate salts poses a serious hazard,[11] in particular when performed on a multigram scale. Because the procedure described herein also applies to the preparation of a host of other donor substituted allenes and related cumulated ylides, it is expected to spur the systematic exploration of the fascinating chemical and physical properties of such electron rich compounds, an area of research which was largely dormant until very recently.

304

1. Max-Planck-Institut für Kohlenforschung, Kaiser-Wilhelm-Platz 1, D-45470 Mülheim, Germany.
2. Department of Chemistry, University of Pittsburgh, Pittsburgh, PA 15260, USA; pwipf@pitt.edu.
3. (a) Tonner, R.; Frenking, G. *Chem. Eur. J.* **2008**, *14*, 3260. (b) Kantlehner, W. in *Science of Synthesis* (de Meijere, A., Ed.), Thieme, Stuttgart, 2005, Vol. 24, 571. (c) Saalfrank, R. W.; Maid, H. *Chem. Commun.* **2005**, 5953. (d) Zimmer, R.; Reissig, H.-U. in *Modern Allene Chemistry* (Krause, N.; Hashmi, A. S. K., Eds.), Wiley-VCH, Weinheim, 2004, 425.
4. Dyker, C. A.; Lavallo, V.; Donnadieu, B.; Bertrand, G. *Angew. Chem. Int. Ed.* **2008**, *47*, 3206.
5. Lavallo, V.; Dyker, C. A.; Donnadieu, B.; Bertrand, G. *Angew. Chem. Int. Ed.* **2008**, *47*, 5411.
6. Viehe, H. G.; Janousek, Z.; Gompper, R.; Lach, D. *Angew. Chem. Int. Ed. Engl.* **1973**, *12*, 566.
7. Fürstner, A.; Alcarazo, M.; Goddard, R.; Lehmann, C. W. *Angew. Chem. Int. Ed.* **2008**, *47*, 3210.
8. Taylor, M. G.; Surman, P. W. J.; Clark, G. R. *J. Chem. Soc. Chem. Commun.* **1994**, 2517.
9. Tonner, R.; Frenking, G. *Angew. Chem. Int. Ed.* **2007**, *46*, 8695.
10. Janousek, Z.; Viehe, H. G. *Angew. Chem. Int. Ed. Engl.* **1971**, *10*, 574.
11. Ragan, J. A.; McDermott, R. E.; Jones, B. P.; am Ende, D. J.; Clifford, P. J.; McHardy, S. J.; Heck, S. D.; Liras, S.; Segelstein, B. E. *Synlett* **2000**, 1172.
12. Gompper, R.; Schneider, C. S. *Synthesis* **1979**, 213.

Appendix
Chemical Abstracts Nomenclature; (Registry Number)

Dichloromethylen-dimethyliminium chloride; (33842-02-3)

Dimethyl acetamide; (127-19-5)

1,3-Dichloro-1,3-bis(dimethylamino)propenium chloride; (34057-61-9)

Dimethylamine; (124-40-3)

1,1,3,3-Tetrakis(dimethylamino)propenium tetrafluoroborate; (125254-01-5)

Sodium amide; (7782-92-5)

Tetrakis(dimethylamino)allene; (42928-64-3)

Alois Fürstner was born in Bruck/Mur, Austria, in 1962. He obtained his training and PhD degree from the Technical University of Graz, Austria, working on carbohydrates under the supervision of Hans Weidmann. After postdoctoral studies with the late Prof. W. Oppolzer in Geneva, he finished his Habilitation in Graz before joining the Max-Planck-Institut für Kohlenforschung, Mülheim/Ruhr, Germany, in 1993, where he is presently the Managing Director. His research interests focus on homogeneous catalysis, metathesis, and organometallic chemistry as applied to natural product synthesis.

Manuel Alcarazo was born in 1978 in Alcalá de Guadaira, Spain. His training as a chemist started at the University of Sevilla where he was graduated in 2000. He obtained his PhD in 2005 at the Instituto de Investigaciones Químicas (CSIC-USe) under the supervision of Prof. Rosario Fernández and Dr. José M. Lassaletta. There he worked on the synthesis and applications of N-heterocyclic carbenes derived from hydrazines. After that he joined Prof. A. Fürstner's research group as a postdoctoral associate at the Max-Plank-Institut für Kohlenforschung where he was involved in various projects ranging from ligand design to natural product total synthesis. He is currently an independent junior scientist at the Max-Planck-Institut für Kohlenforschung working on the design and synthesis of ligands based on low valent p-block elements.

Helga Krause was born in 1952 in Oberhausen, Germany. She started her education in 1969 as a laboratory assistant at the Max-Planck-Institute for Coal Research. After completing these studies in 1972 she worked in the group of Professor Roland Köster in boron-organic chemistry. In 1994 she joined the group of Professor Alois Fürstner working on various projects in the field of organic and organometallic chemistry.

Carl Deutsch was born in Dinslaken (Germany) and studied chemistry in Dortmund, where he obtained his Diploma in 2004 and his PhD in 2008 under the supervision of Prof. N. Krause on the CuH-catalyzed synthesis of allenes. Parts of his research, for which he received a DSM Science & Technology Award in 2008, were carried out with Prof. B. H. Lipshutz (Santa Barbara, CA, U.S.A.) and, as JSPS fellow with Prof. M. Murakami (Kyoto, Japan). He is currently a postdoctoral researcher in the group Prof. P. Wipf, Pittsburgh, U.S.A., sponsored by the DFG, and works in the area of bioactive natural product synthesis.

EFFICIENT ONE-POT SYNTHESIS OF BIS(4-*TERT*-BUTYLPHENYL)IODONIUM TRIFLATE

Submitted by Marcin Bielawski and Berit Olofsson.[1]
Checked by Katja Krämer and Mark Lautens.

1. Procedure

A single-necked, 250-mL round-bottomed flask equipped with a magnetic stirring bar is charged with iodine (2.30 g, 9.05 mmol) (Note 1). Dichloromethane (100 mL) (Note 2) is added, and the mixture is stirred at room temperature until the iodine is dissolved (approximately 10 min) with a plastic stopper loosely attached to the flask. *m*-Chloroperbenzoic acid (*m*-CPBA, 73%, 6.62 g, 28.0 mmol, 3.1 equiv) (Note 3) is added and the purple solution is stirred an additional 10 min. *t*-Butylbenzene (5.70 mL, 36.8 mmol, 4.1 equiv) (Note 4) is added and the flask is then cooled to 0 °C in an ice bath. Trifluoromethanesulfonic acid (TfOH, 4.00 mL, 45.2 mmol, 5 equiv) (Note 5) is then slowly added via a gas-tight Hamilton syringe over 5 min at 0 °C, resulting in a color change to darker purple/black (Note 6). The mixture is then stirred at room temperature for 20 min with a color change to grey and formation of a precipitate (Note 7).

The reaction mixture is subsequently transferred to a 250-mL separatory funnel containing distilled water (30 mL). The reaction flask is rinsed with dichloromethane (2 x 5 mL), and the rinses are added to the separatory funnel. After thorough mixing, the aqueous layer is separated and the organic phase is washed with additional distilled water (30 mL) (Note 8). The combined organic extracts are poured into a 250 mL round bottomed flask and evaporated under reduced pressure (45 °C, ~200 mmHg) using a rotary evaporator to leave 16.7 g of an orange residue (Note 9).

A stirring bar is added to the flask and diethyl ether (30 mL) is added, causing precipitation of the product as a white solid. After 20 min stirring at 0 °C, the solid is collected by suction filtration using a sintered glass filter

Org. Synth. **2009**, *86*, 308-314
Published on the Web 6/16/2009

funnel and washed with cold (0 °C) diethyl ether (2 x 30 mL) (Note 10).

The solid is transferred to a pre-weighed 50-mL round-bottomed flask and dried on a vacuum line (22 °C, <1 mmHg) for 14 h to afford bis(4-*tert*-butylphenyl)iodonium triflate (**1**) as a white solid (7.68 g, 78%) (Notes 11, 12, 13).

2. Notes

1. Iodine (≥ 99%) was purchased from Sigma-Aldrich and used as received.

2. The checkers used dichloromethane (A.C.S. reagent) purchased from ACP, which was used as received. The submitters purchased dichloromethane (puriss) from VWR and used as received. The reaction is run without precaution to avoid moisture or air, *i.e.* without inert gas or dried solvent. The submitters found that dry conditions did not improve the yield.

3. *m*-CPBA (≤ 77%) was purchased from Sigma-Aldrich. It is dried in a round-bottomed flask at room temperature under reduced pressure until the flask is no longer cold. The time required depends on the amount of *m*-CPBA and the pressure; 5 g takes about 1.5 h at 10 mmHg. The percentage of active oxidizing agent is determined by iodometric titration and amounts to 73% in the checker's case. The submitters found the percentage of oxidizing agent to range from 76 to 82 % in different batches. The dried *m*-CPBA can be stored for prolonged time in the refrigerator.[2]

4. *t*-Butylbenzene (99%) was purchased from Sigma-Aldrich and was used as received.

5. Trifluoromethanesulfonic acid (TfOH, ≥98%) was purchased from Fluka. TfOH is highly corrosive, and a glass syringe must be used in the addition. Plastic syringes are rapidly destroyed by TfOH, leading to safety hazards if used for this procedure. The use of 4.1 equiv TfOH results in slightly decreased yield, whereas <4 equiv TfOH gives long reaction times and poor yield.

6. Complete addition of TfOH in a single step results in an exothermic reaction causing the solvent to boil, thus mandating the described slow addition.

7. The precipitate is *m*-chlorobenzoic acid (*m*-CBA); the product is soluble in dichloromethane.

8. The workup removes TfOH and facilitates easier precipitation of the product. *m*-CBA dissolves during the workup, if a large volume of

CH_2Cl_2 is used. The grey slurry may change color to a transparent orange. The use of drying agents, such as Na_2SO_4, should be avoided as partial anion exchange takes place, causing a melting point increase to 163–165 °C.

9. The submitters evaporated under reduced pressure (30 °C, ~150 mmHg) to obtain about 15.5 g of an orange residue.

10. The product is somewhat soluble in diethyl ether, therefore a minimum amount should be used for washing.

11. Running the reaction on half the reported scale afforded a comparable yield of bis(4-*tert*-butylphenyl)iodonium triflate (**1**) as a white solid (3.91 g, 79%).

12. A second crop of precipitate can be obtained by concentration of the filtrate under reduced pressure and addition of chloroform. This precipitates *m*-CBA, which is filtered off. After concentration of the filtrate, diethyl ether is added to precipitate a second crop of product **1** in the same manner as described above. This procedure usually contributes little to the overall yield, but can be employed if the first precipitation is low yielding.

13. Product **1** is stable to air and can be stored at room temperature. Analytical data: mp: 161–162 °C; A lower melting point, attributed to a faulty melting point apparatus, has been reported previously.[3] ^{1}H NMR (400 MHz, $CDCl_3$) δ: 1.29 (s, 18 H), 7.45 (app dt, J = 8.8 Hz, 4 H), 7.92 (app dt, J = 8.8 Hz, 4 H); ^{13}C NMR (100 MHz, $CDCl_3$) δ: 31.0, 35.2, 109.5, 120.3 (q, J = 318 Hz, $CF_3SO_3^-$), 129.5, 134.9, 156.4; IR (film): 2962, 2906, 2871, 1582, 1482, 1464, 1396, 1365, 1277, 1237, 1222, 1196, 1160, 1125, 1107, 1060, 1026, 993, 836, 820, 758, 714 cm^{-1}; HRMS (ESI): calcd for $C_{20}H_{26}I$ ([M – TfO]$^+$): 393.1074; found 393.1073: Anal. Calcd. for $C_{21}H_{26}F_3IO_3S$: C, 46.50; H, 4.83. Found: C, 46.52; H, 4.89.

Safety & Waste Disposal Information

All hazardous materials should be handled and disposed of in accordance with "Prudent Practices in the Laboratory"; National Academy Press; Washington, DC, 1995.

3. Discussion

Diaryl-λ^3-iodanes, also called diaryliodonium salts, constitute the most well known compounds among the iodine(III) reagents with two carbon ligands.[4] Due to their highly electron-deficient nature and hyperleaving group ability, they serve as versatile arylating agents in α-arylation of carbonyl compounds[5] and copper- or palladium-catalyzed cross-coupling reactions.[6] Diaryliodonium salts are also used as photo initiators in polymerizations[7] and to generate benzynes.[8]

Synthetic routes to diaryliodonium salts typically involve 2-3 steps.[4,9] Preformed inorganic iodine(III) reagents can be employed to make symmetric salts.[10] Reported one-pot reactions from arenes and iodine or iodoarenes suffer from narrow substrate scope, need excess reagents and long reaction times.[11]

The one-pot synthesis of diaryliodonium triflates from arenes and iodine reported herein is fast, high yielding and operationally simple.[3] Alternatively, aryl iodides and arenes can be employed as starting materials, which gives access also to unsymmetric salts.[3] We have also developed a route to electron rich diaryliodonium salts[12] and a general, regiospecific route employing arylboronic acids.[13]

The scope of the reaction employing iodine and arenes is shown in Figure 1. The reaction time and temperature is varied depending on the reactivity of the arene.[3] Limitations include the use of very electron-poor arenes, which have low reactivity and very electron-rich arenes, which give byproduct formation. Substituted arenes can give a mixture of *ortho-* and *para*-iodination, *e.g.* toluene, which gives a 3:1 mixture of regioisomeric salts.

Figure 1. Scope of the one-pot synthesis from iodine and arenes.

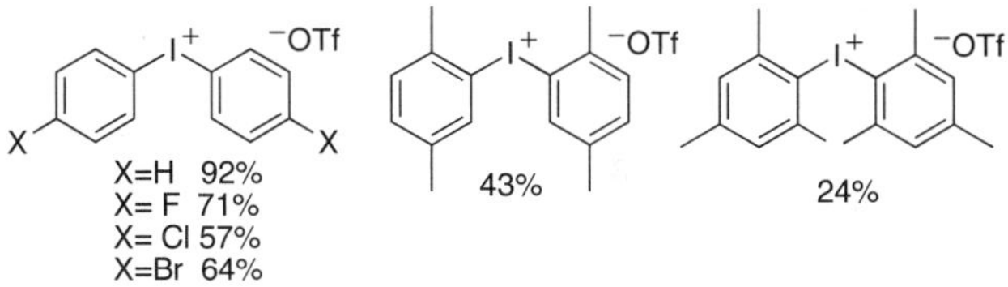

X=H 92%
X= F 71%
X= Cl 57%
X=Br 64%

43%

24%

The scope of the reaction is considerably widened when aryl iodides are employed instead of iodine (Table 1). Both symmetric and unsymmetric salts are efficiently obtained in short reaction times. Salts containing both electron poor and electron rich aryl moieties can be synthesized. The reaction time and temperature needed varies with the reactivity of the substrates.[3] When high temperature is used, the reaction can generally be performed with less trifluoromethanesulfonic acid. The limitations include the synthesis of symmetric electron rich salts and symmetric electron poor salts.

Table 1. Scope of the one-pot synthesis from aryl iodides and arenes.[a]

$$Ar^1\text{-I} \ + \ Ar^2\text{-H} \xrightarrow[\text{CH}_2\text{Cl}_2, \ 10 \ \text{min-22 h}]{\substack{m\text{-CPBA (1.1 equiv),} \\ \text{TfOH (2-3 equiv)}}} Ar^1\text{-}\overset{+}{\text{I}}\text{-}Ar^2 \ ^{-}\text{OTf}$$

Entry	Salt	Yield (%)[b]	Entry	Salt	Yield (%)[b]
1		92	9		91
2		85	10		58
3		85	11		84
4		86	12		85
5		78	13		63
6		83	14		73
7		87	15		60
8		82	16		53

[a]Reactions performed on a 0.2-0.3 mmol scale. [b]Isolated yield.

1. Department of Organic Chemistry, Arrhenius Laboratory, Stockholm University, SE-106 91 Stockholm, Sweden. E-mail: berit@organ.su.se.

2. Vogel, A. I.; Furniss, B. S.; Hannaford, A. J.; Rogers, V.; Smith, P. W. G.; Tatchell, A. R., *Vogel's Textbook of Practical Organic Chemistry.* 1978; p 1280 pp.

3. (a) Bielawski, M.; Olofsson, B., *Chem. Commun.* **2007**, 2521-2523. (b) Bielawski, M.; Zhu, M.; Olofsson, B., *Adv. Synth. Catal.* **2007**, *349*, 2610-2618.

4. Zhdankin, V. V.; Stang, P. J., *Chem. Rev.* **2008**, *108*, 5299-5358.

5. (a) Aggarwal, V. K.; Olofsson, B., *Angew. Chem., Int. Ed.* **2005**, *44*, 5516-5519. (b) Gao, P.; Portoghese, P. S., *J. Org. Chem.* **1995**, *60*, 2276-2278. (c) Ryan, J. H.; Stang, P. J., *Tetrahedron Lett.* **1997**, *38*, 5061-5064.

6. (a) Deprez, N. R.; Sanford, M. S., *Inorg. Chem.* **2007**, *46*, 1924-1935. (b) Phipps, R. J.; Grimster, N. P.; Gaunt, M. J., *J. Am. Chem. Soc.* **2008**, *130*, 8172-8174. (c) Becht, J.-M.; Drian, C. L., *Org. Lett.* **2008**, *10*, 3161-3164. (d) Phipps, R. J.; Gaunt, M. J.: *Science* **2009**, 323, 1593-1597.

7. Toba, Y., *J. Photopolym. Sci. Technol.* **2003**, *16*, 115-118.

8. Kitamura, T.; Yamane, M.; Inoue, K.; Todaka, M.; Fukatsu, N.; Meng, Z.; Fujiwara, Y., *J. Am. Chem. Soc.* **1999**, *121*, 11674-11679.

9. (a) Kitamura, T.; Todaka, M.; Fujiwara, Y., *Org. Synth.* **2002**, *78*, 104-112. (b) Lucas, H. J.; Kennedy, E. R., *Org. Synth.* **1942**, *22*, 52-53.

10. Zefirov, N. S.; Kasumov, T. M.; Koz'min, A. S.; Sorokin, V. D.; Stang, P. J.; Zhdankin, V. V., *Synthesis* **1993**, 1209-1210.

11. (a) Hossain, M. D.; Kitamura, T., *Tetrahedron* **2006**, *62*, 6955-6960. (b) Hossain, M. D.; Ikegami Y., Kitamura, T., *J. Org. Chem.* **2006**, *71*, 9903-9905.

12. Zhu, M.; Jalalian, N.; Olofsson, B., *Synlett* **2008**, 592-596.

13. Bielawski, M.; Aili, D.; Olofsson, B., *J. Org. Chem.* **2008**, *73*, 4602-4607.

Appendix
Chemical Abstracts Nomenclature; (Collective Index Number); (Registry Number)

Bis(4-*tert*-butylphenyl)iodonium triflate; (84563-54-2)
Iodine; (8, 9); (7553-56-2)
tert-Butylbenzene; (98-06-6)

m-Chloroperbenzoic acid; Peroxybenzoic acid, *m*-chloro- (8);
Benzocarboperoxoic acid, 3-chloro- (9); (937-14-4)
Trifluoromethanesulfonic acid; HIGHLY CORROSIVE, Methanesulfonic
acid, trifluoro- (8, 9); (1493-13-6)

Berit Olofsson was born in 1972 in Sundsvall, Sweden. She got her M Sc in 1998 from Lund University, and finished her PhD in asymmetric synthesis at KTH, Stockholm (with P. Somfai) in 2002. She then moved to Bristol University, UK for a post doc in the field of methodology and natural product synthesis (with V. K. Aggarwal). Returning to Sweden, she became assistant supervisor at Stockholm University in the group of J.-E. Bäckvall. In 2006 she was appointed to Assistant Professor, and was promoted to Associate Professor in 2008. Her research interests include the synthesis and application of hypervalent iodine compounds in asymmetric synthesis.

Marcin Bielawski was born in 1981 in Augustów, Poland and moved to Sweden at the age of three. He obtained his M. Sc. in Chemistry at Lund University, and subsequently moved to Stockholm. He started his Ph. D studies at Stockholm University in 2006, under the supervision of Berit Olofsson. He has developed several synthetic routes to diaryliodonium salts, and is currently working with their application in organic synthesis.

Katja Krämer was born in 1979 in Saarbrücken, Germany. She studied Chemistry at the Saarland University, where she received her Ph. D. in 2008 under the supervision of Prof. Uli Kazmaier. After working on the synthesis of non-natural amino acids via palladium-catalyzed allylic alkylation, she is now pursuing post-doctoral research in Prof. Mark Lautens' group at the University of Toronto. Her research focuses on desymmetrization of strained alkenes.

Pd(0)-CATALYZED DIAMINATION OF *TRANS*-1-PHENYL-1,3-BUTADIENE WITH DI-*TERT*-BUTYLDIAZIRIDINONE AS NITROGEN SOURCE

A.

B.

C.

Submitted by Haifeng Du, Baoguo Zhao, and Yian Shi.*[1]
Checked by Eric E. Buck and Peter Wipf.[2]

1. Procedure

A. Di-tert-butylurea (1). A 5-L Erlenmeyer flask equipped with a mechanical stirrer and an internal thermometer is charged with *tert*-butylamine (85.2 mL, 791 mmol, 6.6 equiv), triethylamine (127 mL, 900 mmol, 7.5 equiv), and toluene (2000 mL) (Note 1). The mixture is cooled to an internal temperature of 5 °C in an ice bath and triphosgene (36.3 g, 120 mmol, 1.0 equiv) (Note 2) is added in small portions over a period of 20 min (Note 3). The mixture is warmed to room temperature and stirred for 24 h. The reaction mixture is quenched with water (500 mL), transferred to a separatory funnel, and the aqueous phase is discarded. The organic layer is washed with water (4 x 500 mL) and the white solids are collected by suction filtration (water aspirator) (Note 4 and 5). The combined white solids are washed with water (2000 mL) (Note 6), and dried at 65 °C under vacuum (74 mmHg) for 12 h to give 49.7 g (80%) of di-*tert*-butylurea as a white solid (Notes 7 and 8).

B. Di-tert-butyldiaziridinone (2).[3,4] A 500-mL, one-necked, round-bottomed flask equipped with a Teflon-coated magnetic stir bar (length:

3.3 cm) and a rubber septum is charged with di-*tert*-butylurea (30.0 g, 174 mmol, 1.0 equiv) and Et$_2$O (200 mL) (Note 9). The reaction vessel is protected from light and *tert*-butyl-hypochlorite (20.8 g, 192 mmol, 1.1 equiv) (Note 10) is added to the slurry over a period of 10 min. The resulting pale yellow solution is stirred for an additional 30 min, cooled to 5 °C in an ice bath, and treated with potassium *tert*-butoxide (25.4 g, 226 mmol, 1.3 equiv) (Note 11) in small portions over a period of 20 min (Note 12). The resulting solution is warmed to room temperature and stirred for 5 h (Note 13). The reaction mixture is diluted with hexanes (200 mL), transferred to a 2-L separatory funnel, and washed with water (3 x 150 mL). The organic phase is dried over anhydrous K$_2$CO$_3$ (30.0 g) (Note 14) by stirring over 5 h, filtered, and concentrated by rotary evaporation (25 °C, 18 mmHg) (Note 15). The crude material is purified by fractional distillation under vacuum (7 mmHg) using a vigreux column (length 10.5 cm) fitted with a short path distillation head (Note 16). The oil bath temperature is gradually increased from 24 °C to 70 °C to give 22.6 g (76%) of di-*tert*-butyldiaziridinone as a colorless liquid (Notes 17, 18, 19, and 20).

C. *trans-1,3-Di-tert-butyl-4-phenyl-5-vinyl-imidazolidin-2-one (3)*. A 25-mL, one-necked, round-bottomed flask equipped with a Teflon-coated magnetic stir bar (length: 1.3 cm) and a rubber septum is charged with freshly distilled (*E*)-1-phenyl-1,3-butadiene (2.60 g, 20.0 mmol, 1.0 equiv) (Note 21) and tetrakis(triphenylphosphine)palladium (0.462 g, 0.400 mmol, 0.02 equiv) (Notes 22 and 23). The flask is evacuated and back-filled with argon three times. The mixture is placed in a preheated (65 °C) oil bath and di-*tert*-butyl-diaziridinone (3.75 g, 22.0 mmol, 1.1 equiv) is added over 3 h via syringe pump (Note 24). The resulting red mixture is stirred for an additional 1 h (Note 13), cooled to room temperature, and diluted with 5 mL of a hexanes/Et$_2$O (5:1) solution. The solution is loaded onto a wet-packed silica gel column (305 g SiO$_2$, diameter: 6.5 cm, height: 22 cm, pretreated with hexanes/diethyl ether, 5:1 (Note 25). The product is eluted with hexanes/diethyl ether, 5:1) to give 5.78 g (96%) of *trans*-1,3-di-*tert*-butyl-4-phenyl-5-vinyl-imidazolidin-2-one as a yellow oil (Notes 26 and 27).

2. Notes

1.   *tert*-Butylamine (98%, Alfa Aesar), triethylamine (99%, Fisher Chemicals), and toluene (99.9%, Fisher Chemicals, ACS grade) were used as received.

2. Triphosgene (98%) was purchased from Alfa Aesar and used as received.

3. The addition was carried out at such a rate that an internal temperature of 15 °C was not exceeded.

4. The solids are suspended in the organic phase.

5. The filtrate was concentrated to give an additional 2.35 g of di-*tert*-butylurea.

6. It is imperative to wash the solids extensively with water to remove any residual Et$_3$N•HCl, which may decrease the yield for the subsequent step.

7. Working at 50% scale, the checkers obtained 23.1 g (76%).

8. The product has the following physicochemical properties: White solid, mp 245–246 °C; ^{1}H NMR (300 MHz, DMSO-d_6) δ: 1.18 (s, 18H), 5.45 (s, 2H); ^{13}C NMR (75 MHz, DMSO-d_6) δ: 29.3, 48.8, 157.0; IR (ATR) 3349, 1633 cm^{-1}; MS (EI+) *m/z* 172 (M$^+$, 26), 157 (67), 57 (100); HRMS (EI+) *m/z* calcd for C$_9$H$_{20}$N$_2$O 172.1576, found 172.1569; Anal. Calcd for C$_9$H$_{20}$N$_2$O C, 62.75; H, 11.70; N, 16.26; Found C, 63.02; H, 11.94; N, 16.16.

9. Diethyl ether (ACS grade) was purchased from Fisher Chemicals and dried by distillation over sodium/benzophenone under an argon atmosphere.

10. *tert*-Butyl hypochlorite was prepared according to a literature procedure: Teeter, H. M.; Bell, E. W. *Org. Synth.* **1952**, *32*, 20.

11. Solid potassium *tert*-butoxide (97%) was purchased from Alfa Aesar and used as received.

12. Upon complete addition the reaction vessel does not need to be protected from light.

13. The progress of the reaction can be monitored by ^{1}H NMR analysis of aliquots taken directly from the reaction mixture.

14. Anhydrous K$_2$CO$_3$ (ACS grade) was purchased from Fisher Chemicals and used as received.

15. The *tert*-butanol generated in the reaction should be completely removed to prevent interference with the subsequent distillation.

16. The dimensions of the distillation head are as follows. Single piece construction with inlet for vacuum/inert gas, 10/18 thermometer joint on top, 14/20 joints for distillation and collection flasks, approx. 35 mm length of condenser x 65 mm height (head).

17. Di-*tert*-butyldiaziridinone is collected at 47–49 °C (7 mmHg) and stored away from light.

18. The submitters reported a bp of di-*tert*-butyldiaziridinone of 60–64 °C (7 mmHg).

19. Working at 50% scale, the checkers obtained 11.1 g (75%).

20. The product has the following physicochemical properties: bp 47–49 °C /7 mmHg; ^{1}H NMR (300 MHz, benzene-d_6) δ: 1.07 (s, 18 H); ^{13}C NMR (75 MHz, benzene-d_6) δ: 27.2, 59.3, 159.3; IR (ATR) 1929, 1875, 1856 cm^{-1}; MS (EI+) *m/z* 170 (M$^+$, 5), 157 (67), 131 (15), 84 (40), 57 (100); HRMS (EI+) *m/z* calcd for $C_9H_{18}N_2O$ 170.1419, found 170.1420; Anal. Calcd for $C_9H_{18}N_2O$ C, 63.49; H, 10.66; N, 16.45; Found C, 63.61; H, 10.79; N, 16.17.

21. (*E*)-1-Phenyl-1,3-butadiene was prepared according to a literature procedure: Grummitt, O.; Becker, E. I. *Org. Synth.* **1950**, *30*, 75; or Wittig, G.; Schoellkopf, U. *Org. Synth.* **1960**, *40*, 66. Similar results were obtained for step C using the diene prepared from either protocol.

22. Tetrakis(triphenylphosphine)palladium (9% min. palladium) was purchased from Pressure Chemical Co. and used as received.

23. Reaction is performed neat.

24. The submitters added di-*tert*-butyl-diaziridinone via a 10-mL addition funnel. The checkers observed more consistent results using a syringe pump.

25. Silica gel 40-63 D (60 Å) (Silicycle, Quebec City, Canada) was used.

26. Working at 50% scale, the checkers obtained 2.86 g (95%).

27. The product has the following physicochemical properties: ^{1}H NMR (300 MHz, benzene-d_6) δ: 1.36 (s, 9 H), 1.39 (s, 9 H), 3.55 (dm, J = 8.4 Hz, 1 H), 4.05 (d, J = 1.2 Hz, 1 H), 4.86 – 4.95 (m, 2 H), 5.92 (ddd, J = 17.1, 10.2, 8.4 Hz, 1 H), 7.07-7.16 (m, 3 H), 7.27 (dm, J = 6.9 Hz, 2 H); ^{13}C NMR (75 MHz, benzene-d_6) δ: 29.1, 29.2, 53.7, 53.9, 63.8, 65.5, 115.6, 126.5, 129.4, 141.8, 145.1, 159.2; IR (ATR) 1682 cm^{-1}; MS (EI+) *m/z* 300 (M$^+$, 55), 285 (100), 229 (81), 132 (60); HRMS (EI+) *m/z* calcd for $C_{19}H_{28}N_2O$ 300.2202, found 300.2191; Anal. Calcd for $C_{19}H_{28}N_2O$ C, 75.96; H, 9.39; N, 9.32; Found C, 76.01; H, 9.60; N, 9.31.

Safety and Waste Disposal Information

All hazardous materials should be handled and disposed of in accordance with "Prudent Practices in the Laboratory"; National Academy Press; Washington, DC, 1995.

3. Discussion

The diamination of olefins provides an attractive and efficient approach to biologically and chemically important vicinal diamines.[5] To date, various metal-mediated[6] or catalyzed[7] diaminations have been developed. Recently, we reported that conjugated dienes and trienes can be regio- and stereoselectively diaminated using Pd(0)[8] or Cu(I)[9] as the catalyst and di-*tert*-butyldiaziridinone (**2**) or related compounds as a convenient nitrogen source. When Pd(0) is used as the catalyst, the diamination occurs regioselectively at the internal double bond (Scheme 1).[8a]

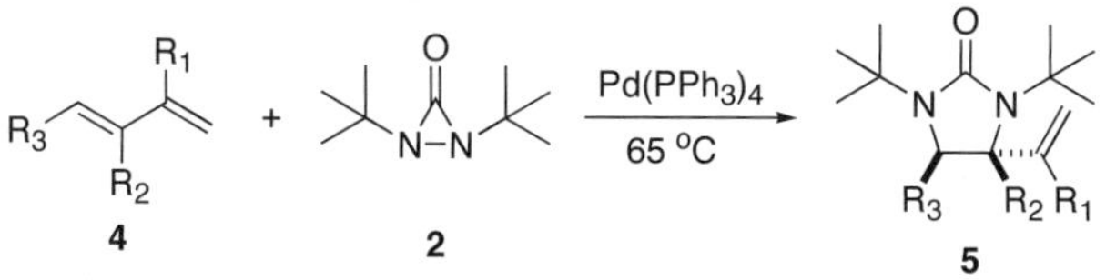

Scheme 1

This diamination can be applied to various alkyl and aryl-substituted dienes as well as electron-rich and electron-deficient dienes (Table 1).[8a] When the diamination is carried out under solvent-free conditions as described herein, the amount of Pd(PPh$_3$)$_4$ catalyst can be further reduced.[10]

Table 1. Pd(0)-Catalyzed Diamination of Conjugated Dienes and Trienes[a]

Entry	Substrate (4)	Product (5)	Yield (%)
1	R = Me		94
2	R = C₅H₁₁		91
3	R = CH₂OBn		76
4	R = p-MeOPh		94
5	R = 2-Furyl		78
6	R = OMe		95
7	R = CO₂Me		62
8			86
9	OTMS		90
10			86

[a] All reactions were carried out with di-*tert*-butyldiaziridinone (2) (1.0 equiv), diene or triene (1.2 equiv), and Pd(PPh₃)₄ (0.1 equiv) in benzene-d_6 in an NMR tube at 65 °C under argon for 0.25 to 5 h.

A possible catalytic cycle for this diamination is shown in Scheme 2.[8a] Oxidative insertion of Pd(0) into the N-N bond of diaziridinone **2** generates four-membered Pd(II) species **6**. This intermediate then reacts with diene **4** to form π-allyl Pd complex **8**, which undergoes a reductive elimination to release **5** and regenerate the Pd(0) catalyst.

Org. Synth. **2009**, *86*, 315-324

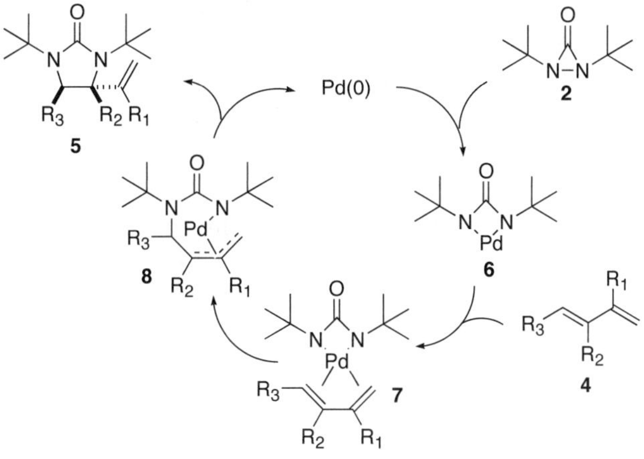

Scheme 2

When a chiral ligand is used, the diamination proceeds with high enantioselectivity. Chiral diamines can be obtained upon deprotection and the double bonds contained in the diamination products can be elaborated into other functionalities.[8d]

1. Department of Chemistry, Colorado State University, Fort Collins, CO 80523; yian@lamar.colostate.edu.

2. Department of Chemistry, University of Pittsburgh, Pittsburgh, PA 15260, USA; pwipf@pitt.edu.

3. Greene, F. D.; Stowell, J. C.; Bergmark, W. R. *J. Org. Chem.* **1969**, *34*, 2254.

4. For a leading review on diaziridinones, see: Heine, H. W. In *The Chemistry of Heterocyclic Compounds*; Hassner, A., Ed.; John Wiley & Sons, Inc: New York, 1983; pp 547.

5. For leading reviews, see: (a) Lucet, D.; LeGall, T.; Mioskowski, C. *Angew. Chem., Int. Ed.* **1998**, *37*, 2580. (b) Mortensen, M. S.; O'Doherty, G. A. *Chemtracts: Org. Chem.* **2005**, *18*, 555. (c) Kotti, S. R. S. S.; Timmons, C.; Li, G. *Chem. Biol. Drug Des.* **2006**, *67*, 101.

6. For metal-mediated diaminations, see: (a) Gomez Aranda, V.; Barluenga, J.; Aznar, F. *Synthesis* **1974**, 504. (b) Chong, A. O.; Oshima,

K.; Sharpless, K. B. *J. Am. Chem. Soc.* **1977**, *99*, 3420. (c) Bäckvall, J.-E. *Tetrahedron Lett.* **1978**, 163. (d) Barluenga, J.; Alonso-Cires, L.; Asensio, G. *Synthesis* **1979**, 962. (e) Becker, P. N.; White, M. A.; Bergman, R. G. *J. Am. Chem. Soc.* **1980**, *102*, 5676. (f) Fristad, W. E.; Brandvold, T. A.; Peterson, J. R.; Thompson, S. R. *J. Org. Chem.* **1985**, *50*, 3647. (g) Muñiz, K.; Nieger, M. *Synlett* **2003**, 211. (h) Muñiz, K. *Eur. J. Org. Chem.* **2004**, 2243. (i) Muñiz, K.; Nieger, M. *Chem. Commun.* **2005**, 2729. (j) Zabawa, T. P.; Kasi, D.; Chemler, S. R. *J. Am. Chem. Soc.* **2005**, *127*, 11250. (k) Zabawa, T. P.; Chemler, S. R. *Org. Lett.* **2007**, *9*, 2035.

7. For metal-catalyzed diaminations, see: (a) Li, G.; Wei, H.-X.; Kim, S. H.; Carducci, M. D. *Angew. Chem. Int. Ed.* **2001**, *40*, 4277. (b) Bar, G. L. J.; Lloyd-Jones, G. C.; Booker-Milburn, K. I. *J. Am. Chem. Soc.* **2005**, *127*, 7308. (c) Streuff, J.; Hövelmann, C. H.; Nieger, M.; Muñiz, K. *J. Am. Chem. Soc.* **2005**, *127*, 14586. (d) Muñiz, K.; Streuff, J.; Hövelmann, C. H.; Núñez, A. *Angew. Chem., Int. Ed.* **2007**, *46*, 7125. (e) Muñiz, K. *J. Am. Chem. Soc.* **2007**, *129*, 14542. (f) Muñiz, K.; Hövelmann, C. H.; Streuff, J. *J. Am. Chem. Soc.* **2008**, *130*, 763.

8. (a) Du, H.; Zhao, B.; Shi, Y. *J. Am. Chem. Soc.* **2007**, *129*, 762. (b) Xu, L.; Du, H.; Shi, Y. *J. Org. Chem.* **2007**, *72*, 7038. (c) Du, H.; Yuan, W.; Zhao, B.; Shi, Y. *J. Am. Chem. Soc.* **2007**, *129*, 7496. (d) Du, H.; Yuan, W.; Zhao, B.; Shi, Y. *J. Am. Chem. Soc.* **2007**, *129*, 11688. (e) Xu, L.; Shi, Y. *J. Org. Chem.* **2008**, *73*, 749. (f) Du, H.; Zhao, B.; Shi, Y. *J. Am. Chem. Soc.* **2008**, *130*, 8590.

9. (a) Yuan, W.; Du, H.; Zhao, B.; Shi, Y. *Org. Lett.* **2007**, *9*, 2589. (b) Zhao, B.; Yuan, W.; Du, H.; Shi, Y. *Org. Lett.* **2007**, *9*, 4943. (c) Zhao, B.; Du, H.; Shi, Y. *Org. Lett.* **2008**, *10*, 1087.

10. The detailed studies will be described elsewhere.

Appendix
Chemical Abstracts Nomenclature; (Registry Number)

Di-*tert*-butylurea: Urea, *N,N'*-bis(1,1-dimethylethyl)-; (5336-24-3)
tert-Butylamine: 2-Propanamine, 2-methyl-; (75-64-9)
Triethylamine: Ethanamine, *N,N*-diethyl-; (121-44-8)
Triphosgene: Methanol, 1,1,1-trichloro-, 1,1'-carbonate; (32315-10-9)
Di-*tert*-butyldiaziridinone: 3-Diaziridinone, 1,2-bis(1,1-dimethylethyl)-;
(19656-74-7)

tert-Butyl-hypochlorite: Hypochlorous acid, 1,1-dimethylethyl ester; (507-40-4)

Potassium *tert*-butoxide: 2-Propanol, 2-methyl-, potassium salt (1:1); (865-47-4)

trans-1,3-Di-*tert*-butyl-4-phenyl-5-vinyl-imidazolidin-2-one: 2-Imidazolidinone, 1,3-bis(1,1-dimethylethyl)-4-ethenyl-5-phenyl-, (4*R*,5*R*)-rel-; (927902-91-8)

(*E*)-1-Phenyl-1,3-butadiene: Benzene, (1*E*)-1,3-butadienyl-: (16939-57-4)

Tetrakis(triphenylphosphine)palladium; (14221-01-3)

Yian Shi was born in Jiangsu, China in 1963. He obtained his B.Sc. degree from Nanjing University in 1983, M.Sc. degree from University of Toronto with Professor Ian W.J. Still in 1987, and Ph.D. degree from Stanford University with Professor Barry M. Trost in 1992. After a postdoctoral study at Harvard Medical School with Professor Christopher Walsh, he joined Colorado State University as assistant professor in 1995 and was promoted to associate professor in 2000 and professor in 2003. His current research interests include the development of new synthetic methods, asymmetric catalysis, and synthesis of natural products.

Haifeng Du was born in Jilin, China in 1974. He received his B.Sc. degree in 1998 and M.Sc. degree in 2001 from Nankai University. He then moved to Shanghai Institute of Organic Chemistry, Chinese Academy of Sciences, and obtained his Ph.D. degree in 2004 under the supervision of Professor Kuiling Ding. In the fall of 2004, he joined the Department of Chemistry at Colorado State University as a postdoctoral fellow with Professor Yian Shi. His research interests include the development of novel synthetic methodology and asymmetric synthesis.

Baoguo Zhao was born in Hubei, China in 1973. He received his B.Sc. degree from Wuhan University in 1996 and M.Sc. degree from Nanjing University in 2002 under the supervision of Professor Jianhua Xu. After completing his Ph.D. degree under the supervision of Professor Kuiling Ding at Shanghai Institute of Organic Chemistry, Chinese Academy of Science in 2006, he joined Department of Chemistry at Colorado State University as a postdoctoral fellow with Professor Yian Shi. His current research interests include the development of novel synthetic methodology and asymmetric synthesis.

Eric Buck was born in 1985 in Fairmont, West Virginia. He graduated from the University of Minnesota with his B.Sc. in 2006. During his freshman year he entered the laboratory of Professor Thomas R. Hoye where he pursued the synthesis of petromyzonamine disulfate analogs with a focus on 5-β petromyzonamine disufate. While attending the U of M he was supported by the David A. and Merece H. Johnson Scholarship. In 2007 he moved to Pittsburgh, Pennsylvania where he is currently a graduate student under the direction of Professor Peter Wipf. His current research interests include asymmetric synthesis and the synthesis of natural products.

SYNTHESIS OF ETHYL
2-ETHANOYL-2-METHYL-3-PHENYLBUT-3-ENOATE

Submitted by Taisuke Fujimoto,[1] Kohei Endo,[2] Masaharu Nakamura,[3] and Eiichi Nakamura.[1]
Checked by Mark Webster and John A. Ragan.

1. Procedure

Ethyl 2-ethanoyl-2-methyl-3-phenylbut-3-enoate. A flame-dried, 10-mL Schlenk tube (Notes 1 and 2) connected to a vacuum/argon manifold through a glass stopcock and fitted with a glass stopper is equipped with a 1-cm Teflon-coated magnetic stirring bar. In(OTf)$_3$ (64.0 mg, 0.11 mmol) (Notes 3 and 4) is placed in the Schlenk tube and connected to the vacuum line (1.0–1.5 mmHg). The Schlenk tube is immersed in an oil bath. The oil-bath temperature is gradually increased to 180 °C over 1 h and then kept for 30 min at that temperature (Note 5). The Schlenk tube is cooled to ambient temperature and charged with argon. The glass stopper is replaced with a rubber septum, and acetonitrile (4 mL) (Note 6) is introduced into the Schlenk tube via a syringe under argon to obtain a 0.025-M acetonitrile solution of In(OTf)$_3$.

A 50-mL, 3-necked, round-bottomed flask (Note 1), connected to a vacuum/argon manifold through a glass stopcock vacuum adaptor, is equipped with a 2-cm Teflon-coated magnetic stirring bar. A solution of 0.025 M In(OTf)$_3$ in acetonitrile (2.0 mL, 0.050 mmol) is introduced into the flask via a syringe under argon, and the remaining two necks of the flask are equipped with glass stoppers. The solution is stirred under reduced pressure (1.0-1.5 mm Hg) at room temperature for 1 h to remove acetonitrile. The flask is flushed with argon, and one of the glass stoppers is replaced with a rubber septum. Ethyl 2-methyl-3-oxobutanoate (14.48 g, 14.2 mL, 0.1 mol) (Notes 7 and 8) and phenylacetylene (12.26 g, 13.2 mL, 0.12 mol) (Note 9) are introduced into the flask via syringe under argon, and the septum is replaced with the glass stopper. The resulting clear yellow solution is stirred and the flask is immersed in an oil bath (140 °C).

After stirring for 3 h at 140 °C, the reaction mixture is cooled to ambient temperature (Note 10). The vacuum adaptor is replaced with a distillation head. Distillation of the reaction mixture under reduced pressure (1.0–1.5 mmHg) at 150–160 °C (oil bath temperature) gives the title compound as a pale yellow liquid (22.2–23.0 g, 0.090–0.093 mol) in 90-93% yield (>99.9% purity, non-calibrated GC area ratio) (Notes 11 and 12).

2. Notes

1. All glassware was dried in an oven (110 °C), assembled while hot, and allowed to cool to room temperature under argon atmosphere.

2. The checkers used the mass of In(OTf)$_3$ prior to drying for determination of the concentration of the resulting acetonitrile solution.

3. In(OTf)$_3$ (complexiometric EDTA specification of 19.2-21.7% indium, 19.5% for the batch used by the checkers, theory for In(OTf)$_3$ = 20.4%) was purchased from Aldrich and used as received. Alternatively, the submitters report that it can be made from In$_2$O$_3$ and TfOH in boiling water.[4] In$_2$O$_3$ was purchased from Kanto Kagaku. TfOH was purchased from Wako Pure Chemical Industries, Ltd. Both reagents were used as received. In order to dispose of trifluoromethanesulfonic acid, the acid should be carefully introduced into water in a dropwise fashion.

4. In(OTf)$_3$ is extremely hygroscopic.

5. Rapid increase of the oil bath temperature causes the decomposition of hydrated In(OTf)$_3$.

6. Anhydrous acetonitrile (<50 ppm water) was purchased from EMD Chemicals (Merck KGaA) and used as received.

7. Ethyl 2-methyl-3-oxobutanoate was purchased from Acros and was purified by silica gel flash column chromatography (5% ethyl acetate in hexane as eluent) and distillation (bp 80 °C/20 mmHg) before use. The chromatographic purification is necessary to remove ethyl 3-oxobutanoate.

8. The exact mass of keto ester was determined by weight of the syringe before and after the addition.

9. Phenylacetylene was purchased from Aldrich Inc. and purified by vacuum distillation before use (bp 60 °C/20 mmHg). The exact mass of the

charge was determined as in Note 8.

10. The reaction was monitored by GC/MS on an HP-1 capillary column (0.2 mm x 12 m, 0.33μm) at 30 °C to 290 °C raised at 30 °C /min. Typical retention time of product is 5.7 min. The submitters monitored the reaction by TLC on glass plates coated with 0.25 mm of 230-400 mesh silica gel containing a fluorescent indicator (Merck #1.05715.0009). Plates were visualized with UV light (254 nm) and/or by immersion in an acidic staining solution of *p*-anisaldehyde followed by heating on a hot plate.

11. The purity was determined by GC/MS analysis, per Note 10.

12. The product displays the following physicochemical properties: ^{1}H NMR (400 MHz, CDCl$_3$): δ 1.20 (t, J = 7.0 Hz, 3 H), 1.50 (s, 3 H), 2.28 (s, 3 H), 4.16 (q, J = 7.0, 2 H), 5.26 (s, 1 H), 5.41 (s, 1 H) 7.16–7.30 (m, 5 H); ^{13}C NMR (100 MHz, CDCl$_3$): δ 14.1, 21.5, 27.6, 61.8, 66.1, 119.0, 127.8, 128.0 (2C), 128.3 (2C), 140.6, 148.2, 172.0, 205.5; IR (neat) cm^{-1}; 1710, 1355, 1092, 1021, 915, 776; Anal. Calcd. For C$_{15}$H$_{18}$O$_3$: C, 73.15; H, 7.37. Found: C, 73.00; H, 7.22.

Safety and Waste Disposal Information

All hazardous materials should be handled and disposed of in accordance with "Prudent Practices in the Laboratory"; National Academy Press; Washington, DC, 1995.

3. Discussion

The present procedure provides a simple and efficient way to produce α-alkenyl carbonyl compounds through addition of an active methylene compound to an unactivated alkyne or acetylene catalyzed by indium(III) tris(trifluoromethanesulfonate) [In(OTf)$_3$]. The reaction does not necessarily require solvent, and, as the result, the 20-g scale synthesis can be carried out in a 50- to 100-mL flask. When viscous or solid keto esters are used, the reaction mixture can be diluted with a solvent such as toluene. The reaction also represents a method for the construction of a quaternary carbon center.

The reaction takes place smoothly at room temperature if one uses 20 mol % of In(OTf)$_3$, which however consumes the same amount of the alkyne. For the reaction to be performed with low catalyst loading (0.05

mol % catalyst), higher temperature is needed, but the reaction still shows good functional group tolerance as shown in Table 1.[5] Acid sensitive compounds such as benzyl propargyl ether (entry 5) and ethyl 2-allylacetoacetate (entry 9) require the presence of triethylamine as a base. The reaction of ethynylsilane requires In(OTf)₃ (5 mol %) and DBU (6 mol %) at 100 °C for 16 hours (entry 8). The addition of DBU is essential to prevent the desilylation of the vinylsilane product. The *trans*-stereochemistry of the double bond in the product indicates the *cis*-addition of an indium(III) enolate intermediate to the triple bond.

1,3-Diketones also take part in the reaction under slightly modified conditions. The addition of 3-methyl-2,4-pentanedione to phenylacetylene takes place in the presence of In(OTf)₃ (5 mol %), Et₃N (5 mol %), and *n*-BuLi (5 mol %) in 32 hours at 100 °C to give the desired product in 88% yield (entry 11). The presence of Et₃N and *n*-BuLi suppresses the formation of side products. As suggested by the data in Table 1, the present reaction is the most suited for creation of a quaternary carbon center, where there is no possibility of enolization. However, the creation of a tertiary center may be achieved in some limited examples of 3,3-diprotio-2,4-pentanedione congeners.[5a,d,e]

The present reaction can also be used for the creation of chiral quaternary carbon stereocenters (eq. 1).[5g] An indium(III) enamide intermediate bearing a chiral auxiliary undergoes highly diastereoselective addition to an alkyne. Intramolecular version of the reaction shows remarkable generality, creating six to fifteen membered rings in good to excellent yields (eq. 2).[5e,h,i]

Table 1. Addition of Active Methylene Compounds to Alkynes

entry	substrate	alkyne (eq)	product (% yield)[a]
1	**1**	X—⬡—≡ (1.2)	**2**, X = -H (99)
2			X = -CO₂CH₃ (97)
3[b]			X = -OCH₃ (98)
4	**1**	R—≡ (2.0–5.0)	R = C₆H₁₃ (2.0) (99)
5[c]			R = -CH₂OBn (3.0) (90)
6[d]			R = -(CH₂)₄NPht (2.0) (89)
7[e]			R = 1-cyclohexene (5.0) (94)
8[f]	**1**	C₆H₅-Si-≡ (2.0)	(94)
9		**2** (1.2)	(93)
10[g]		acetylene (balloon)	(100)
11[h]		**2** (5.0)	(88)
12	C₂H₅O— —OC₂H₅	**2** (1.2)	(99)

a) Isolated yield. b) In(OTf)₃ (1 mol%) was used at 100 °C in toluene (1 M) for 2 h. c) In(OTf)₃ (5 mol%) and Et₃N (5 mol%) were used at 80 °C for 22 h. d) NPht is the abbreviation of phthalamide. In(OTf)₃ (1 mol%) was used at 100 °C in toluene (2 M) for 10 h. e) In(OTf)₃ (2 mol%) was used at 60 °C for 4 h. f) In(OTf)₃ (5 mol%) and 1,8-diazabicyclo[5. 4. 0]undec-7-ene, DBU (6 mol%) were used at 100 °C for 16 h. g) In(OTf)₃ (20 mol%), DBU (20 mol%), and MS 3A were used in toluene (1 M) at 100 °C. h) In(OTf)₃ (5 mol%), Et₃N (5 mol%), and n-BuLi (5 mol%) were used at 100 °C for 32 h.

In summary, the present procedure is useful for the introduction of an alkenyl group, featuring such synthetically convenient attributes as 1) perfect regioselectivity as to the alkyne acceptor, 2) good functional group compatibility, 3) high catalytic performance, 4) requirement of no solvent and 5) good atom economy.

1. Department of Chemistry, The University of Tokyo, Hongo, Bunkyo-ku, Tokyo, 113-0033, Japan.
2. Department of Chemistry and Biochemistry School of Advanced Science and Engineering, Waseda University, Ohkubo, Shinjuku, Tokyo 169-8555, Japan
3. International Research Center for Elements Science, Institute for Chemical Research, Kyoto University, Gokasho, Uji, Kyoto, 611-0011, Japan.
4. Koshima, H.; Kubota, M. *Synth. Commun.* **2003**, *33*, 3983–3988.
5. (a) Nakamura, M.; Endo, K.; Nakamura, E. *J. Am. Chem. Soc.* **2003**, *125*, 13002–13003. (b) Nakamura, M.; Endo, K.; Nakamura, E. *Org. Lett.* **2005**, *7*, 3279–3281. (c) Nakamura, M.; Endo, K.; Nakamura, E. *Adv. Synth. Catal.* **2005**, *347*, 1681–1686. (d) Endo, K.; Hatakeyama, T.; Nakamura, M.; Nakamura, E. *J. Am. Chem. Soc.* **2007**, *129*, 5264–5271. (e) Tsuji, H.; Yamagata, K.; Ito, Y.; Endo, K.; Nakamura, M.; Nakamura, E. *Angew. Chem. Int. Ed.* **2007**, *46*, 8060–8062; highlighted by *Synfact*, **2008**, 0077. (f) Tsuji, H.; Fujimoto, T.; Endo, K.; Nakamura, M.; Nakamura, E. *Org. Lett.* **2008**, *10*, 1219–1221. (g) Fujimoto, T.; Endo, K.; Tsuji, H.; Nakamura, M.; Nakamura, E. *J. Am. Chem. Soc.* **2008**, *130*, 4492–4496. (h) Ito, Y.; Yamagata, K.-i.; Tsuji, H.; Endo, K.; Nakamura, M.; Nakamura, E. *J. Am. Chem. Soc.* **2008**, *130*, 17161–17167. (i) Tsuji, H.; Tanaka, I.; Endo, K.; Yamagata, K.-i.; Nakamura, M.; Nakamura, E. *Org. Lett.* **2009**, *11*, 1845–1847.

Appendix
Chemical Abstracts Nomenclature (Registry Number)

Ethyl 2-methylacetoacetate: Butanoic acid, 2-methyl-3-oxo-, ethyl ester; (609-14-3)

Phenylacetylene: Benzene, ethynyl-; (536-74-3)

Indium(III) tris(trifluoromethanesulfonate): Methanesulfonic acid, trifluoro-, indium(3+) salt; (128008-30-0)

Eiichi Nakamura received his Ph.D. degree from Tokyo Institute of Technology in 1978. He became assistant professor in the same institute in 1980 after two-year post doc at Columbia University, and rose to the rank of professor. Since 1995, he has been professor of chemistry in the University of Tokyo. He is currently directing JST Nakamura Functional Carbon Cluster ERATO project. He received the Chemical Society of Japan Award (2003) and the Humboldt Research Award (2006), and is elected Fellow of the American Association for the Advancement of Science (1998), Fellow of the Royal Society of Chemistry (2005) and Honorary Foreign Member of the American Academy of Arts and Sciences (2008). His research focuses on physical organic chemistry, organic synthesis, material science and use of electron microscopy in chemistry.

Taisuke Fujimoto was born in Sendai, Japan in 1980. Under the direction of Prof. Eiichi Nakamura, he received his bachelor's degree in 2004 and his Ph. D. degree in 2008 from The University of Tokyo, where he worked on the development of Zn or In-mediated reactions. Currently he works for Fujifilm Corporation as organic synthetic chemist, engaging in the syntheses of functional dyes and medicinal chemicals.

Kohei Endo was born in Yokohama, Kanagawa in 1979. He received his bachelor's degree in 2001 from Tokyo Institute of Technology. Under the direction of Eiichi Nakamura he received his Ph. D. degree from The University of Tokyo in 2006. He joined the Noyori laboratory at Nagoya University as COE postdoctoral researcher in the same year. In 2007, he became an assistant professor at Waseda University. His research interest includes development of multi-functionalized molecular catalyst and catalysis system.

Masaharu Nakamura was born in Asagaya, Tokyo in 1967. He received his bachelor's degree in 1991 from Science University of Tokyo. Under the direction of Eiichi Nakamura he received his Ph. D. degree from Tokyo Institute of Technology in 1996. He became an assistant professor at The University of Tokyo in the same year. After promotions to a lecturer (2002) and an associate professor (2004) he moved to Kyoto. Since 2006, he has been a professor of Institute for Chemical Research at Kyoto University, where his research focuses on the development of future molecular/material transformations toward full utilization of chemical resources.

Mark Webster was born in 1956 in Newport, Kentucky. He received his B.S. degree from Northern Kentucky University in 1979 and began working at Hilton-Davis Chemical Company. While working at Hilton-Davis as a process chemist he attended Xavier University and received his M.S. degree in 1986. He began working in the pharmaceutical industry in 1987 at Merrell Dow Pharmaceuticals. He then worked at Procter and Gamble before joining Pfizer's Chemical Research and Development group where he currently works as a process chemist.

SYNTHESIS OF 4,5-DIMETHYL-1,3-DITHIOL-2-ONE

A.

1

B.

Submitted by Perumalreddy Chandrasekaran and James P. Donahue.[1]
Checked by Daniel M. Bowles and John A. Ragan.[2]

1. Procedure

A. O-Isopropyl S-3-Oxobutan-2-yl Dithiocarbonate (**1**). A 1-L, 3-necked flask, equipped with a 125-mL pressure equalizing addition funnel and nitrogen inlet is oven-dried, assembled hot, and cooled to ambient temperature under nitrogen vented to an external bubbler. The center neck is equipped with an overhead stirrer and the third neck is fitted with an internal temperature probe. The apparatus is cooled to ambient temperature and is charged with anhydrous K_2CO_3 (65 g, 470 mmol, 1.4 equiv) (Note 1) followed by freshly distilled 2-butanone (500 mL, 403 g, 5.58 mol, 16 equiv) (Note 2). The resulting suspension is cooled to below –75 °C with an acetone/dry ice bath, and a solution of bromine (17.7 mL, 55.2 g, 0.345 mol, 1.0 equiv) (Note 3) in 100 mL CH_2Cl_2 (Note 4) is added dropwise via the addition funnel over 30 min, maintaining the internal pot temperature below –70 °C (Note 15, photograph A). The resulting orange-colored reaction mixture is removed from the cold bath and is allowed to warm to 18 °C over 4 h. During this time, the color of the mixture changes from orange to colorless. The reaction mixture is stirred for an additional 4 h at ambient temperature. Potassium *O*-isopropylxanthate (Note 5) (60.4 g, 0.346 mol, 1.0 equiv) is then charged via a solids addition funnel in a single portion under nitrogen.

The resulting reaction mixture is heated to 60 °C (internal temperature) with a heating mantle for 6 h. The progress of the reaction can be conveniently monitored by silica gel TLC (9:1 heptanes/EtOAc; potassium O-isopropylxanthate R_f=0.0; desired product R_f=0.25). The heterogeneous mixture is cooled to 18 °C and then vacuum filtered in the open air through a packed 2 inch celite pad on a medium porosity, 3.5" diameter glass frit. The filter contents are washed with CH_2Cl_2 (3 × 100 mL), and the combined filtrates are concentrated to dryness at 50 °C and 20 mmHg in a 1-L, round-bottomed flask on a rotary evaporator to afford 59.6–61.7 g (0.289–0.299 mol, 83–86% yield based on potassium O-isopropylxanthate) of O-isopropyl S-3-oxobutan-2-yl dithiocarbonate **1** as an orange oil that can be carried forward without further purification. Analytically pure material is obtained by vacuum distillation from a 150-mL, single-necked, round-bottomed flask with a 10-cm Vigreaux column at 79–82 °C and 0.1 mmHg, which yields 48.9–50.1 g (0.237–0.243 mol, 68–70%) of **1** as a pale yellow, pungent smelling oil (Notes 6 and 7).

  B. 4,5-Dimethyl-1,3-dithiol-2-one (**2**). A 250-mL, 3-necked, round-bottomed flask equipped with a magnetic stirring bar, 250-mL pressure-equalizing addition funnel, internal temperature probe and glass stopper is charged with previously distilled O-isopropyl S-3-oxobutan-2-yl dithiocarbonate **1** (53.0 g, 0.257 mol), and the neat oil is cooled to 0 °C. Ice-cooled 15 M aqueous H_2SO_4 (100 mL, 1.5 mol, Note 8) is added dropwise via addition funnel over 30 min, maintaining the pot temperature below 10 °C. The reaction mixture is maintained between 0 °C and 10 °C for 1 h, warmed to room temperature, and then stirred for an additional hour. During this time, the reaction mixture darkens progressively to a brown color. Reaction progress is conveniently monitored by silica gel TLC analysis (9:1 heptane/EtOAc; SM R_f=0.25; product R_f=0.35). The reaction is quenched by slowly pouring the contents over 500 mL of ice in 10 mL portions over a 15 min period. The resulting two-phase aqueous–organic mixture is transferred to a 1-L separatory funnel and is extracted with Et_2O (3 × 200 mL) (Note 9). The combined Et_2O extracts are dried by stirring over $MgSO_4$ (75 g) for 2 h. The $MgSO_4$ is removed by vacuum filtration on a coarse, sintered glass filter, and the cake is washed with 2 × 200 mL Et_2O. The filtrates are concentrated on a rotary evaporator (25 °C, 20 mmHg) to produce a crude, dark, waxy solid that is purified by short-path vacuum distillation from a 100-mL, single-necked, round-bottomed flask (Note 10 and Note 15, photograph B) at 60–64 °C and 0.1 mmHg to afford 31.6–32.5 g (0.216–0.222 mol, 84–87%) of 4,5-dimethyl-1,3-dithiol-2-one as a pale yellow oil that solidifies upon standing (Notes 11 and 13).

Org. Synth. **2009**, *86*, 333-343

Highly crystalline material is prepared by dissolving 5.0 g of freshly distilled 4,5-dimethyl-1,3-dithiol-2-one in 25 mL warm hexanes followed by cooling to –25 °C to afford 4.10 g (82%) of off-white needle crystals that are isolated by decanting the mother liquor and drying at 18 °C in open air. Crystallization of 5.0 g crude (non-distilled) **2** in the same manner produces a lower recovery (2.1 g, 42%) of brown needles that are relatively clean by ^{1}H and ^{13}C NMR (Note 15, photograph C). Compound **2** is stable to air, moisture and light and is soluble in common organic solvents such as pentane, hexane, diethyl ether, dichloromethane, benzene, toluene, acetonitrile and methanol (Note 14).

2. Notes

1. Anhydrous potassium carbonate (ReagentPlus grade, 99%) from Sigma-Aldrich was used as received. The use of K_2CO_3 is necessary to scavenge the HBr generated during the bromination reaction. In one instance, the checkers noted a poor yield (21%) resulting from using older, 'wet' potassium carbonate.

2. Reagent grade 2-butanone (methyl ethyl ketone) was obtained from Sigma Aldrich (99+%) and distilled from $CaSO_4$ at ambient pressure using a fractionating column. A forerun is set aside. The 2-butanone used in the procedure is collected when the distillation is steady at ~80 °C. The 2-butanone is used both as solvent and substrate in this reaction. The use of smaller volumes of 2-butanone results in heterogeneous mixtures that are difficult to stir.

3. Reagent grade bromine from Sigma Aldrich (99.5%) was used as received.

4. Reagent grade dichloromethane from JT Baker was used without additional purification.

5. Potassium *O*-isopropylxanthate is available commercially from a variety of suppliers (e.g. $44.10 / 25 g from TCI America, used as received). The checkers noted that one-portion addition of this material resulted in an average exotherm of 20 °C at this scale. Care should be taken on larger scales to insure proper jacket cooling is available.

6. Pure *O*-isopropyl *S*-3-oxobutan-2-yl dithiocarbonate may be stored in the open air at ambient temperature without noticeable decomposition for a period of months. The material obtained after rotary evaporation of the CH_2Cl_2 and 2-butanone solvent is generally of sufficient purity for the next

step. This determination is made on the basis of ^{1}H NMR spectroscopy and by the observation of only slightly diminished overall yields using undistilled **1**.

7.   The spectroscopic and analytical properties of *O*-isopropyl *S*-3-oxobutan-2-yl dithiocarbonate (**1**) are as follows: bp: 79–82 °C at 0.1 mmHg. R_f: 0.25 on silica (9:1 heptane:EtOAc), visualization by UV-vis. ^{1}H NMR (CDCl$_3$, 400 MHz, 25 °C) δ: 1.359 (d, J = 6.4 Hz, 3 H), 1.363 (d, J = 6.4 Hz, 3 H), 1.43 (d, J = 7.6 Hz, 3 H), 2.28 (s, 3 H), 4.34 (q, J = 7.2 Hz, 1 H), 5.69 (sep, J = 6.4 Hz, 1 H). ^{13}C NMR (CDCl$_3$, 100 MHz, 25 °C) δ: 15.0, 20.69, 20.74, 26.9, 53.1, 78.3, 203.9, 210.8. FTIR (neat oil, ATR, cm^{-1}): 2981, 1717, 1386, 1374, 1356, 1242, 1157, 1143, 1091, 1038, 900, 631, 537. CI-MS: *m/z* 207 ([M + H]$^+$, 46), 165 (M − ^{i}Pr, 77), 147 (M − O^{i}Pr, 26), 105 (100).   HRMS: (ESI+) *m/z*: 207.05080 [calcd for C$_8$H$_{15}$O$_2$S$_2$ (M+H) 207.05090].   Anal. Calcd. for C$_8$H$_{14}$O$_2$S$_2$: C, 46.57; H, 6.84; S, 31.08. Found: C, 46.53; H, 6.80; S, 31.15.

8.   An 80 mL volume of concentrated H$_2$SO$_4$ is diluted to 100 mL by slowly pouring it into a 20 mL volume of ice cold water, resulting in a 15 *M* solution.   The checkers noted a poor yield (34%) when sulfuric acid was added too quickly to the pot mixture without proper cooling.

9.   The aqueous acidic phase should be brought to a neutral pH with saturated aqueous K$_2$CO$_3$ and transferred to an aqueous waste storage bottle.

10.   The condenser of the short-path distillation apparatus should not be cooled with cold water. During the course of the vacuum distillation, 4,5-dimethyl-1,3-dithiol-2-one solidifies within the distillation condenser and requires mild, continuous heating with a heat gun to ensure that it remains melted until collected in the distillation receiver. Alternatively, the checkers found that circulation of warm (55 °C) fluid through the short-path distillation head is highly effective.

11. The submitters obtained an additional 2.8% of product by chromatographic purification of the dark distillation pot residue as follows. Following the distillation of 4,5-dimethyl-1,3-dithiol-2-one obtained from the cyclization of 62.3 g (0.302 mol) of undistilled *O*-isopropyl *S*-3-oxobutan-2-yl dithiocarbonate, an additional 1.25 g (8.53 mmol, 2.8% yield) of product is recovered. The column is packed as a silica slurry (Note 12) in hexanes (1 in diameter × 10 in height) and is loaded by evaporation of a concentrated CH$_2$Cl$_2$ solution of the distillation pot residues onto 10 g silica followed by transferal of this dry coated silica directly onto the column. The column is then eluted with a gravity drip. The 4,5-dimethyl-1,3-dithiol-2-one

readily and completely resolves from the other components of the mixture as the leading band and is completely eluted within a 150 mL volume of hexanes after loading.

12. The silica used for the column was 70-230 micron and was obtained from EMD Chemicals (Merck), Gibbstown NJ, Catalog Number 7734-3.

13. The checkers found that using undistilled **1** in Step B (61.7 g, 0.299 mol) provided 25.9 g of **2** (0.177 mol, 59% yield).

14. Spectroscopic and analytical data for 4,5-dimethyl-1,3-dithiol-2-one are as follows: mp: 46–48 °C. bp: 59–64 °C at 0.1 mmHg. R_f: 0.35 on silica (9:1 heptane:EtOAc), visualization by UV-vis. ^{1}H NMR (CDCl$_3$, 400 MHz, 25 °C) δ: 2.13 (s). ^{13}C NMR (CDCl$_3$, 100 MHz, 25 °C) δ: 13.7, 122.8, 192.1. FTIR (neat oil, ATR, cm^{-1}): 3243, 2942, 2851, 1753, 1650, 1596, 1433, 1373, 1189, 1092, 1017, 941, 866, 756. CI-MS: m/z 147 ([M + H]$^+$, 100). HRMS: [M + H] calcd for C$_5$H$_7$O$_1$S$_2$: 146.99329. Found: 146.99333. Anal. Calcd. for C$_5$H$_6$OS$_2$: C, 41.07; H, 4.14; S, 43.86. Found: C, 41.11; H, 4.04; S, 44.75.

15. Photographs of the procedure:

(A) Step A during bromine addition.

(B) Step B product distillation

(C) Comparison of Step B product (**2**) obtained by recrystallization of distilled (left two samples) or undistilled (right two samples) material

Scheme 1

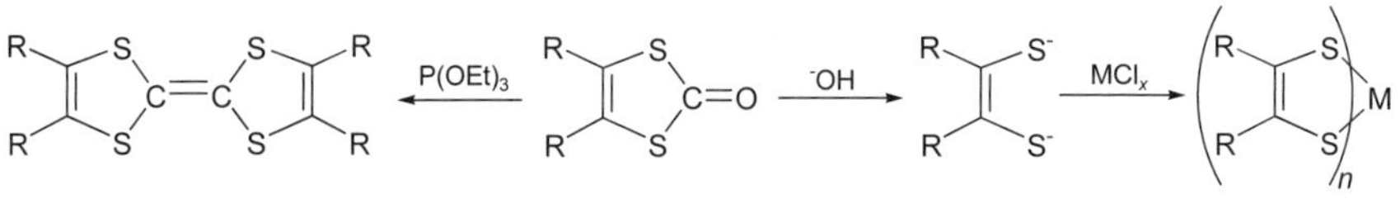

3. Discussion

The 1,3-dithiol-2-ones constitute a class of molecules with demonstrated synthetic utility for the synthesis of tetrathiafulvalenes and related materials (Scheme 1). Oligo- and polytetrathiafulvalenes have been subjected to intense investigation for a broad variety of applications ranging from conducting organic materials[3-9] to solar energy conversion systems[10] to sensors[3,11] and switching devices.[7,11] A key reaction for the synthesis of many kinds of elaborated tetrathiafulvalenes is the phosphite mediated coupling of 1,3-dithiol-2-chalcogenones to form the tetrathio-substituted olefinic bond. First reported by Corey[12] and later extended by Miles,[13] this reaction type has become ubiquitous for the synthesis of all manner of tetrathiafulvalenes. The 1,3-dithiol-2-ones are also useful synthetically as protected forms of dithiolene ligands that are readily unmasked by simple base hydrolysis (Scheme 1).[14-15] The rich electrochemical and photophysical behavior associated with metal dithiolene complexes has also stimulated a large body of research aimed at applications such as superconducting materials,[16] sensing devices,[17] photocatalysts,[18] and reversibly bleachable dyes for Q-switching in near IR lasers.[19]

Scheme 2

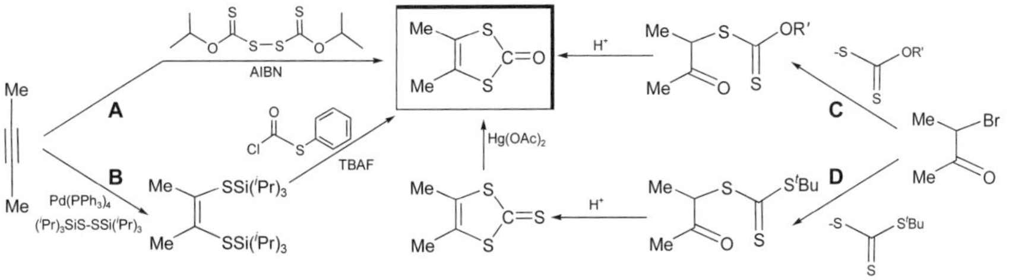

One of the more simple, symmetrically substituted 1,3-dithiol-2-ones is the 4,5-dimethyl form. Despite the amenability of this compound to further derivatization, e.g., by bromination of the methyl groups and subsequent nucleophilic substitution,[6,20] its use has been more limited than would be expected. A considerable number of molybdenum and tungsten dithiolene complexes with the dimethyl-substituted ligand have been reported as structural and functional analogues of the catalytic sites of molybdo-and tungstoenzymes,[21-22] but in all instances the ligand is derived ultimately from 3-hydroxy-2-butanone/P$_4$S$_{10}$ rather than through a more direct approach proceeding through 4,5-dimethyl-1,3-dithiole-2-one. A possible reason for the surprisingly little use that has been made of 4,5-dimethyl-1,3-dithiole-2-one as a protected dithiolene ligand is an underappreciation of the ease by which it may be prepared in large, multigram quantities.

The preparation of this useful molecule can be accomplished by several distinct routes (Scheme 2). One approach involves the radical-initiated reaction of diisopropyl xanthogen disulfide with 2-butyne to form the 1,3-dithiol-2-one directly (Path **A**).[23] A related synthesis describes the palladium-mediated coupling of bis(triisopylsilyl)disulfide with 2-butyne to yield 2,3-bis(triisopropylsilanylsulfanyl)but-2-ene, which is subsequently treated with phenyl chlorothiolformate and TBAF to form the 1,3-dithiol-2-one (Path **B**).[24] Another method begins with the reaction between 3-bromo-2-butanone and an *O*-alkylxanthate salt to afford an *O*-alkyl *S*-3-oxobutan-2-yl dithiocarbonate intermediate (Path **C**) that is then subjected to acid-catalyzed ring closure.[20,25-26] The 4,5-dimethyl-1,3-dithiol-2-one product is also available via mercuric acetate mediated chalcogen exchange[27] from the corresponding trithiocarbonate (Path **D**).[28] The route to 4,5-dimethyl-1,3-dithiol-2-one that begins with 3-bromo-2-butanone and potassium *O*-isopropylxanthate avoids the use of 2-butyne and its associated flammability and expense. While this reaction sequence has been previously reported,[20,25-26] we describe procedural modifications that enable the cost effective synthesis of 4,5-dimethyl-1,3-dithiol-2-one on a scale of tens of grams. In contrast, the largest reported scale on which it has previously been reported is seven grams.[27]

The first point we emphasize is that the α-bromo ketone that is condensed with *O*-isopropylxanthate can be generated *in situ* from 2-butanone/Br$_2$ and used immediately. The 2-butanone can in fact be used in excess as both solvent and reagent. Previously published procedures have described the use of α-bromo ketone purchased from commercial sources or have implied a need to prepare and isolate it in a discrete step. The ease and selectivity with which 2-butanone is brominated in the 3-position and the

340

intrinsic reactivity of the resulting α-bromoketone render *in situ* generation and immediate use a greatly preferable approach. The second procedural improvement we note for the synthesis of 4,5-dimethyl-1,3-dithiol-2-one is that, although a solid at ambient temperature and pressure, it has a low melting point and is sufficiently volatile at reduced pressure as to be amenable to vacuum distillation. A simple short-path distillation can yield a large volume of pure 4,5-dimethyl-1,3-dithiol-2-one without the need to resort to column chromatography as a purification method, as has been reported earlier.[20] Together, these improvements should make this useful molecule much more accessible than has heretofore been recognized.

1. Department of Chemistry, Tulane University, 6400 Freret Street, New Orleans, LA 70118-5698. Email: donahue@tulane.edu.
2. Pfizer Global Research & Development, Eastern Point Road, Groton, CT, 06340.
3. Sarhan, Abd El-Wareth A. O. *Tetrahedron* **2005**, *61*, 3889.
4. Iyoda, M.; Hasegawa, M.; Miyake, Y. *Chem. Rev.* **2004**, *104*, 5085.
5. Jeppesen, J. O.; Nielsen, M. B.; Becher, J. *Chem. Rev.* **2004**, *104*, 5115.
6. Gorgues, A.; Hudhomme, P.; Sallé, M. *Chem. Rev.* **2004**, *105*, 5151.
7. Rovira, C. *Chem. Rev.* **2004**, *104*, 5289.
8. Fourmigué, M.; Batail, P. *Chem. Rev.* **2004**, *104*, 5379.
9. Dressel, M.; Drichko, N. *Chem. Rev.* **2004**, *104*, 5689.
10. Martín, N.; Sánchez, L.; Herranz, M. A.; Illescas, B.; Guldi, D. M. *Acc. Chem. Res.* **2007**, *40*, 1015.
11. Moonen, N. N. P.; Flood, A. H.; Fernández, J. M.; Stoddart, J. F. *Top. Curr. Chem.* **2005**, *262*, 99.
12. Corey, E. J.; Märkl, G. *Tetrahedron Lett.* **1967**, 3201.
13. Miles, M. G.; Wilson, J. D.; Dahm, D. J.; Wagenknecht, J. H. *J. Chem. Soc., Chem. Commun.* **1974**, 751.
14. Robertson, N.; Cronin, L. *Coord. Chem. Rev.* **2002**, *227*, 93.
15. Rauchfuss, T. B. *Prog. Inorg. Chem.* **2004**, *52*, 1.
16. Faulmann, C.; Cassoux, P. *Prog. Inorg. Chem.* **2004**, *52*, 399.
17. Pilato, R. S.; Van Houten, K. A. *Prog. Inorg. Chem.* **2004**, *52*, 369.
18. Cummings, S. D.; Eisenberg, R. *Prog. Inorg. Chem.* **2004**, *52*, 315
19. Mueller-Westerhoff, U. T.; Vance, B.; Yoon, D. I. *Tetrahedron* **1991**, *47*, 909.

20. Crivillers, N.; Oxtoby, N. S.; Mas-Torrent, M.; Veciana, J.; Rovira, C. *Synthesis* **2007**, 1621.

21. McMaster, J.; Tunney, J. M.; Garner, C. D. *Prog. Inorg. Chem.* **2004**, *52*, 539.

22. Enemark, J. H.; Cooney, J. J. A.; Wang, J.-J.; Holm, R. H. *Chem. Rev.* **2004**, *104*, 1175.

23. Gareau, Y.; Beauchemin, A. *Heterocycles* **1998**, *48*, 2003.

24. Gareau, Y.; Tremblay, M.; Gauvreau, D.; Juteau, H. *Tetrahedron* **2001**, *57*, 5739.

25. Rae, I. D. *Int. J. Sulf. Chem.* **1973**, *8*, 273.

26. Bhattacharya, A. K.; Hortmann, A. G. *J. Org. Chem.* **1974**, *39*, 95.

27. Schulz, R.; Schweig, A.; Hartke, K.; Köster, J. *J. Am. Chem. Soc.* **1983**, *105*, 4519.

28. Haley, N. F.; Fichtner, M. W. *J. Org. Chem.* **1980**, *45*, 175.

Appendix
Chemical Abstracts Nomenclature; (Registry Number)

O-Isopropyl *S*-3-Oxobutan-2-yl Dithiocarbonate: Carbonodithioic acid, O-(1-methylethyl) S-(1-methyl-2-oxopropyl) ester; (958649-73-5)
2-Butanone; (78-93-3)
Potassium *O*-isopropylxanthate: Carbonodithioic acid, O-(1-methylethyl) ester, potassium salt (1:1); (140-92-1)
4,5-Dimethyl-1,3-dithiol-2-one; (49675-88-9)

James P. Donahue was born in 1968 in Mishawaka, IN. He graduated in 1991 from M.I.T. with a B.S. in chemistry and joined the laboratory of Richard H. Holm of Harvard University as an NDSEG predoctoral fellow. His doctoral work concerned synthetic analogues of the active sites of molybdo- and tungstoenzymes. In 1998, he received his Ph.D. and moved to Texas A&M University to do postdoctoral work with F. A. Cotton with the support of an NIH postdoctoral fellowship. In 2004, he accepted a position at Tulane University in New Orleans, where he is currently an assistant professor of chemistry.

Chandrasekaran was born in 1979 in Kakkanur, Tamil Nadu, India. He obtained his Bachelors degree in chemistry from Sacred Heart College (Tirupattur) and Masters from Pachaiyappa's College at Chennai. He then moved to Indian Institute of Technology Bombay for his Ph.D. studies under the guidance of Prof. Maravanji S. Balakrishna. His doctoral research was focused on organometallic chemistry of cyclic phosphorus-nitrogen compounds. He is currently a post-doctoral researcher with Prof. James P. Donahue at Tulane University.

Daniel M. Bowles was born in Charleston, West Virginia and received a B.S. in Chemistry from West Virginia Wesleyan College in 1996. He earned a Ph.D. in Organic Chemistry from the University of Kentucky under Dr. John Anthony in 2000 for his work on multi-step synthesis of a series of polycyclic aromatic oligomers. He joined C. Edgar Cook at RTI in Research Triangle Park, North Carolina for postdoctoral study in steroid research, then took a position with Pfizer Chemical Research and Development in Holland, Michigan in late 2001. He moved with Pfizer to the Ann Arbor site in 2003, and again to Groton, CT in 2007, where he currently works as a Principal Scientist in the RAPI (early candidate development) group.

B-PROTECTED HALOBORONIC ACIDS
FOR ITERATIVE CROSS-COUPLING

Submitted by Steven G. Ballmer, Eric P. Gillis, and Martin D. Burke.[1]
Checked by Daniel Morton and Huw M. L. Davies.

1. Procedure

A. N-Methyliminodiacetic acid (MIDA, **2**) (Note 1). A 1000-mL,
three-necked, round-bottomed flask equipped with a PTFE-coated magnetic
stir bar is charged with iminodiacetic acid (**1**) (100.5 g, 755.2 mmol, 1
equiv) (Note 2) and formalin (84.5 mL, 92.1 g, 1.13 mol, 1.50 equiv) (Note
3) to give an off-white suspension. The flask is then fitted with a water-
cooled Friedrichs condenser in the center neck, a 125-mL addition funnel
(Note 4) containing formic acid (57.0 mL, 69.5 g, 1.51 mol, 2.00 equiv)

344

(Note 5) in a side neck, and a thermometer in the other side neck. The stirred reaction mixture is brought to reflux (90 °C) using a heating mantle, and then maintained at reflux with stirring for 30 min. After 30 min the formic acid is added dropwise over 20 min (approx 3 mL/min) (Note 6). During this time, the reaction mixture becomes clear and yellow. The addition funnel stop-cock is then closed and the solution is allowed to reflux for one hour. At the end of one hour the heat source is removed and the solution is allowed to cool with stirring to 23 °C over one hour. The Friedrichs condenser, addition funnel, and thermometer are then removed and the reaction mixture is poured into a 4000-mL Erlenmeyer flask equipped with a large PTFE-coated magnetic stir bar. Deionized water (2 x 25 mL) is used to quantitatively transfer the contents of the reaction mixture to the Erlenmeyer flask. To the stirred reaction mixture is then added absolute ethanol (750 mL) dropwise over one hour (approx 12.5 mL/min) (Note 7) leading to the precipitation of a colorless, crystalline powder. The precipitate is collected by vacuum filtration. The 4000-mL Erlenmeyer flask is then rinsed with absolute ethanol (4 x 200 mL), with each washing being poured over the collected precipitate. The precipitate is then washed with absolute ethanol (200 mL). The precipitate is allowed to air dry under vacuum suction for 10 min. The solid is then transferred to a tared 500-mL round-bottomed flask and residual solvent is removed under reduced pressure (23 °C, 1 mmHg) for 12 h to give the title compound as a free-flowing, air-stable, white powder (98.30 g, 668.1 mmol, 88% yield) (Note 8).

B. 4-Bromophenylboronic MIDA ester (**4**). A 500-mL, single-necked, round-bottomed flask equipped with a PTFE-coated magnetic stir bar is charged with 4-bromophenylboronic acid (**3**) (24.99 g, 124.4 mmol, 1.0 equiv) (Note 9) and methyliminodiacetic acid (**2**) (18.31 g, 124.4 mmol, 1.0 equiv). To the flask is then added a freshly prepared 5% (v/v) solution of dimethyl sulfoxide in toluene (125 mL) (Note 10) to afford a white solid suspended in a clear, colorless solution. The flask is then fitted with a toluene-filled Dean-Stark trap topped with a water-cooled Friedrichs condenser (Note 11) and the stirred reaction mixture is brought to reflux using a heating mantle. The reaction mixture is allowed to reflux with stirring for 6 h, during which time the reaction remains heterogeneous, but darkens in color, giving a tan solid suspended in a clear, colorless solution. Approximately 2.1 mL of water was collected in the Dean-Stark trap. The heating mantle is removed, and the reaction is allowed to cool to 23 °C with stirring for one hour. The Dean-Stark trap and magnetic stir bar are removed (Note 12) and the reaction mixture is concentrated on a rotary evaporator (40 °C, 15 mmHg) (Note 13) to afford the crude product as a tan, chunky solid.

Acetone (15 mL) (Note 14) is then added and the flask is swirled vigorously to afford a white solid suspended in a clear tan solution. To this mixture is added diethyl ether (150 mL) (Note 15) in 25-mL portions. After the addition of each portion the flask is swirled gently causing the precipitation of additional white solid. The white solid is collected via vacuum filtration on a 150-mL medium-porosity glass frit. The collected product is then washed with diethyl ether (3 x 50 mL) and is allowed to dry for 5 min under vacuum suction (Note 16). The product is then transferred to a tared 250-mL round-bottomed flask and residual solvent is removed at reduced pressure (23 °C, 1 mmHg) for 4 h to give 4-bromophenylboronic acid MIDA ester as a free-flowing, air-stable, white powder (36.30 g, 116.4 mmol, 94% yield) (Note 17).

C. 4-(p-Tolyl)-phenylboronic acid MIDA ester (**5**). An oven-dried, 500-mL Schlenk flask equipped with a magnetic stir bar is charged with palladium (II) acetate (361 mg, 1.61 mmol, 0.020 equiv) and (2-biphenyl)-dicyclohexylphosphine (1.160 g, 3.310 mmol, 0.041 equiv) (Note 18) and then quickly sealed with a rubber septum and placed under an inert atmosphere through five cycles of evacuation (1 mmHg) and purging with dry argon (Note 19). Tetrahydrofuran (400 mL) is then cannulated into the 500-mL Schlenk flask resulting in a clear, pale yellow-orange solution (Note 20). Under positive argon pressure the rubber septum is removed and replaced with an oven-dried water-cooled Graham condenser topped with an oven-dried hose barb adapter. Argon is allowed to flow through the system for 60 seconds at which point an argon inlet is attached to the hose barb and the Schlenk valve is closed. The reaction vessel is lowered into an oil bath preheated to 70 °C and the solution is allowed to reflux with stirring for 20 min, during which time the catalyst solution turns colorless. The heating bath is then removed and the catalyst solution is allowed to cool to 23 °C over 10 min. The Schlenk valve is reopened and under a positive argon pressure the Graham condenser is replaced with a rubber septum and the head space is purged for 60 sec through a 20G (1.5 inch) vent needle.

In parallel with the catalyst preparation, an oven-dried, 2000-mL, three-necked, round-bottomed flask equipped with a magnetic stir bar and fitted with an oven-dried water-cooled Graham condenser topped with a hose-barb adapter in the center neck and a rubber septa on each side neck is charged with **4** (25.00 g, 80.15 mmol, 1 equiv), *p*-tolylboronic acid (16.39 g, 120.6 mmol, 1.50 equiv) (Note 21), and freshly-ground, anhydrous potassium phosphate (51.06 g, 240.5 mmol, 3.00 equiv) (Note 22). The reaction flask is then quickly placed under inert atmosphere (through the condenser top) through five cycles of evacuation (1 mmHg) and purging with dry argon. Tetrahydrofuran (400 mL) is then added into the 2000-mL flask through one of the side necks (Note 23) by syringe, resulting in a white suspension.

The catalyst solution is then cannulated into the 2000-mL reaction flask with stirring resulting in a yellow suspension. The reaction flask is lowered into an oil bath preheated to 70 °C and allowed to reflux with stirring for 6 h. The heat source is then removed and the reaction mixture is allowed to cool for 20 min with stirring. The Graham condenser and septa are removed from the flask necks and the reaction is quenched with 1000 mL saturated ammonium chloride (Note 24) giving a biphasic mixture comprised of a clear colorless bottom layer and a clear yellow top layer. This mixture is poured into a 4000-mL separatory funnel. The reaction mixture is quantitatively transferred to the separatory funnel with a freshly prepared THF:diethyl ether (1:1) solution (2 x 200-mL). The layers are separated, and the aqueous layer is extracted with THF:diethyl ether (1:1) solution (400 mL). The combined organic layers are washed with saturated aqueous sodium chloride (150 mL) (Note 25), dried over anhydrous $MgSO_4$, and filtered through Celite. The solvent is removed via rotary evaporation (40 °C, 20 mmHg). Residual solvent is removed at reduced pressure (23 °C, 1 mmHg) to afford the crude product as a yellow solid. Acetone (120 mL) is added to the crude product and the resulting mixture is swirled vigorously to give a yellow slurry. Diethyl ether (800 mL) is then added in 4 equal portions with swirling to give an off-white precipitate in a clear yellow solution. The precipitate is collected via vacuum filtration and allowed to air dry over vacuum for 5 minutes. The solid is transferred to a tared 100-mL round-bottomed flask and residual solvent is removed at reduced pressure (23 °C, 1 mmHg) to give the title compound as a free-flowing, air stable, off-white powder (19.86 g, 61.46 mmol, 77% yield).

To this solid is then added acetone (100 mL) and the thin slurry is heated with stirring at 60 °C until the volume is reduced to 60 mL. Diethyl ether (400 mL) is then added in 4 equal portions with swirling to give a white precipitate in a clear light yellow solution. The precipitate is collected via vacuum filtration and allowed to air dry over vacuum for 5 min. The solid is then transferred to a tared 100-mL round-bottomed flask and residual solvent is removed under reduced pressure (23 °C, 1 mmHg) to give the title compound as a free-flowing, air stable, white powder (17.30 g, 53.53 mmol, 67% yield) (Note 26).

D. 4-(p-Tolyl)-phenylboronic acid (**6**). A 1000-mL, single-necked, round-bottomed flask equipped with a PTFE-coated magnetic stir bar is charged with **5** (10.11 g, 31.27 mmol, 1 equiv), tetrahydrofuran (220 mL), and an aqueous solution of 1 M sodium hydroxide (93.5 mL, 93.5 mmol, 2.99 equiv) to give a biphasic system consisting of a clear colorless bottom layer and a cloudy white top layer. The flask is covered with a polypropylene cap and the reaction is stirred vigorously at 23 °C for 10 min to give a biphasic system consisting of a clear colorless bottom layer and a clear yellow top layer. A saturated aqueous solution of ammonium chloride (250 mL) is then added and the reaction is allowed to stir vigorously for five min. The resulting mixture is poured into a 1000-mL separatory funnel and the reaction vessel is rinsed with diethyl ether (4 x 50 mL), each washing being poured into the separatory funnel. The layers are separated and the aqueous layer is extracted with a freshly prepared tetrahydrofuran:diethyl ether (1:1) solution (400-mL). The combined organic layers are dried over anhydrous MgSO$_4$, filtered through Celite, and concentrated via rotary evaporation (40 °C, 20 mmHg). The residual solvent is removed via three azeotropic cycles with acetonitrile on a rotary evaporator (3 x 50 mL, 40 °C, 20 mmHg) and then at reduced pressure (23 °C, 1 mmHg) for 12 h to afford the title compound as a fine, off-white powder (6.24 g, 29.4 mmol, 94% yield) (Notes 27 and 28).

2. Notes

1. MIDA is a commercially available reagent. However, the synthesis described here is highly convenient and very inexpensive (estimated cost including all reagents and solvents is < 10 cents/gram of product).

2. Iminodiacetic acid (98%) was obtained from Alfa Aesar (Lot No. A13R006) and used as received.

3. Formalin (37 wt % formaldehyde) was obtained from Sigma-Aldrich (Lot No. 06010EH) and used as received.

4. Failure to condense the formaldehyde also results in significant formation of paraformaldehyde on the condenser and addition funnel.

Paraformaldehyde can be easily removed with an alcoholic sodium hydroxide solution. All ground glass joints were sealed with Apiezon H high temperature vacuum grease and secured with Keck clips.

5.   Formic acid (98+%) was obtained from Acros (Lot No. A0254874) and used as received.

6.   Addition of formic acid results in effervescence of CO_2, which can become vigorous if the addition is performed too quickly.

7.   Ethyl alcohol (200 Proof, absolute, anhydrous, ACS/USP grade) was obtained by the submitters from Pharmco-Aaper, and by the checkers from Decon Labs, Inc., and used as received.

8.   The physical and spectral data for **2** are as follows: mp 215–216 °C dec, uncorrected; ^{1}H NMR (400 MHz, D_2O) δ: 2.98 (s, 3 H), 3.96 (s, 4 H); ^{13}C NMR (100 MHz, 95:5 DMSO-d_6:D_2O w/ TMS) δ: 41.7, 56.7, 170.0; IR (thin film): 723, 886, 903, 958, 982, 1018, 1065, 1126, 1172, 1223, 1328, 1380, 1477, 1682, 2955, 2998 cm^{-1}; LRMS (ESI+) m/z (rel. intensity) 219.1 (24%), 148.1 (M$^+$, 100%), 102.1 (8%). HRMS (ESI+) for $C_5H_{10}NO_4$ [M+H$^+$] calcd 148.0604; Found: 148.0603 Anal. calcd. for $C_5H_9NO_4$: C, 40.82; H, 6.17; N, 9.52; found: C, 40.55; H, 6.13; N, 9.40.

9.   4-Bromophenylboronic acid (containing varying amounts of anhydride) was obtained from Aldrich (Lot No. 78396DJ) and used as received. To the best of the authors' knowledge, the amount of boroxine present in the starting boronic acid has no effect on the complexation reaction.

10.   Toluene (certified ACS) was obtained from Fisher Scientific (Lot No. 072584) and used as received. Dimethyl sulfoxide (certified ACS) was obtained from Fisher Scientific (Lot No. 066635) and used as received.

11.   All ground glass joints were sealed with Apiezon H high vacuum grease and secured with Keck clips. The arm of the Dean-Stark trap was wrapped in two layers of aluminum foil to facilitate refluxing.

12.   The submitters removed the magnetic stir bar with forceps and rinsed with toluene in order to flush particulates back into the reaction flask, the checkers used a magnetic retriever and rinsed with toluene.

13.   The reaction is concentrated to remove toluene. There is no need to remove the DMSO at this point.

14.   Acetone (certified ACS) was obtained from Fisher Scientific and used as received.

15. Diethyl ether was obtained from a solvent delivery system, with solvent purified via passage through packed dry neutral alumina columns as described by Pangborn and coworkers.[0]

16. It should be noted that the tan filtrate may be concentrated and the tan solid purified via silica gel column chromatography to obtain a nearly quantitative yield of the title product.

17. In an experiment on 50% scale (12.5 g of 4-bromophenylboronic acid), the checkers obtained a yield of 83%. The physical and spectral data for **4** are as follows: mp 238–240 °C, uncorrected; ^{1}H NMR (400 MHz, CD_3CN) δ: 2.50 (s, 3 H), 3.89 (d, J = 16.0 Hz, 2 H), 4.07 (d, J = 16.0 Hz, 2 H), 7.41 (d, J = 8.0 Hz, 2 H), 7.55 (d, J = 8.0 Hz, 2 H); ^{13}C NMR (100 MHz, CD_3CN) δ: 48.2, 62.5, 124.1, 131.6, 135.2, 169.2; ^{11}B NMR (100 MHz, CD_3CN) δ: 12.0; IR (thin film, acetone): 707, 812, 867, 995, 1037, 1187, 1216, 1237, 1294, 1339, 1459, 1584, 1745, 3012 cm^{-1}. LRMS (EI^+) m/z (rel. intensity) 314.0 (97%), 313.0 (20%), 312.0 ([M^++H], 100%), 311.0 (23%), 283.0 (16%), 255.9 (16%). HRMS (EI+) for $C_{11}H_{12}BBrNO_4$ [M+H]$^+$ calcd: 312.0037, found: 312.0035; Anal. calcd. for $C_{11}H_{11}BBrNO_4$: C, 42.36; H, 3.55; N, 4.49; Found: C, 42.42; H, 3.61; N, 4.67.

18. Palladium (II) acetate (98%) was obtained from Sigma-Aldrich (Lot No. 09417MH) and used as received. (2-biphenyl)-dicyclohexylphosphine (97%) was obtained from Sigma-Aldrich (Lot No. 12209BH) and used as received. Both compounds were massed out at a bench-top balance open to air.

19. Although it is possible to perform this type of selective cross-coupling reaction without the use of a glovebox, it is very important that rigorous Schlenk techniques are utilized to exclude water. Failure to exclude water can result in hydrolysis of the MIDA boronate ester.

20. The submitters obtained tetrahydrofuran from a solvent delivery system, with solvent passage through packed dry neutral alumina columns as described by Pangborn and coworkers.[2] It was dispensed directly from the system into an oven-dried, 1000-mL single-necked, round-bottomed flask which was quickly sealed with a rubber septum. Immediately following, the head space of the flask was purged with dry argon for 60 sec. The checkers distilled tetrahydrofuran from sodium and benzophenone.

21. *p*-Tolylboronic acid was obtained from Oakwood Products, Inc. (Lot No. A30J) and used as received.

22. K_3PO_4 (anhydrous, 97%) was obtained from Alfa Aesar (Lot No. A23R022) and finely-ground just prior to use. It is very important that the

K$_3$PO$_4$ is finely ground and that it remains anhydrous throughout this process. This can be achieved using a glove box. Alternatively, a convenient way to achieve this without the use of a glove box is as follows: ~10% excess of the desired amount of K$_3$PO$_4$ is massed out on a benchtop balance and quickly poured into a hot mortar (removed from a 60 °C oven just prior to use) and finely ground quickly using a hot pestle (removed from a 60 °C oven just prior to use). The ground base is massed quickly on a benchtop balance and transferred to the reaction vessel. This was the method used by the checkers.

23. The submitters obtained tetrahydrofuran and diethyl ether from a solvent delivery system, with solvent purified via passage through packed dry neutral alumina columns as described by Pangborn and coworkers.[2] The checkers distilled the tetrahydrofuran from sodium and benzophenone.

24. Ammonium chloride (99.5% ACS reagent) was obtained from Sigma-Aldrich and added to deionized water until saturated.

25. Sodium chloride (ReagentPlus ≥ 99.5%) was obtained from Sigma-Aldrich and added to deionized water until saturated.

26. In an experiment on 50% scale (12.5 g of 4-bromophenylboronic acid MIDA **4**), the checkers obtained a yield of 70%. The physical and spectral data for **5** are as follows: mp 214–216 °C dec, uncorrected; ^{1}H NMR (400 MHz, CD$_3$CN) δ: 2.31 (s, 3 H), 2.54 (s, 3 H), 3.91 (d, J = 16.0, 2 H), 4.09 (d, J = 16.0 Hz, 2 H), 7.28 (d, J = 8.0 Hz, 2 H), 7.56 (d, J = 8.0 Hz, 4 H), 7.64 (dt, J = 8.0, 2.0 Hz, 2 H); ^{13}C NMR (100 MHz, CD$_3$CN) δ: 20.6, 48.0, 62.3, 126.6, 127.2, 130.0, 133.6, 137.9, 138.1, 142.0, 169.1; ^{11}B NMR (100 MHz, CD$_3$CN) δ: 12.4; IR (thin film): 800, 985, 1035, 1236, 1299, 1341, 1745, 3022 cm^{-1}; LRMS (EI$^+$) m/z (rel. intensity) 325.1 (18%), 324.1 (M+H, 100%), 323.1 (23%), 323.0 (15%); HRMS (EI+) for C$_{18}$H$_{19}$BNO$_4$ [M+H]$^+$ calcd: 324.1402, found: 324.1406; The submitters found: Anal. calcd. for C$_{18}$H$_{18}$BNO$_4$: C, 66.90; H, 5.61; N, 4.33; found: C, 66.72; H, 5.56; N, 4.47. An accurate microanalysis was not achieved by the checkers. The solid, free flowing product **5** was first dried under reduced pressure (23 °C, 1 mmHg), for a period of 5 h; found: C, 65.77; H, 5.56; N, 4.21. The solid was then re-precipitated from acetone and diethyl ether as described in the procedure and dried under reduced pressure (23 °C, 1 mmHg), for 24 h; found: C, 65.92; H, 5.70; N, 4.27. A sample that was precipitated twice and dried extensively under reduced pressure while being heated (35 °C, 1 mmHg), for 48 h, was further away than before; found: C, 63.90; H, 5.21; N, 4.23.

27. In an experiment on 50% scale (5.05 g of 4-(p-tolyl)-phenylboronic acid MIDA **5**), the checkers obtained a yield of 91%. The physical and spectral data for **6** are as follows: mp 136–138 °C dec, uncorrected; ^{1}H NMR (400 MHz, 95:5 DMSO-d_6:D$_2$O) δ: 2.29 (s, 3 H), 7.22 (d, J = 8.0 Hz, 2 H), 7.53 (d, J = 8.0 Hz, 2 H), 7.56 (d, J = 8.0 Hz, 2 H), 7.80 (d, J = 8.0 Hz, 2 H); ^{13}C NMR (100 MHz, 95:5 DMSO-d_6:D$_2$O) δ: 20.9, 125.7, 126.8, 129.8, 135.0, 137.2, 137.4, 141.8; ^{11}B NMR (95:5 DMSO-d_6:D$_2$O w/ TMS) δ: 36.3; IR (thin film, acetone): 740, 806, 1003, 1093, 1154, 1339, 1530, 1607, 3331 cm^{-1}; LRMS (EI$^+$) m/z (rel. intensity) 212.1 (M+, 8%), 185.1 (16%), 170.1 (14%), 169.1 (M-BO$_2$, 100%); HRMS (EI+) for C$_{13}$H$_{13}$BO$_2$ [M$^+$] calcd: 212.1003, found: 212.1006. Due to the unpredictable composition of boronic acids and their corresponding boroxines, elemental analysis does not provide an accurate measure of purity. Based on the ^{1}H NMR the checkers afforded the title compound in 91–93% purity. Based on ^{1}H NMR, the submitters afforded the title compound in 82–87% purity (Note 28).

28. The ^{1}H NMR spectrum was obtained of a freshly prepared solution of product **6** in DMSO-d_6:D$_2$O 95:5 with tetramethylsilane (TMS) added as an internal reference. The following procedure was followed by the checkers to establish the purity of product **6**: The integration for the methyl resonance at 2.29 ppm was normalized to 3.00 and the two minor resonances at 2.30 and 2.25 ppm were integrated. The sum integration for the minor resonances was 0.21, which represented 7% of the total integration area in that region. Similarly, the aryl resonances at 7.80, 7.56, 7.53, and 7.22 ppm were integrated as were the eight minor resonances at 7.84, 7.74, 7.72, 7.65, 7.49, 7.41, 7.39, and 7.16 ppm. The sum integration for the minor resonances was 0.79, which represented 9% of the total integration area in that region. The combined calculations suggest the presence of 7–9% impurity (91–93% purity). The following procedure was followed by the submitters to establish the purity of product **6**: The integration for the methyl resonance at 2.35 ppm was normalized to 3.00 and the two minor resonances at 2.36 and 2.31 ppm were integrated. The sum integration for the minor resonances was 0.46, which represented 13% of the total integration area in that region. Similarly, the aryl resonances at 7.87, 7.62, 7.59, and 7.28 ppm were integrated as were the eight minor resonances at 7.91, 7.82, 7.77, 7.70, 7.55, 7.45, 7.34, and 7.23 ppm. The sum integration for the minor resonances was 1.90, which represented 18% of the total integration area in

that region. The combined calculations suggest the presence of 13–18% impurity (82–87% purity).

Safety and Waste Disposal Information

All hazardous materials should be handled and disposed of in accordance with "Prudent Practices in the Laboratory"; National Academy Press; Washington, DC, 1995.

3. Discussion

Figure 1. Analogous strategies for the synthesis of peptides and small molecules.

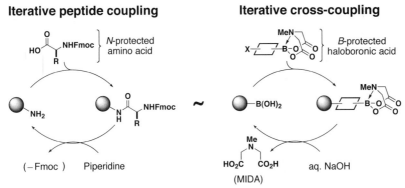

Inspired by the powerful simplicity of peptide coupling (Fig. 1, left), we recently reported an analogous strategy for the synthesis of small molecules involving the iterative cross-coupling (ICC) of B-protected "haloboronic acids" (Fig. 1, right).[3] In an ideal ICC pathway, only stereospecific cross-coupling reactions are utilized to assemble a collection of building blocks having all of the required functional groups preinstalled in the correct oxidation state and with the desired stereochemical relationships. This type of pathway is inherently modular and flexible, and thus highly amenable to analog and/or library synthesis. The approach is also well-suited for automation, a goal which is currently being pursued.

To avoid random oligomerization, this strategy necessitates methodology for reversibly attenuating the reactivity of one end of the bifunctional haloboronic acids, analogous to an Fmoc protective group routinely utilized to control the iterative coupling of amino acids. In this vein, we discovered that the trivalent heteroatomic ligand MIDA can pyramidalize and thereby inhibit the reactivity of a boronic acid under anhydrous cross-coupling conditions.[3] This methodology enables the selective cross-coupling of B-protected haloboronic acids, and is general for aryl, heteroaryl, alkenyl, and alkyl derivatives (Table 1 and Fig. 2).[3] Critical

Table 1. Cross-coupling and Deprotection of B-Protected Haloboronic Acids

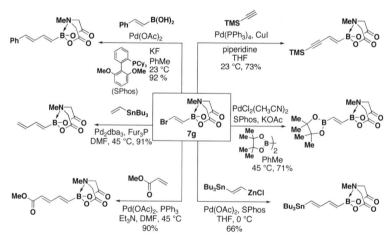

Entry	7	Protected product	% Yield	Deprotected product	% Yield
1	7a	**8a** (*para*)	87[a]	**9a** (*para*)	86
2	7b	**8b** (*meta*)	85	**9b** (*meta*)	92
3	7c	**8c** (*ortho*)	80	**9c** (*ortho*)	97[b]
4[c]	7d	**8d**	81	**9d**	88
5	7e	**8e**	82	**9e**	83
6	7f	(±) **8f**	94	(±) **9f**	91

[a] The same yield was observed whether this reaction was set up in the glovebox or in the air.

[b] B-Deprotection was achieved via treatment with saturated aq. NaHCO$_3$/MeOH, 23 °C, 6 h, (85%).

[c] 2-(Dicyclohexylphosphino)-2',4',6'-triisopropyl-1,1'-biphenyl was used as ligand.

Figure 2. Diverse Uses of B-Protected Haloboronic Acids

for applications in complex small molecule synthesis, this ligand can be removed under very mild aqueous basic conditions, including 1N aqueous NaOH or even saturated aqueous NaHCO$_3$ in methanol (Table 1).[3]

Our early studies have additionally demonstrated the feasibility of the ICC concept with the synthesis of a variety of small molecule natural products (for example, see Schemes 1 and 2).[3] We herein describe methods for translating some of this chemistry to the decagram scale.

Scheme 1

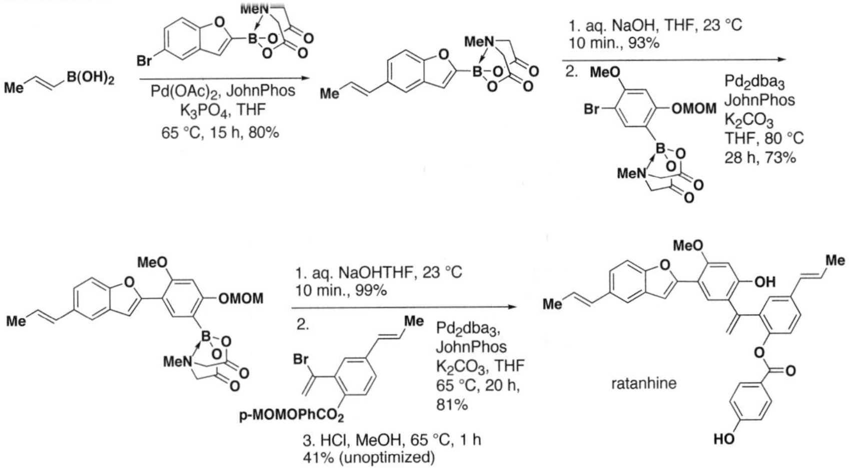

Scheme 2

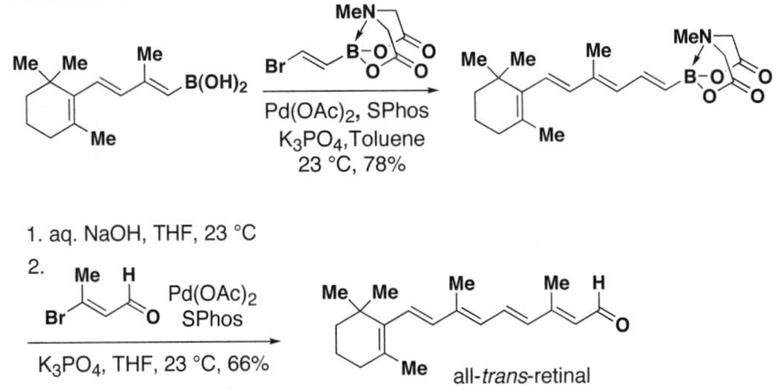

Importantly, there are many features of this chemistry that make it very well-suited for execution on scale. First, MIDA is non-toxic, biodegradable,[4] and can be conveniently prepared in analytically pure form using the procedure described herein for very low cost (estimated cost including all solvents and reagents is < 10 cents/gram). In addition, the MIDA ligand is indefinitely stable on the benchtop under air. The low cost of this procedure is in large part because the key starting material, iminodiacetic acid, is a commodity chemical used in the preparation of herbicides (>60,000 metric tons of iminodiacetic acid is synthesized worldwide each year).[5] The only other reagents are formic acid and formaldehyde.

A previously reported synthesis of MIDA (2) proceeds in moderate yield from the reaction of methylamine and chloroacetic acid.[6] A more effective preparation utilizes formalin and formic acid to reductively methylate iminodiacetic acid in good yield (the Eschweiler-Clarke conditions).[7] The simple method described herein is an optimized version of this reaction, and is viewed by the submitters to be considerably more convenient and effective than syntheses reported previously.

The synthesis of MIDA boronates were first reported by Mancilla and coworkers.[8] We herein describe a simple and efficient procedure for preparing substantial quantities of B-protected haloboronic acids using a minimum volume of DMSO (employed to solubilize the MIDA ligand). Specifically, using a standard Dean-Stark apparatus, the complexation between p-bromophenylboronic acid and MIDA proceeds in excellent yield with purification via simple precipitation from acetone/Et_2O. Moreover, the selective Suzuki-Miyaura cross-coupling between p-tolylboronic acid and this B-protected haloboronic acid[3a] was achieved on the decagram scale without the use of a glove box. In order to avoid MIDA hydrolysis during the course of this reaction, it is critical to utilize rigorous Schlenk techniques and avoid the introduction of water. Freshly grinding anhydrous K_3PO_4 using a hot mortar and pestle (recently removed from a 60 °C oven) is a convenient way to maintain anhydrous conditions during the reaction setup. Alternatively, a glove box can be used. Finally, we demonstrate that the hydrolysis of the MIDA ligand can conveniently be performed on scale to yield the corresponding boronic acid.

In addition to the capacity for iterative cross-coupling, there are numerous enabling features that make MIDA boronates highly attractive intermediates for organic synthesis. In contrast to their boronic acid

356

counterparts, these compounds are invariably monomeric and highly crystalline free-flowing solids. They have also proven to be extremely stable to benchtop storage under air and universally compatible with silica gel chromatography.[3] As demonstrated herein, the synthesis, selective cross-coupling, and deprotection of MIDA boronates is also scalable. Moreover, we have recently discovered that the MIDA boronate functional group is stable to a wide range of common synthetic reagents, thereby enabling complex boronic acid building blocks to be reliably prepared from simple B-containing starting materials via multistep synthesis.[9] In addition, we have found that MIDA boronates can serve directly as cross-coupling partners under aqueous basic conditions via the in situ release of the corresponding boronic acids.[10] Collectively, these features suggest that MIDA boronates represent a superior platform for the preparation, purification, storage, and utilization of organoboranes in organic synthesis.

1. Department of Chemistry, University of Illinois at Urbana-Champaign, Urbana, IL 61801-3602, USA; E-mail: burke@scs.uiuc.edu; Sigma-Aldrich is gratefully acknowledged for generous gifts of 4-bromophenylboronic acid, (2-biphenyl)dicyclohexylphosphine, and palladium (II) acetate. We also thank the NSF (CAREER 0747778), Dreyfus Foundation, and the Arnold and Mabel Beckman Foundation for financial support. EPG is a Seemon H. Pines Graduate Fellow.

2. Pangborn, A. B.; Giardello, M. A.; Grubbs, R. H.; Rosen, R. K.; Timmers, F. J. *Organometallics* **1996**, *15*, 1518-1520.

3. (a) Gillis, E. P.; Burke, M. D. *J. Am. Chem. Soc.* **2007**, *129*, 6716-6717. (b) Lee, S. J., Gray, K.C., Paek, J. S., Burke, M. D. *J. Am. Chem. Soc.* **2008**, *130*, 466-468. For a related strategy for oligoarene synthesis, see: Noguchi, H.; Hojo, K.; Suginome, M. *J. Am. Chem. Soc.* **2007**, *129*, 758-759.

4. Warren, C. B.; Malec, E. J. *Science* **1972**, *176*, 277-279.

5. Yangong, Z. *China Chemical Reporter* **2005** p. 16.

6. Berchet, G. J. *Org. Syn.* **1938**, *18*, 56.

7. (a) Childs, A. F.; Goldsworthy, L. J.; Harding, G. F.; King, F. E.; Nineham, A. W.; Norris, W. L.; Plant, S. G. P.; Selton, B.; Tompsett, A. L. L. *J. Chem. Soc.* **1948**, 2174-2177. (b) Chase, B. H.; Downes, A. M. *J. Chem. Soc.* **1953**, 3874-3877.

8. (a) Mancilla, T.; Contreras, R.; Wrackmeyer, B. *J. Organomet. Chem.* **1986**, *307*, 1-6. (b) Mancilla, T.; Zamudio-Rivera, L. S.; Beltrán, H., I.; Santillan, R.; Farfán, N. *ARKIVOC* **2005**, 366.

9. (a) Gillis, E. P.; Burke, M. D. *J. Am. Chem. Soc.* **2008**, *130*, 14084-14085. (b) Uno, B. E.; Gillis, E. P.; Burke, M. D. *Tetrahedron* **2009**, *65*, 3130-3138. (c) Gillis, E.P.; Burke, M.D. *Aldrichimica Acta* **2009**, *42*, 17-27.

10. Knapp, D. M.; Gillis, E. P.; Burke, M. D. *J. Am. Chem. Soc.* **2009**, *131*, 6961-6963.

Appendix
Chemical Abstracts Nomenclature; (Registry Number)

N-Methyliminodiacetic acid;(4408-64-4)

Iminodiacetic acid; (142-73-4)

Formalin; (50-00-0)

Formic acid; (64-18-6)

4-Bromophenylboronic acid; (5467-74-3)

Palladium (II) acetate; (3375-31-3)

(2-Biphenyl)dicyclohexylphosphine; (247940-06-3)

p-Tolylboronic acid; (5720-05-8)

 Marty Burke completed his undergraduate education at Johns Hopkins in 1998 and then moved to Harvard Medical School as a PhD/MD student in the Health Sciences and Technology program. He completed his thesis research under the direction of Prof. Stuart L. Schreiber and graduated from medical school in 2005. That same year, he began his independent career at the University of Illinois at Urbana-Champaign. His research program focuses on the synthesis and study of small molecules that perform protein-like functions. To enable these studies, Marty's group is developing a synthesis strategy, dubbed *iterative cross-coupling*, that aims to make the process of complex small molecule making as simple, efficient, and flexible as possible.

Steve Ballmer was born in Toledo, Ohio and attended Wright State University in Dayton, Ohio, where he majored in chemistry and received his bachelor's of science degree in 2007. There he performed undergraduate research under the supervision of Prof. Daniel M. Ketcha. He is currently pursuing a PhD under the supervision of Prof. Martin Burke.

Eric Gillis grew up in Portland, ME and received his undergraduate education at Grinnell College in Grinnell, IA. While at Grinnell he worked with Professor T. Andrew Mobley on the synthesis and characterization of tungsten-tin complexes. In 2005 he began his doctoral studies under the direction of Martin Burke at the University of Illinois, Urbana-Champaign. His current research focuses on the development MIDA boronate esters as a platform to enable the simple, efficient, and flexible synthesis of small molecules.

Daniel Morton obtained his MChem from the University of East Anglia in 2001, where he remained for his PhD (supervised by Dr. Rob Stockman and Prof. Rob Field) which concerned the synthesis of chiral aziridines. In 2005, he joined the group of Prof. Adam Nelson at the University of Leeds, as a post-doctoral research associate, where he worked on the total synthesis of hemibrevetoxin B and the diversity-oriented synthesis of natural product-like molecules. In 2008 he moved to the group of Huw Davies, at Emory University, Atlanta, where he is currently exploring the use of C-H functionalization in the generation of molecular complexity.

RHODIUM-CATALYZED ENANTIOSELECTIVE ADDITION OF ARYLBORONIC ACIDS TO *IN SITU* GENERATED *N*-BOC ARYLIMINES. PREPARATION OF (*S*)-*TERT*-BUTYL (4-CHLOROPHENYL)(THIOPHEN-2-YL)METHYLCARBAMATE

A.

B.

Submitted by Morten Storgaard and Jonathan A. Ellman.[1]
Checked by Jason A. Bexrud and Mark Lautens.

1. Procedure

A. tert-Butyl phenylsulfonyl(thiophen-2-yl)methylcarbamate (1). In a 250-mL, round-bottomed flask benzenesulfinic acid sodium salt (13.13 g, 80.0 mmol, 2.0 equiv) (Note 1) is dissolved in H_2O (105 mL) (Note 2). *tert*-Butyl carbamate (4.69 g, 40.0 mmol, 1.0 equiv) (Note 3) is added, but does not dissolve. 2-Thiophene-carboxaldehyde (5.50 mL, 6.73 g, 60.0 mmol, 1.5 equiv) (Note 4) is added forming a yellow emulsion. Formic acid (3.10 mL, 3.68 g, 80.0 mmol, 2.0 equiv) (Note 5) is added. The flask is loosely fitted with a rubber septum and the white, opaque, biphasic mixture is stirred

Org. Synth. **2009**, *86*, 360-373
Published on the Web 7/9/2009

vigorously at room temperature (23 °C). After a couple of hours the water phase becomes clear. The product **1** is formed as yellow chunks, which become more dispersed in the water phase as the reaction proceeds. After 3 days (Note 6) of stirring the suspension is vacuum filtered (Note 7). The yellow chunks are crushed with a spatula, and the product is triturated with H_2O (2 × 10 mL) and Et_2O (2 × 10 mL) (Note 8). After each trituration the solvent is removed by vacuum filtration. Finally, it is dried for an hour under high vacuum (Note 9) to give the imine precursor **1** as a white solid (8.21 g, 58%) (Notes 10 and 11).

B. *(S)-tert-Butyl (4-chlorophenyl)(thiophen-2-yl)methylcarbamate (2).* An oven-dried (Note 12), 250-mL, three-necked round-bottomed flask with a magnetic stir bar is equipped with a vacuum adaptor in the middle neck and glass stoppers in the two other necks (one of which is loosely fitted to allow an outflow of nitrogen gas). The adaptor is connected to a nitrogen gas line (Note 13) and the flask is purged with nitrogen as it is allowed to cool to ambient temperature (23 °C). The flask is then charged with [RhCl(cod)]₂ (247 mg, 0.50 mmol, 0.025 equiv) (Note 14) and (*R,R*)-deguPHOS (583 mg, 1.1 mmol, 0.055 equiv) (Note 15) by removing one of the glass stoppers. A septum is used to seal the flask and the other glass stopper is exchanged with an adaptor equipped with a thermometer. The flask is then purged with nitrogen for 5 min and a positive nitrogen flow is thereafter maintained to ensure an oxygen-free atmosphere inside the flask (Note 16). Dry dioxane (80 mL) (Note 17) is added through the septum via a syringe and the flask is submerged into an oil bath (70 °C), and the mixture is stirred for 1 h (internal temperature: 65 °C, reached after 20 min). Initially, the precatalyst is not fully soluble in dioxane, but as the preincubation proceeds it completely dissolves. The solution of the active catalyst is clear and dark orange.

Meanwhile (Note 18), a 500-mL, oven-dried, three-necked round-bottomed flask (Note 19) with a magnetic stir bar is equipped with a vacuum adaptor in the middle neck and glass stoppers in the two other necks (one of which is loosely fitted to allow outflow of nitrogen gas). The adaptor is connected to a nitrogen gas line and the flask is purged with nitrogen as it is allowed to cool to ambient temperature (23 °C). The flask is then charged with *tert*-butyl phenylsulfonyl(thiophen-2-yl)methylcarbamate (**1**) (7.07 g, 20.0 mmol, 1.0 equiv), 4-chlorophenylboronic acid (6.26 g, 40.00 mmol, 2.0 equiv) (Notes 20 and 21), K_2CO_3 (16.58 g, 120.0 mmol, 6.0 equiv) (Note 22) and 4Å powdered molecular sieves (32 g) (Note 23) by removing one of the glass stoppers. A septum is used to seal the flask, and the other glass stopper is exchanged with an adaptor equipped with a thermometer. The flask is then purged with nitrogen for 5 min, and a positive nitrogen inflow is maintained to ensure an oxygen-free atmosphere inside the flask. Dry dioxane (240 mL) is added through the septum via a syringe immediately before the preincubation is complete (described above). Additionally, dry triethylamine (4.20 mL, 3.04 g, 30.00 mmol, 1.5 equiv) (Note 24) is added via a syringe. The white suspension is stirred vigorously at room temperature (23 °C) while adding the preincubated solution of catalyst and ligand via cannula transfer (Note 25) resulting in a yellow suspension. The reaction flask is submerged into an oil bath (70 °C), and the yellow suspension is stirred vigorously for 16 h (internal temperature: 70 °C) (Note 26). The yellow suspension is allowed to cool to ambient temperature (23 °C) over the course of one hour and vacuum filtered through a plug of Celite^{TM} (Note 27), which is rinsed with EtOAc (300 mL) (Note 28). The combined yellow filtrates are evaporated *in vacuo* (Note 29) to give a yellow solid (Note 30). The crude product is purified by flash chromatography (6.5 × 20 cm, 270 g silica gel) (Note 31) using a gradient of 5 to 15% EtOAc in hexanes and fractions of 50 mL. The column is eluted with 500 mL of 1:19 EtOAc:hexanes (Note 32), 500 mL of 1:12 EtOAc:hexanes, 1500 mL of 1:9 EtOAc:hexanes, 500 mL of 1:7 EtOAc:hexanes and finally with 500 mL of 1:5 EtOAc:hexanes. Fractions 32–65 (Note 33) are combined, evaporated *in vacuo* and dried overnight under high vacuum affording the title compound **2** as a white solid (4.92 g, 76%) (Notes 34 and 35) with 93% ee (Notes 36 and 37).

2. Notes

1.   Benzenesulfinic acid sodium salt (98%) was purchased from Sigma-Aldrich and used without further purification.

2.   Deionized water (H_2O) was used in all cases where the procedures call for water.

3.   *tert*-Butyl carbamate (98%) was purchased from Sigma-Aldrich and was used without further purification.

4.   2-Thiophene-carboxaldehyde (98%) was purchased from Sigma-Aldrich and was used without further purification.

5.   Formic acid (HCOOH) (reagent grade, >95%) was purchased from Sigma-Aldrich and was used without further purification.

6. This reaction was originally published by Wenzel and Jacobsen[2] giving only a 44% yield of product **1** when MeOH:H$_2$O (1:2) was used as the solvent and with a 3 day reaction time. We attempted to increase the yield by heating the reaction mixture to 50 °C, but this resulted in product decomposition. Increasing the reaction concentration resulted in only a slight increase in the yield of **1**. Reducing the amount of MeOH resulted in the most significant increase in yield. Ultimately, running the reaction in pure H$_2$O gave the reported yield. Lower yields were achieved with a reaction time of only 1 day (44%), while a further increase in the yield can be achieved after 5 days (74%).

7. Wilmad Labglass sintered glass funnel, 60 mL, size M, was used.

8. Diethyl ether (Et$_2$O), anhydrous HPLC grade, stabilized, was purchased from Fisher Scientific Chemicals and was used without further purification.

9. High vacuum refers to 0.025 mmHg at 23 °C.

10. *tert*-Butyl phenylsulfonyl(thiophen-2-yl)methylcarbamate (**1**) exhibits the following properties: mp 160–162 °C (decomp.). ^{1}H NMR (400 MHz, CDCl$_3$) δ: 1.26 (s, 9 H), 5.62 (d, J = 10.0 Hz, 1 H), 6.18 (d, J = 10.8 Hz, 1 H), 7.05–7.09 (m, 1 H), 7.26–7.28 (m, 1 H), 7.41–7.43 (m, 1 H), 7.52–7.57 (m, 2 H), 7.63–7.65 (m, 1 H), 7.90–7.94 (m, 2 H). ^{13}C NMR (100 MHz, CDCl$_3$) δ: 28.2, 70.4, 81.7, 127.6, 128.0, 129.3, 129.6, 129.8, 131.7, 134.3, 136.7, 153.4. IR (neat) 3347, 2955, 1699, 1510, 1306, 1150 cm^{-1}. Anal. calcd for C$_{16}$H$_{19}$NO$_4$S$_2$: C, 54.37; H, 5.42; N, 3.96; found: C, 54.39; H, 5.40; N, 3.87.

11. A second run by the checkers provided 10.57 g (75%) of **1** with a melting point range of 159–161 °C. The submitters reported a yield of 8.46 g (60%) with a melting point range of 162–164 °C.

12. Oven-dried refers to drying of flasks, glass stoppers, adaptors and magnetic stir bars in an oven (150 °C) overnight before use. The glassware was assembled while still hot and cooled to ambient temperature (23 °C) under high vacuum. The submitters cooled the glassware under high vacuum.

13. The nitrogen gas line was a standard dual manifold with multiple ports with stopcocks that allow vacuum or nitrogen to be selected without the need for placing the flask on a separate line. One manifold was connected to a source of nitrogen dried through a tube of Drierite® (>98% CaSO$_4$, >2% CoCl$_2$), while the other was connected to a high-vacuum Fisher Scientific Maxima® C Plus Model M8C pump (0.025 mm Hg). The nitrogen

gas line was vented through an oil bubbler that was connected to the manifold through a valve (making it possible to disconnect the bubbler during cannula transfer), while solvent vapors were prevented from contaminating the pump through a dry ice/isopropanol cold trap.

14. [RhCl(cod)]$_2$ (chloro(1,5-cyclooctadiene)rhodium(I) dimer), 98%, which is air stable, was purchased from Strem and was used without further purification.

15. (R,R)-deguPHOS ((3R,4R)-1-benzyl-3,4-bis-diphenylphosphanyl-pyrrolidine), 98%, was purchased from Strem and was used without further purification.

16. The active catalyst is very sensitive to air. It is important to introduce a nitrogen atmosphere to the flask and maintain a positive pressure of nitrogen throughout the preincubation and the reaction to prevent catalyst decomposition.

17. 1,4-Dioxane, HPLC grade, was purchased from Fisher Scientific Chemicals and passed through a column of dry, activated, basic alumina under a nitrogen atmosphere. The solvent is transferred to the flask via a syringe without exposure to air.

18. This part of the procedure can be performed while stirring the precatalyst and ligand, but the dioxane and Et$_3$N should not be added until just immediately before the active catalyst is ready (1 h at 70 °C). This is to avoid premature hydrolysis of the in situ generated imine, which results in a decreased yield of the title compound 2.

19. A 500-mL flask was used instead of a 1000-mL to reduce the risk of catalyst decomposition – minimizing unoccupied volume reduces the risk of oxygen contamination.

20. 4-Chlorophenylboronic acid (95%) was purchased from Sigma-Aldrich and recrystallized from H$_2$O before use as described in Note 21.

21. Commercially available arylboronic acids contain boroximes (anhydride trimers) that do not add efficiently to the in situ generated imine. Therefore, to maximize formation of the title compound 2, we found it very important to recrystallize and dry the arylboronic acid before use. This was carried out as follows: in a 1000-mL conical flask was added 4-chlorophenylboronic acid (10 g) (Note 20) and H$_2$O (400 mL) and the flask was covered with a watch glass. The suspension was heated to boiling over the course of 25 minutes on a heating plate (115 °C) under vigorous stirring with a magnetic stir bar. The boiling point was maintained for 5 minutes to fully dissolve the boronic acid. The hot solution was filtered through filter

paper using gravity filtration to remove insoluble particles. The colorless solution was cooled to ambient temperature (23 °C) overnight and then was cooled in an ice bath for 1 hour (internal temperature: 5 °C). During the cooling process the aryboronic acid precipitated and was isolated by vacuum filtration and dried by continuing the vacuum filtration for an additional 15 minutes. To remove further amounts of water the boronic acid was dried in high vacuum at room temperature (23 °C) until ^{1}H NMR analysis in DMSO-d_6 showed a composition of no more than approx. 5% boroxime and 30% H_2O, at which point the mass of product was 8.0–8.6 g of white microplates. The drying procedure is important because reaction of pure boroxime will cause a reduction in the yield of the title compound **2** down to 52%. Depending on the initial amount of water in the recrystallized batch and the vacuum pump capacity, the time of drying may vary. Usually we were able to obtain the above-mentioned composition requirements within 5–15 minutes of drying in high vacuum. It is highly recommended to dry the arylboronic stepwise, e.g. 5 minutes at a time and then analyze the arylboronic by NMR. In DMSO-d_6 (dried prior to use over 4Å molecular sieves, 3.2 mm pellets) 4-chlorophenylboronic acid exhibits the following chemical shifts (300 MHz): δ 8.16 (s, broad), 7.79 (d, J = 8.3, 2H), 7.39 (d, J = 8.3, 2H), while the corresponding boroxime exhibits these shifts: δ 7.86 (d, J = 8.1, 2H), 7.42 (d, J = 8.1, 2H). The composition can be determined using the integrals directly if the DMSO is water-free. Occasionally, we found it difficult to remove the excess of water without increasing the amount of boroxime to strictly more than 5%. In such cases the batch should be recrystallized again.

22. Potassium carbonate (K_2CO_3), anhydrous, was purchased from EM Science (an affiliate of Merck KGaA) and was dried overnight before use under high vacuum at 100 °C in a thermostatically controlled oil bath.

23. Molecular sieves, 4Å, <5 microns, powdered, were purchased from Sigma-Aldrich and activated under high vacuum at 230–260 °C overnight. Heating was achieved by a Glas-Col® heating mantle, 2/3 filled with sand and connected to a Powerstat® variable autotransformer (in: 120 V, 50/60, ~1 PH, out: 0–140 V, 10 A, 1.4 KVA). The transformer was adjusted to approx. 250 °C as measured with a thermometer placed directly into the sand.

24. Triethylamine was purchased from Fisher Scientific Chemicals and was freshly distilled from CaH_2 under a nitrogen atmosphere before use.

25. Cannulation technique (Figure 1) was used to conveniently transfer the active catalyst solution (A) to the mixture of starting materials, bases and molecular sieves (B) through a cannula (C) without exposure to air. Before inserting the cannula into the flasks, an extra oil bubbler (D) was attached to flask B via a needle through the septum. The cannula (C) was then inserted into flask A and after a minute the other end of C was inserted into flask B. To cannulate the catalyst solution (A), the nitrogen inlet to flask B (E) and the Schlenk line oil bubbler (F) were both closed making the extra oil bubbler (D) the only outlet from the system. After complete cannulation E and F were both opened again, and the extra bubbler (D) and cannula (C) were removed.

26. It is not convenient to monitor the progress of the reaction by TLC because the diagnostic compound (the imine formed *in situ* from **1**) is unstable and does not elute without decomposition on TLC. Therefore, we ran a number of reactions on small scale (0.250 mmol) at different reaction times. We found that the amount of the title product **2** reaches a 65–67% NMR yield after 10 hours. Neither product decomposition nor an increase in yield are observed with prolonged reaction times, e.g., 40 hours at 70 °C. For convenience we chose a 16 h reaction time.

27. Celite^{TM} powder, 545 filter aid, not acid washed, was purchased from Fisher Scientific and was used without further purification. The filter plug was prepared by mixing Celite^{TM} (20 g) with EtOAc (80 mL) and filtered through a Kimax$^®$ sintered glass funnel, 150 mL – 60F.

28. Ethyl acetate (EtOAc), HPLC grade, was purchased from Fisher Chemicals and was used without further purification.

29. Evaporation *in vacuo* was carried out on a Büchi Rotavapor R-114 at 45 mmHg with a Büchi Waterbath B-480 at 35 °C, unless otherwise stated.

30. The ^{1}H NMR (CDCl$_3$) spectrum of the crude product was recorded to determine if the reaction proceeded as expected. Besides EtOAc, dioxane and the peaks corresponding to the title compound **2** (Note 34), the crude product also contains 2-thiophene-carboxyaldehyde (9.68 ppm), and other decomposition compounds: δ 7.86, 7.75, 4.57 and 1.28 ppm. If the reaction has been performed correctly there should be only trace amounts of the the *in situ* generated imine (9.05 and 1.57 ppm) in the crude product, and the crude product should be a yellow solid rather than an oil.

31. Silica gel 60 (0.040–0.063 mm), 230–400 mesh ASTM, was purchased from Merck KGaA and used without further purification.

32. Hexanes, HPLC grade, was purchased from Fisher Scientific and used without further purification.

33. TLC of fractions is performed using Dynamic Adsorbents, Inc. glass plates coated with 250 mm F-254 silica gel. 15% EtOAc in hexanes is used as the eluent. Visualization is achieved with UV (Spectroline®, Model EF-140C, short wave ultraviolet 254 nm) and subsequently with PMA staining (10 g phosphomolybdic acid + 100 mL absolute EtOH) by immersion and heating with a heat gun. The title compound **2** is visible by UV and stains dark brown with PMA at an R_f = 0.41. Trace amounts of 2-thiophene-carboxaldehyde, which is generated by decomposition of *tert*-butyl phenylsulfonyl(thiophen-2-yl)methylcarbamate (**1**), elutes at R_f = 0.35. This aldehyde is only visible by UV and does not stain with PMA. Fractions containing both thiophene-carboxaldehyde and **2** are collected because this aldehyde is easy to remove under vacuum (boils at 75 – 77 °C at 11 mmHg). Fractions containing an impurity with R_f = 0.22 (visible by UV and stains brown with PMA) were not collected.

34. The title compound (**2**) exhibits the following properties: mp 138–140 °C. ^{1}H NMR (400 MHz, CDCl$_3$) δ: 1.43 (s, 9 H), 5.20 (broad s, 1 H), 6.10 (broad s, 1 H), 6.77–6.80 (m, 1 H), 6.90–6.94 (m, 1 H), 7.22–7.34 (m, 5 H). ^{13}C NMR (100 MHz, CDCl$_3$) δ: 28.6, 54.2, 80.5, 125.5, 125.9, 127.1, 128.5, 129.0, 133.7, 140.6, 145.9, 154.9. IR (neat) 3347, 2979, 2921, 2361, 1686, 1515, 1233, 1169 cm^{-1}. [α]$_D^{20}$ +11.0 (c = 0.5, EtOH). MS (ESI+) m/z 346 (M$^+$ + Na, 100%), 347 (17%), 348 (40%). Anal. calcd. for C$_{16}$H$_{18}$ClNO$_2$S: C, 59.34; H, 5.60; N, 4.33; found: C, 59.50; H, 5.63; N, 4.23.

35. The checkers also performed the reaction at half-scale and isolated pure product in a 65% yield at 96% ee. The submitters report a full-scale reaction to provide product in 65% yield at 95-99% ee.

36. The absolute configuration was shown by anomalous dispersion to be (*S*) using X-ray crystallography. This configuration is consistent with prior additions of this type [see Reference 13].

37. Enantiomeric excess is determined by chiral HPLC using an Agilent 1100 series instrument and a Chiralpak® AS-H column (amylose tris[(*S*)-α-methylbenzyl-carbamate] coated on 5 mm silica gel), L = 250 mm, I.D. = 4.6 mm, from Danicel Technologies, LTD. 1% EtOH in hexanes is used as the eluent (isochratic) with a flow rate of 1.00 mL/min (max. 70 bar) for 25 minutes. For optimal performance the column is equilibrated with the solvent system for at least 45 minutes before running the sample. A sample

is prepared by dissolving approx. 1 mg compound in 1 mL of 1% EtOH in hexanes and filtering through a 4 mm nylon syringe filter (0.45 mm) purchased from National Scientific. 5.0 mL of this solution is used for injection. To determine the retention times for both enantiomers, a racemate of **2** (synthesized with dppBenz as ligand) can be analyzed: (*R*)-enantiomer (minor): 11.1 minutes and (*S*)-enantiomer (major): 13.5 minutes. Samples are analyzed at the following wavelengths: 222, 230, 250 and 254 nm each of which gave similar %ee.

Waste Disposal Information

All hazardous materials should be handled and disposed of in accordance with "Prudent Practices in the Laboratory"; National Academy Press; Washington, DC, 1995.

3. Discussion

Synthesis of enantiomerically pure functionalized amines is of great importance because such compounds are widely used in drugs. The rhodium-catalyzed enantioselective addition of arylboronic acids to *in situ* generated *N*-Boc aromatic imines is a general and easy method for the preparation of *N*-Boc protected diaryl methanamines. The first reported example of addition of arylboronic acids to an imine was the addition to *N*-sulfonyl aldimines published in 2000 by Miyaura and co-workers.[3] A number of enantioselective variants were later developed using chiral ligands, such as *N*-Boc-L-valine amidomonophosphanes,[4] (1*R*,4*R*)-bicyclo[2.2.2]-octadienes,[5] (*S*)-ShiP,[6] monodentate phosphoramidites,[7,8] binaphtholic phosphites,[8] tetrahydropentalenes,[9] and (*R*,*R*)-deguPHOS.[10] Most of the methods are limited to aromatic imines, but more recently enantioselective catalytic additions to aliphatic imines have also been reported.[11,12]

However, these methods suffer from a number of drawbacks. For example, all of the methods utilize unstable imine substrates, and many of the methods necessitate the use of very harsh conditions to remove the *N*-substituent present in the addition products. Some of these problems were previously solved by the Ellman group using *N*-Boc aromatic imines generated *in situ* from easily prepared and stable α-carbamoyl sulfones in an enantioselective addition with arylboronic acids (Table 1).[13] Commercially available (*R,R*)-deguPHOS was used as the chiral ligand to obtain enantiomeric excesses up to 99%. However, Rh(acac)(coe)$_2$ was used as the precatalyst, and it is currently not commercially available. Moreover, Rh(acac)(coe)$_2$ is highly air-sensitive necessitating that the reactions be set up in a nitrogen-filled glovebox,[13] which is inconvenient for most research laboratories.

Table 1. Synthesis of various *N*-Boc amines.[13]

5% Rh(acac)(coe)$_2$
5.5% (*R,R*)-deguPHOS
Ar^{2}B(OH)$_2$ (2 equiv)
$\xrightarrow{\hspace{3cm}}$
K$_2$CO$_3$ (6 equiv), Et$_3$N (1.5 equiv)
4Å sieves, dioxane, 70 °C, 20 h

BocHN—Ar^1 (SO$_2$Ph) → BocHN—Ar^1 (Ar^2)

Entry	Ar^1	Ar^2	Yield (%)^a	ee (%)^b
1	Ph	4-ClC$_6$H$_4$	76	98^c
2	Ph	4-MeC$_6$H$_4$	70	96
3	Ph	4-MeOC$_6$H$_4$	76	93^c
4	Ph	4-CF$_3$C$_6$H$_4$	51	95^c
5	Ph	3-ClC$_6$H$_4$	55	99
6	Ph	3-MeC$_6$H$_4$	66	95
7	Ph	3-AcC$_6$H$_4$	52	94
8	Ph	2-MeC$_6$H$_4$	62	93
9	4-MeC$_6$H$_4$	Ph	71	90
10	3-MeC$_6$H$_4$	Ph	70	95
11	2-MeC$_6$H$_4$	Ph	63	97
12	4-BrC$_6$H$_4$	Ph	59	90
13	2-thienyl	Ph	71	96
14	4-MeOC$_6$H$_4$	Ph	76	96^c
15	4-CF$_3$C$_6$H$_4$	Ph	69	79^c

^{a} Isolated yields after chromatography. ^{b} Determined by chiral HPLC analysis. ^{c} Absolute configuration established by comparison of the optical rotation of amine obtained upon Boc cleavage to literature values.[14]

Optimization of this chemistry was therefore revisited. We found that the inexpensive and air-stable precatalyst, [RhCl(cod)]$_2$, performed equally well to Rh(acac)(coe)$_2$. Unfortunately, an enantiomeric excess of only 40% was achieved with this precatalyst. To improve the enantiomeric excess, we therefore performed a series of preincubation experiments whereby [RhCl(cod)]₂ and (*R,R*)-deguPHOS were stirred at 70 °C in dioxane prior to adding the starting materials, bases and molecular sieves. We found that one hour of preincubation resulted in a dramatic improvement in the enantiomeric excess to at least 95% ee. Shorter preincubations gave lower enantiomeric excess, whereas longer incubations were not beneficial.

Furthermore, we discovered that the presence of significant quantities of the boroxime (cyclic anhydride) in the boronic acid resulted in a decreased yield of the title compound **2**. Decreased yields may occur because the boroxime adds only slowly to the *in situ* generated imine, which competitively hydrolyzes under the reaction conditions. Commercially available boronic acids contain varying amounts of boroxime and therefore should be recrystallized from water before use. To avoid too much water in the reaction mixture, the boronic acid should also be dried prior to use. Boronic acids should not contain more than 5% boroxime and preferentially no more than 30% water as determined by ¹H NMR in dry DMSO-*d₆*.

Figure 1: Cannulation technique

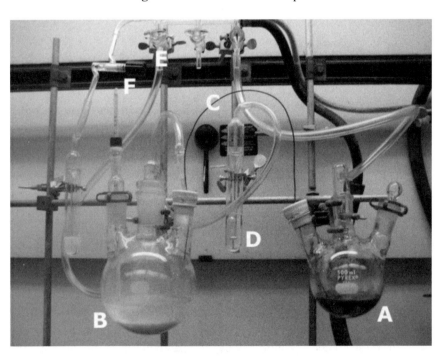

To expand the usability of the chemistry and to make it easier to perform on larger scale the reaction was set up using Schlenk techniques. This reaction set up provides for efficient reactions on both small and large scale, but it is important to transfer the active catalyst solution by cannulation technique to completely avoid exposure to air.

In conclusion, the title product **2** has been prepared in good yield and with high enantioselectivity. We believe that these optimized conditions should be compatible with the same range of different α-carbamoyl sulfones and arylboronic acids reported previously (Table 1).[13] This method, which utilizes the commercially available (*R,R*)-deguPHOS chiral ligand and the commercially available and air stable [RhCl(cod)]₂ precatalyst, does not require the use of a glovebox and represents a straightforward and general method for the enantioselective synthesis of *N*-protected diaryl methanamines.

1. Department of Chemistry, University of California, Berkeley, California 94720-1460 (email: jellman@berkeley.edu).
2. Wenzel, A. G.; Jacobsen, E. N. *J. Am. Chem. Soc.* **2002**, *124*, 12964 – 12965.
3. Ueda, M.; Saito, A.; Miyaura, N. *Synlett* **2000**, 1637 – 1639.
4. Kuriyama, M.; Soeta, T.; Hao, X., Chen, Q.; Tomioka, K. *J. Am. Chem. Soc.* **2004**, *126*, 8128 – 8129.
5. Tokunaga, N.; Otomaru, Y.; Okamoto, K.; Ueyama, K.; Shintani, R.; Hayashi, T. *J. Am. Chem. Soc.* **2004**, *126*, 13584 – 13585.
6. Duan, H.-F.; Jia, Y.-X.; Wang, L.-X.; Zhou, Q.-L. *Org. Lett.* **2006**, 8, 2567 – 2569.
7. Jagt, R. B. C.; Toullec, P. Y.; Geerdink, D.; Vries, J. G. d.; Feringa, B. L.; Minnaard, A. J. *Angew. Chem. Int. Ed.* **2006**, *45*, 2789 – 2791.
8. Marelli, C.; Monti, C.; Gennari, C.; Piarulli, U. *Synlett* **2007**, 2213 – 2216.
9. Wang, Z.-Q., Feng, C.-G., Xu, M.-H., Lin, G.-Q. *J. Am. Chem. Soc.* **2007**, *129*, 5336 – 5337.
10. Weix, D. J.; Shi, Y., Ellman, J. A., *J. Am. Chem. Soc.* **2005**, *127*, 1092 – 1093.
11. Trincado, M.; Ellman, J. A. *Angew. Chem. Int. Ed.* **2008**, 47, 5623 – 5626.
12. For diastereoselective arylboronic acid additions to *N-tert*-butanesulfinyl aldimines using achiral ligands and rhodium catalysts see: (a) See

reference 10; (b) Bolshan, Y.; Batey, R. *Org. Lett.* **2005**, *7*, 1481 – 1484;
(c) Beenen, M. A.; Weix, D. J.; Ellman, J. A. *J. Am. Chem. Soc.* **2006**,
128, 6304 – 6305.

13. Nakagawa, H.; Rech, J. C.; Sindelar, R. W.; Ellman, A. J. *Org. Lett.*
2007, *9*, 5155 – 5157.

14. (a) Hayashi, T.; Ishigedani, M. *J. Am. Chem. Soc.* **2000**, *122*, 976 – 977;
(b) Plobeck, N.; Powell, D. *Tetrahedron: Asymmetry*, **2002**, *13*, 303 –
310.

Appendix
Chemical Abstracts Nomenclature; (Registry Number)

tert-Butyl phenylsulfonyl(thiophen-2-yl)methylcarbamate: Carbamic acid,
  N-[(phenylsulfonyl)-2-thienylmethyl]-, 1,1-dimethylethyl ester;
  (479423-34-2)
tert-Butyl carbamate: Carbamic acid, 1,1-dimethylethyl ester; (4248-19-5)
2-Thiophene-carboxaldehyde; (98-03-3)
Benzenesulfinic acid sodium salt; (873-55-2)
[RhCl(cod)]$_2$; (12092-47-6)
(*R,R*)-deguPHOS: Pyrrolidine, 3,4-bis(diphenylphosphino)-1-
  (phenylmethyl)-, (3*R*,4*R*)-; (99135-95-2)
4-Chlorophenylboronic acid: Boronic acid, B-(4-chlorophenyl)-; (1679-18-
  1)

Morten Storgaard was born in Denmark in 1980. He graduated
from Technical University of Denmark in 2006 with a M.Sc.
degree in chemistry and in 2007 he continued as a Ph.D.
student under the supervision of professor David Tanner and
Dr. Bernd Peschke from Novo Nordisk. His research has
mainly been focusing on palladium catalyzed coupling
reactions towards the synthesis of biologically active
compounds. In the summer and fall of 2008 he visited the
group of Jonathan A. Ellman at University of California at
Berkeley and carried out research on the rhodium-catalyzed
enantioselective synthesis of amines.

 Jason Bexrud received his B.Sc. from Simon Fraser University in 2003. After which, he began doctoral work at the University of British Columbia with Laurel Schafer. His thesis focused on the development of titanium and zirconium-catalyzed hydroamination and C-H functionalization reactions.

MAGNESIATION OF WEAKLY ACTIVATED ARENES USING tmp$_2$Mg·2LiCl: SYNTHESIS OF *TERT*-BUTYL ETHYL PHTHALATE

Submitted by Christoph J. Rohbogner,[1] Andreas J. Wagner,[1] Giuliano C. Clososki,[2] and Paul Knochel.[*1]

Checked by Jeremy P. Olson and Huw M.L. Davies.

1. Procedure

A. Preparation of tmpMgCl·LiCl. A dried and nitrogen-flushed 1-L Schlenk flask (Note 1), equipped with a magnetic stirring bar and rubber septum, is charged with *i*PrMgCl·LiCl (792 mL, 1.2 M in THF,[3] 950 mmol) (Note 2) then 2,2,6,6-tetramethylpiperidine (141.3 g, 1.00 mol) (Note 3) is added dropwise within 5 min via syringe. The mixture is stirred until gas evolution ceases (24–48 h) (Note 4). Titration prior to use shows a concentration of 0.95 M (Notes 4 and 5).[4,5]

B. Preparation of tmp$_2$Mg·2LiCl and synthesis of tert-butyl ethyl phthalate. A flame-dried and nitrogen-flushed 500-mL Schlenk flask (Note 1), equipped with a magnetic stirring bar and rubber septum, is charged with 100 mL of dry THF (Note 7) cooled in a –40 °C cooling bath (Note 8) and stirred for 15 min at this temperature. Then *n*-BuLi (45.5 mL, 2.22 M in hexanes, 100 mmol, 1.1 equiv) (Note 9) is added at once via syringe. After stirring for 15 min at –40 °C, 2,2,6,6-tetramethylpiperidine (14.1 g, 100 mmol, 1.1 equiv) (Note 3) is added at once via syringe. The resulting mixture is stirred at –40 °C for 5 min and the flask is then transferred to an ice bath (0 °C) and is stirred for 30 min. Then, tmpMgCl·LiCl (105 mL, 0.95 M in THF, 100 mmol, 1.1 equiv.) is added dropwise via syringe in one portion (addition time < 1 min.). The mixture is stirred in a 0 °C bath for 30 minutes and at 25 °C for another 60 minutes. The rubber septum is removed and the flask is stoppered with a glass stopper. The solvents are removed *in*

Org. Synth. **2009**, *86*, 374-384

Published on the Web 7/16/2009

vacuo using a rotary vane vacuum pump and liquid nitrogen cooled cooling trap (25 °C at 1.5 mmHg). The stopper is replaced by a septum and the resulting pale brown solid is redissolved in 112 mL of dry THF (Note 11) and stirred for 10 minutes at 25 °C. Ethyl benzoate (13.5 g, 90.0 mmol, 1.0 equiv) (Note 12) is added at once via syringe (addition time < 1 min.). The greenish brown solution immediately turns deep red. The reaction mixture is stirred at 25 °C for 30 minutes (Note 13). The red solution is cooled in a -40 °C cooling bath. After stirring for 10 min, Boc$_2$O (28.0 g, 130 mmol, 1.44 equiv) (Note 14) is added in one portion via syringe (addition time < 1 min.). After the addition, the reaction mixture is stirred at 25 °C until a slightly exothermic reaction starts. Then, the flask is put into a water bath (~20 °C) to moderate the exotherm. The reaction is stirred for a further 30 min (Note 15).

At this time, 300 mL of a saturated aqueous NH$_4$Cl solution is added. Large amounts of a precipitate are formed, which are removed by suction filtration (Note 16). The precipitate is washed with approx. 800 mL of diethyl ether until it is colorless. The resulting mixture is brought into a 2-L separatory funnel, rinsing with 50 mL of diethyl ether. The phases are separated and the organic layer is extracted with a saturated aqueous NH$_4$Cl solution (3 x 300 mL). The combined aqueous layers are extracted with ethyl acetate (400 mL). The combined organic layers are washed with 300 mL of brine and dried by stirring for 15 min over 110 g of anhydrous MgSO$_4$ (Note 17). After filtration, the solvents are removed under reduced pressure using a rotary evaporator and vacuum (40 °C, 720 to 50 mmHg). The resulting red oil is loaded on 25.5 g of silica gel, dried for 2 h under high vacuum (Note 18), and purified by column chromatography (Note 19). The solvents are evaporated under reduced pressure using a rotary evaporator and vacuum (40 °C, 720 to 50 mmHg) and dried for 4 h at 25 °C using a rotary vane pump yielding 15.1 g of *tert*-butyl ethyl phthalate (60.3 mmol, 67 %) as a red oil. (Notes 20 and 21)

2. Notes

1.   The glassware was oven-dried overnight and evacuated using high vacuum (1 mmHg) and backfilled with nitrogen (this procedure was repeated three times). All syringes, cannulas and needles were purged with nitrogen prior to use.

2. *i*-PrMgCl·LiCl was purchased from Aldrich and the solution was titrated using iodine (1 mmol) in 2 mL of dry THF.[3] The submitters purchased *i*-PrMgCl·LiCl from Chemetall GmbH (Frankfurt, Germany).

3. 2,2,6,6-Tetramethylpiperidine was purchased from TCI America and the material was distilled before use (120 °C, 75 mmHg). The submitters purchased 2,2,6,6-tetramethylpiperidine from Evonik Industries AG (Essen, Germany).

The amounts of all liquids used in the synthesis were determined by weighing them in a syringe, since this method is more accurate than the determination of the volumes by the scales of the syringes used.

4. Titration of the base was performed in a dried 10-mL Schlenk tube (Note 1). The base was titrated against benzoic acid (1 mmol) using (4-phenylazo)diphenylamine (3 mg, Aldrich) as indicator in 1 mL dry THF. Color change from orange to dark violet indicated the end of the titration (consumption 1.05 mL). The submitters purchased (4-phenylazo)diphenylamine from Acros.

5. The checkers used the base of known concentration in subsequent reactions. The submitters also checked complete formation of the base by quenching an aliquot with benzaldehyde. Absence of 2-methyl-1-phenylpropan-1-ol (detected by GC/MS; $M^+=150$) indicates full consumption of the Grignard reagent. The submitters performed GC/MS-analysis using an HP 6890 Series GC system equipped with an Agilent 5973 Network Mass Selective Detector. Column HP-5MS (J&W Scientific) (15 m x 0.25 mm x 0.25 μm). Oven program for GC/MS-Analysis: Starting temperature 70 °C for 0.5 min; heating to 250 °C at a rate of 50 °C/min; 5 min at 250 °C; heating to 300 °C at a rate of 50 °C/min, 3 min 300 °C. Aliquots of the base were quenched with benzaldehyde (approx. 0.2 mL) then diethyl ether was added (1 mL). The mixture was extracted with a saturated aqueous NH_4Cl solution.

6. It is not mandatory to prepare the tmpMgCl·LiCl on this large scale. It is possible to prepare the base in just the amount needed for this reaction. The description of the preparation refers to a synthesized batch in the submitter's group's usual scale. The submitter has reported that it can be stored at room temperature under nitrogen for more than 6 months without losing efficiency.[5]

7. The checkers purchased all solvents from Fischer and THF (99%) was distilled and then constantly refluxed from sodium/benzophenone. The other solvents were used by the checkers without further purification. The

submitters purchased THF (95 %) from BASF (Ludwigshafen, Germany), distilled and then constantly refluxed from sodium/benzophenone, all other solvents were purchased from Biesterfeld AG (Hamburg, Germany) and distilled before use.

8. The checkers performed constant cooling using a NESLAB CC-100 cooling device and a stirred ethanol bath. The submitters used a ThermoHaake EK90 cooling device and a stirred ethanol bath.

9. n-BuLi was purchased from Aldrich and titrated using menthol (1 mmol) in 1 mL of dry THF and (4-phenylazo)diphenylamine (3 mg) (Aldrich) as indicator.[4] The submitters purchased n-BuLi from Chemtall GmbH (Frankfurt, Germany), and (4-phenylazo)diphenylamine and used the same titration method.

10. Vacuum was 1 mmHg. During the evaporation process the flask is stored in a water bath (~20°C) to maintain the temperature and ensure steady evaporation.

11. The actual quantity of dry THF needed may vary up to 10%.

12. Ethyl benzoate (99%) was purchased from Aldrich and used without further purification The submitters purchased ethyl benzoate (99%) from Merck.

13. The checkers continued with the subsequent addition after the reaction had stirred for 30 minutes. The submitters confirmed full consumption of the starting material by GC analysis of iodolyzed reaction aliquots. The submitters performed GC analysis using an Agilent Technologies 6850 Series equipped with an HP-5 column (J&W Scientific) (15m x 0.25mm x 0.25µm). Oven program for GC analysis: Starting temperature 70 °C for 0.5 min; heating to 250 °C at a rate of 50 °C/min; 5 min at 250 °C. Reaction aliquots were quenched with 0.2 mL of a 0.5 M I_2 solution in dry THF mixed with approx. 1 mL of a sat. aq. NH_4Cl and 1 mL of a sat. aq. $Na_2S_2O_3$ solution and extracted with diethyl ether (1 mL). The conversion was monitored by GC by controlling the production of ethyl 2-iodobenzoate (retention time 3.45 min) and the consumption of ethyl benzoate (retention time: 2.31 min).

14. Boc_2O (99%) was purchased from Aldrich and melted in a water bath (40 °C) to facilitate the weighing and adding processes. The submitters purchased Boc_2O (99%) from Merck.

15. The checkers quenched the reaction after 30 minutes. The submitters followed the progress of the reaction by quenching reaction aliquots with 1 mL of a saturated NH_4Cl solution and extracting with diethyl

ether. GC analysis after 30 minutes showed the complete absence of ethyl benzoate. The retention time of *tert*-butyl ethyl benzoate was 3.75 min.

16. A glass filter with P3 pore size and a volume of 500 mL were used. Filtration was carried out under reduced pressure (300 mmHg).

17. $MgSO_4$ was purchased from Fischer and used as received. The submitters purchased $MgSO_4$ from Grüssing (Filsum, Germany).

18. The checkers loaded the crude product onto 25.5 g of silica. It was added to the oil and dried in high vacuum for 2 h. The submitters loaded the crude product onto 25.5 g of Isolute (Biotage) and dried it in high vacuum for 1 h. Isolute is comparable to loading on silica, but it offers very sharp bands as it plays no part in the chromatographical process.

19. Column chromatography was carried out using a 10-cm diameter column packed with 787 g of silica gel (the checkers used Siliaflash 60, 0.040-0.060 mm; the submitters used Gelduran SI60 from Merck 0.040-0.063 mm) using *n*-pentane:diethyl ether 4:1 as eluent (R_f=0.62, TLC aluminium sheets, silica gel 60 F254, Merck). The product is visible under UV radiation as a bright blue fluorescent band. After 2.6 L of eluent, the band is ready to be collected. Collection is stopped when the fluorescence subsides; approx 1.5 L.

20. The product exhibits the following properties ^{1}H-NMR (400 MHz, $CDCl_3$) δ: 1.38 (t, J = 7.0 Hz, 3 H), 1.58 (s, 9 H), 4.37 (q, J = 7.2 Hz, 2 H), 7.48 (m, 2 H), 7.67 (m, 2 H); ^{13}C-NMR (100 MHz, $CDCl_3$) δ: 13.9, 27.8, 61.3, 81.8, 128.4, 128.8, 130.45, 130.5, 132.3, 133.2, 166.3, 167.8; MS (70 ev, EI) *m/z*: 251 (29), 246 (21), 195 (100), 177 (23) [M^+-CO_2Et], 150 (4), 149 (35); HRMS (ESI) calcd. for $C_{14}H_{19}O_4$ 251.1278 found: 251.1280; IR (Film) $\tilde{v}$ (cm^{-1}): 2979 (w), 1715 (vs), 1599 (w), 1579 (w), 1477 (w), 1447 (w), 1392 (w), 1367 (vs), 1286 (s), 1255 (s), 1172 (S), 1123 (vs), 1072 (vs), 1038 (vs), 1017 (vs), 845 (vs), 784 (w), 737 (s), 705 (vs); CHN Analysis for $C_{14}H_{18}O_4$: calcd.: C, 67.18; H, 7.25; found: C, 67.46; H, 7.48.

21. A ¾-scale run afforded 12.1 g (71%) of analytically pure product.

Safety and Waste Disposal Information

All hazardous materials should be handled and disposed of in accordance with "Prudent Practices in the Laboratory"; National Academy Press; Washington, DC, 1995.

3. Discussion

Directed lithiations are important reactions for the functionalization of aromatics and heterocycles.[6] The high reactivity of the organolithium reagents leads to low functional group tolerance. In contrast, magnesium reagents offer also high reactivity in combination with high tolerance of functional groups such as ester, cyano and keto groups. We have shown that the use of magnesium bases like tmpMgCl·LiCl are suitable for the deprotonation of a wide range of aromatic and heteroaromatic substrates.[5] However, some poorly activated substrates like ethyl benzoate do not undergo metallation at all or the metallation does not lead to satisfactory results. Eaton demonstrated that tmp₂Mg offers high reactivity and functional group tolerance.[7] The limited solubility in common organic solvents, as well as the use in high excess (of the reagent and the electrophile) to achieve full conversion, precluded further use in organic synthesis.[8] During the studies on Grignard reagents we realized that the addition of LiCl could enhance the reactivity and solubility dramatically.[9] Therefore, the addition of two equivalents of LiCl leads to the highly soluble and reactive complex tmp₂Mg·2LiCl. This composition facilitates the metallation of moderately activated aromatic substrates such as ethyl benzoate, benzonitrile, ethyl naphthanoate or derivatives thereof. The metallation of heteroaromatic subtrates like ethyl nicotinate or diethyl pyridine-3,5-dicarboxylate proceeds without any problems.[8] These organomagnesium reagents can now be used in any typical trapping process, for example Negishi cross coupling reactions,[10] copper(I) mediated acylations[11] or iodo- or bromolysis (Table 1), leading to the expected products in good to excellent yields.[8]

Table 1 Products of type **4** obtained by the magnesiation of aromatics and heterocycles with tmp$_2$Mg·2LiCl and reactions with electrophiles.[a]

Entry	Substrate	T [°C],t[h]	E+	Product	Yield[%][b]
1	**2a**	25, 1	PhCOCl	**4a**	93[c]
2	**2b**	0, 3	(BrCl$_2$C)$_2$	**4b** E = Br	83
3	**2b**	0, 3	p-IC$_6$H$_4$CN	**4c** E = p-C$_6$H$_4$CN	81[d]
4	**2c**	–20, 1	I$_2$	**4d**	71
5	**2d**	–30, 3	p-IC$_6$H$_4$CO$_2$Et	**4e**	70[d]
6	**2e** R = t-Bu; X = CH	0, 1	I$_2$	**4f** E = I	94
7	**2e** R = t-Bu; X = CH	0, 1	p-IC$_6$H$_4$CO$_2$Et	**4g** E = p-C$_6$H$_4$CO$_2$Et	88[d]
8	**2f** R = Et; X = N	–40, 3	I$_2$	**4h** E = I	77
9	**2f** R = Et; X = N	–40, 3	p-IC$_6$H$_4$CN	**4i** E = p-C$_6$H$_4$CN	73[d]
10	**2g**	–40, 12	I$_2$	**4j**	66

[a]Reactions performed on 1-2 mmol scale. [b]Isolated yield of analytically pure product. [c]Transmetalation with CuCN·2LiCl (0.2 mol%) was performed. [d]Obtained by palladium-catalyzed cross-coupling after transmetalation with ZnCl$_2$ (1.2 to 1.3 equiv).

1. Prof. Dr. Paul Knochel, Dipl.-Chem. Christoph J. Rohbogner, Dipl.-Chem. Andreas J. Wagner, Department Chemie & Biochemie, Ludwig-Maximilians-Universität München, Butenandtstr. 5-13, 81377 München (Germany) email: paul.knochel@cup.uni-muenchen.de

2. Prof. Dr. Giuliano C. Clososki, Faculdade de Ciências Farmacêuticas de Ribeirão Preto, Universidade de São Paulo, Av. Do Café s/n, 14040-903 Ribeirão Preto-SP (Brazil).

3. Titration of Organomagnesium reagents: Krasovskiy, A.; Knochel, P. *Synlett* **2006**, *5*, 890.

4. Hammett, L. P.; Walden, G. H.; Edmonds, S. M. *J. Am. Chem. Soc.* **1934**, *56*, 1092.

5. Krasovskiy, A.; Krasovskaya, V.; Knochel, P. *Angew. Chem.* **2006**, *118*, 3024; *Angew. Chem. Int. Ed.* **2006**, *45*, 2958; b) Lin, W.; Baron, O.; Knochel, P. *Org. Lett.* **2006**, *8*, 5673.

6. (a) Schlosser, M. *Angew. Chem.* **2005**, *117*, 380; *Angew. Chem. Int. Ed.* **2005**, *44*, 376; (b) Turck, A.; Plé, N.; Mongin, F.; Quéguiner, G. *Tetrahedron* **2001**, *57*, 4489; (c) Mongin, F.; Quéguiner, G. *Tetrahedron* **2001**, *57*, 4059; (d) Schlosser, M. *Eur. J. Org. Chem.* **2001**, *21*, 3975; (e) Hodgson, D. M.; Bray, C. D.; Kindon, N. D . *Org. Lett.* **2005**, *7*, 2305; (f) Plaquevent, J.-C.; Perrard, T.; Cahard, D. *Chem. Eur. J.* **2002**, *8*, 3300; (g) Chang, C.-C.; Ameerunisha, M. S. *Coord. Chem. Rev.* **1999**, *189*, 199; (h) Clayden, J. *Organolithiums: Selectivity for Synthesis* (Ed.: Baldwin, J. E.; Williams, R. M.), Elsevier, **2002**; (i) "*The Preparation of Organolithium Reagents and Intermediates*": Leroux, F.; Schlosser, M.; Zohar, E.; Marek, I. *Chemistry of Organolithium Compounds* (Ed.: Rappoport, Z.; Marek, I.), Wiley, New York, **2004**, Chapt.1, P. 435; (j) Whisler, M. C.; MacNeil, S.; Snieckus, V.; Beak, P. *Angew. Chem.* **2004**, *116*, 2256; *Angew. Chem. Int. Ed.* **2004**, *43*, 2206; (k) Quéguiner, G.; Marsais, F.; Snieckus, V.; Epsztajn, J. *Adv. Heterocycl. Chem.* **1991**, *52*, 187; (l) Veith, M.; Wieczorek, S.; Fries, K.; Huch, V. *Z. Anorg. Allg. Chem.* **2000**, *626*, 1237.

7. (a) Eaton, P. E.; Martin, R. M., *J. Org. Chem.* **1988**, *53*, 2728; (b) Eaton, P.E.; Lee, C. H.; Xiong, Y. *J. Am. Chem. Soc.* **1989**, *111*, 8016; (c) Eaton, P. E.; Lukin, K. A. *J. Am. Chem. Soc.* **1993**, *115*, 11370; (d) Zhang, M.–X.; Eaton, P. E. *Angew. Chem.* **2002***, 114, 2273; Angew.*

Chem. Int. Ed. **2002**, *41*, 2169; (e) Eaton, P. E.; Zhang, M.-X.; Komiya, N.; Yang, C.-G.; Steele, I.; Gilardi, R. *Synlett* **2003**, *9*, 1275.

8. Clososki, G. C.; Rohbogner, C. J.; Knochel, P. *Angew. Chem.* **2007**, *119*, 7825; *Angew. Chem. Int. Ed.* **2007**, *46*, 7681.

9. (a) Krasovskiy, A.; Knochel, P. *Angew. Chem.* **2004**, *116*, 3396; *Angew. Chem. Int. Ed.* **2004**, *43*, 3333; (b) Krasovskiy, A.; Straub, B.; Knochel, P. *Angew. Chem.* **2006**, *118,* 165; *Angew. Chem. Int. Ed.* **2006**, *45*, 159; (c) Shi, L.; Chu, Y.; Knochel, P.; Mayr, H. *Angew. Chem.* **2007**, *119,* 208; *Angew. Chem. Int. Ed.* **2007**, *47*, 202.

10. (a) Negishi, E.; King, A. O.; Okukado, N. *J. Org. Chem.* **1977**, *42*, 1821; (b) Negishi, E.; Valente, L. F.; Kobayashi, M. *J. Am. Chem. Soc.* **1980**, *102*, 3298; (c) Negishi, E.; Kobayashi, M. *J. Org. Chem.* **1980**, *45*, 5223; (d) Negishi, E. *Acc. Chem. Res.* **1982**, *15*, 340.

11. (a) Knochel, P.; Yeh, M. C. P.; Berk, S. C.; Talbert, J. *J. Org. Chem.* **1988**, *53*, 2390; (b) Knochel, P.; Rao, S. A. *J. Am. Chem. Soc.* **1990**, *112*, 6146.

Appendix
Chemical Abstracts Nomenclature; (Registry Number)

*i*PrMgCl·LiCl: isopropylmagnesium chloride lithium chloride complex; (807329-97-1)

2,2,6,6-Tetramethylpiperidine; (768-66-1)

Ethyl benzoate: Benzoic acid, ethyl ester; (93-89-0)

Boc$_2$O: Dicarbonic acid, *C,C'*-bis(1,1-dimethylethyl) ester; (244424-99-5)

tert-Butyl ethyl phthalate: 1,2-Benzenedicarboxylic acid, 1,2-bis[2-[3,5-bis(1,1-dimethylethyl)-4-hydroxyphenyl]ethyl] ester; (259254-67-6)

Paul Knochel was born in 1955 in Strasbourg, France. He completed his undergraduate studies at the University of Strasbourg and his Ph.D. at the ETH Zurich with D. Seebach. He spent 4 years with Prof. J.-F. Normant (Paris) and 1 year with Prof. M. F. Semmelhack (Princeton) as a postdoctoral researcher. After professorships at the University of Michigan and the Philipps-Universität (Marburg), he moved to the Ludwig-Maximilians-Universität (Munich) in 1999. His research interests include the development of new synthetic methods with organometallic reagents, new asymmetric catalysts and natural product synthesis.

Christoph J. Rohbogner was born in 1980 in Munich, Germany. After undergraduate studies at the Ludwig-Maximilans-Universität (Munich), he joined the group of Prof. Knochel where he received his diploma degree in Organic Chemistry in 2006. He stayed in the same group for his Ph.D. studies. His current research focuses on C-H activation on arenes and heteroarenes.

Andreas J. Wagner was born in 1982 in Regensburg, Germany. After undergraduate studies at the Ludwig-Maximilans-Universität (Munich), he joined the group of Prof. Knochel where he received his diploma degree in Organic Chemistry in 2007. He stayed in the same group for his Ph. D. studies. His current research focuses on amination reactions and organocopper chemistry.

Giuliano Cesar Clososki was born in 1974 in Capitão Leônidas Marques-PR, Brazil. He completed his undergraduate studies (1999) and got his Master degree (2000) at the Federal University of Paraná (Brazil) under the supervision of Prof. Fabio Simonelli. He spent 1 year at the University of California working with Prof. Bruce H. Lipshutz (2003) and got his Ph.D degree at the University of São Paulo (2005) under the supervision of Prof. João V. Comasseto. After spending 1 year with Prof. José A. R. Rodrigues (Brazil) and 2 years with Prof. P. Knochel (Germany) as a postdoctoral researcher he got a position at the University of São Paulo - Ribeirão Preto in 2008. His research interests include the development of new synthetic methods with organometallic reagents, biocatalysis and natural product synthesis.

Jeremy Olson was born in 1981 in Plano, Texas. He earned a B.S. degree in chemistry from Armstrong Atlantic State University in 2003. He then joined Prof. Huw Davies' lab at The State University of New York at Buffalo, where he focused on the synthesis of natural products using asymmetric [4+3] cycloaddition methodology.

384

ERRATA   (modification to procedure originally published in *Org. Synth.* **1987**, *65*, 52; Coll. Vol. 8, **1993**, 63.)

1,4-BIS(TRIMETHYLSILYL)BUTA-1,3-DIYNE

Graham E. Jones, David A. Kendrick, and Andrew B. Holmes

$$Me_3Si\text{---}\!\equiv\!\text{---}H \xrightarrow[\text{Me}_2\text{NCH}_2\text{CH}_2\text{NMe}_2]{\text{O}_2,\ \text{CuCl}} Me_3Si\text{---}\!\equiv\!\text{---}\!\equiv\!\text{---}SiMe_3$$

Warning: A serious explosion shattering the reaction flask has been reported to occur on one occasion since publication when carrying out this procedure. The experiment was performed at the described scale and the explosion occurred a few seconds after starting addition of the catalyst solution to the reaction flask through a syringe. The most plausible explanation for the explosion was an ignition of the acetone/ trimethylsilylacetylene/ oxygen gas mixture by a spark directly in the flask. It is believed that the source of the spark was the discharge of static electricity accumulated on a plastic syringe when the metal needle of the syringe contacted the metallic tip of a digital thermometer (connected by a cable to IKA digital hotplate). Users are reminded of the absolute necessity to follow the submitted procedure, especially Notes 6-8. As indicated in Note 6, it is critical that this procedure (as well as any syntheses involving highly flammable compounds in an atmosphere of oxygen) is performed behind a safety shield and that all possible sources of ignition (including static electricity) are carefully examined and eliminated.

CUMULATIVE AUTHOR INDEX FOR VOLUMES 85-86

This index comprises the names of contributors to Volumes **85** and **86**. For authors of previous volumes, see either indices in Collective Volumes I through XI, or the single volume entitled *Organic Syntheses, Collective Volumes I-VIII, Cumulative Indices,* edited by J. P. Freeman.

Dai, P., **86**, 236
Daugulis, O., **86**, 105
Deng, X., **85**, 179
Denmark, S. E., **86**, 274
Do, N., **85**, 138
Donahue, J. P., **86**, 333
Drago, C., **86**, 121
Du, H., **86**, 315
Duchêne, A., **85**, 231
Dudley, M. E., **86**, 172

Ebner, D. C., **86**, 161
Ellman, J. A., **86**, 360
Endo, K., **86**, 325

Fidan, M., **86**, 47
Fleming, M. J., **85**, 1
Franckevičius, V., **85**, 72
Fujimoto, T., **86**, 325
Fujiwara, H., **86**, 130
Fukuyama, T., **86**, 130
Fürstner, A., **85**, 34, **86**, 298

Gálvez, E., **86**, 70, 81
Gillis, E. P., **86**, 344
Glasnov, T. N., **86**, 252
Glorius, F., **85**, 267
Gooßen, L. J., **85**, 196
Goss, J. M., **86**, 236
Greszler, S., **86**, 18
Guichard, G., **85**, 147

Hahn, B. T., **85**, 267
Harada, S., **85**, 118
Hierl, E., **85**, 64
Hill, M. D., **85**, 88
Hodgson, D. M., **85**, 1
Hossain, M. M., **86**, 172
Huard, K., **86**, 59
Humphreys, P. G., **85**, 1
Hwang, S., **86**, 225
Hwang, S. J., **85**, 131

Jackson, R. F. W., **86**, 121
Javed, M. I., **85**, 189
Johnson, J. S., **85**, 278

Kang, H. R., **86**, 225
Kappe, C. O., **86**, 252
Kim, S., **86**, 225
Kitching, M. O., **85**, 72
Kocienski, P. J., **85**, 45
Knauber, T., **85**, 196
Knochel, P., **86**, 374
Kramer, J. W., **86**, 287
Krause, H., **85**, 34, **86**, 298
Krout, M. R., **86**, 181, 194
Kuethe, J. T., **86**, 92
Kwon, O., **86**, 212

Landais, Y., **86**, 1
Langle, S., **85**, 231
Lautens, M., **85**, 172, **86**, 36
La Vecchia, L., **85**, 295
Lazareva, A., **86**, 105
Lebel, H., **86**, 113
Lebel, H., **86**, 59
Lebeuf, R., **86**, 1
Leogane, O., **86**, 113
Ley, S. V., **85**, 72
Linder, C., **85**, 196
List, B., **86**, 11
Longbottom, D. A., **85**, 72
Lou, S., **86**, 236
Lu, C. -D., **85**, 158
Lu, K., **86**, 212

Mani, N. S., **85**, 179
Mans, D. J., **85**, 238, 248
Marin, J., **85**, 147
Matsunaga, S., **85**, 118
Maw, G., **85**, 219
McAllister, G. D., **85**, 15
McDermott, R. E., **85**, 138
McNaughton, B. R., **85**, 27
Meletis, P., **86**, 47
Meyer, H., **85**, 287, 295

CUMULATIVE SUBJECT INDEX FOR VOLUMES 85-86

This index comprises subject matter for Volumes **85** and **86**.. For subjects in previous volumes, see either the indices in Collective Volumes I through XI or the single volume entitled *Organic Syntheses, Collective Volumes I-VIII, Cumulative Indices,* edited by J. P. Freeman.

The index lists the names of compounds in two forms. The first is the name used commonly in procedures. The second is the systematic name according to Chemical Abstracts nomenclature, accompanied by its registry number in parentheses. Also included are general terms for classes of compounds, types of reactions, special apparatus, and unfamiliar methods.

Most chemicals used in the procedure will appear in the index as written in the text. There generally will be entries for all starting materials, reagents, intermediates, important by-products, and final products.

1,1-dimethylethyl ester; (151476-40-3) **86**, 113

tert-Butylamine: 2-Propanamine, 2-methyl-; (75-64-9) **86**, 315

tert-Butylbenzene; (98-06-6) **86**, 308

(*E*)-*tert-Butyl benzylidenecarbamate* **86**, 11

(*E*)-*tert*-Butyl benzylidenecarbamate: Carbamic acid, *N*-(phenylmethylene)-,1,1-dimethyl-ethyl ester, [*N*(*E*)]-; (177898-09-2) **86**, 11

(-)-2-tert-Butyl-(4S)-benzyl-(1,3)-oxazoline: 4,5-Dihydrooxazole, (4S)-benzyl, 2-tert-butyl; (75866-75-0) **85**, 267

tert-Butyl bromoacetate; (5292-43-3) **85**, 10

tert-Butyl *tert*-butyldimethylsilylglyoxylate **85**, 278

(S)-tert-Butyl (4-chlorophenyl)(thiophen-2-yl)methylcarbamate **86**, 360

tert-Butyl (1R)-2-cyano-1-phenylethylcarbamate (126568-44-3) **85**, 219

Butyl di-1-adamantylphosphine; (321921-71-5) **86**, 105

tert-Butyl diazoacetate; (35059-50-8) **85**, 278

tert-Butyldimethylsilyl chloride: Silane, chloro(1,1-dimethylethyl)dimethyl-; (18162-48-6) **86**, 130

tert-Butyldimethylsilyl trifluoromethanesulfonate; (69739-34-0) **85**, 278

tert-Butyl ethyl phthalate: 1,2-Benzenedicarboxylic acid, 1,2-bis[2-[3,5-bis(1,1-dimethylethyl)-4-hydroxyphenyl]ethyl] ester; (259254-67-6) **86**, 374

tert-Butyl (1S)-2-hydroxy-1-phenylethylcarbamate (117049-14-6) **85**, 219

tert-Butyl-hypochlorite: Hypochlorous acid, 1,1-dimethylethyl ester; (507-40-4) **86**,315

n-Butyllithium; (109-72-8) **85**, 1, 45, 53, 158, 238, 248, 295; **86**, 47, 70, 262

tert-Butyllithium: Lithium, (1,1-dimethylethyl)-; (5944-19-4) **85**, 1, 209

tert-Butyl phenyl(phenylsulfonyl)methylcarbamate **86**, 11

tert-Butyl phenyl(phenylsulfonyl)methylcarbamate: Carbamic acid, *N*-[phenyl(phenyl-sulfonyl)methyl]-1,1-dimethylether ester; (155396-71-7) **86**, 11

tert-Butyl phenylsulfonyl(thiophen-2-yl)methylcarbamate: Carbamic acid, N-[(phenylsulfonyl)-2-thienylmethyl]-, 1,1-dimethylethyl ester; (479423-34-2) **86**, 360

(S)-tert-ButylPHOX: Oxazole, 4-(1,1-dimethylethyl)-2-[2-(diphenylphosphino)phenyl]-4,5-dihydro-, (4S)-; (148461-16-9) **86**, 181

Carbamates **86**, 11, 59, 81, 113, 141, 151, 236, 333

Carbamic acid, (hydroxymethyl)-, phenylmethyl ester; (31037-42-0) **85**, 287

Carbamic acid, *N*-[(1*S*,2*S*)-2-methyl-3-oxo-1-phenylpropyl]-1,1-dimethylethyl ester; (926308-17-0) **86**, 11

Carbamic acid, N-[phenyl(phenylsulfonyl)methyl]-, 2-propen-1-yl ester; (921767-12-6) **86**, 236

Carbamic acid, 2-propen-1-yl ester; (2114-11-6) **86**, 236

(Carbethoxymethylene)triphenylphosphorane; (1099-45-2) **85**, 15

Carbon disulfide; (75-15-0) **86**, 70

Carbon monoxide; (630-08-0) **86**, 287

Carbonyl(dihydrido)tris(triphenylphosphine)ruthenium (II); (25360-32-1) **86**, 28

1,1'-Carbonyldiimidazole: Methanone, di-1*H*-imidazol-1-yl-; (530-62-1) **86**, 58

Cbz-L-proline: 1,2-Pyrrolidinedicarboxylic acid, 1-(phenylmethyl) ester, (2*S*)-; (1148-11-4) **85**, 72

Cesium carbonate: Carbonic acid, cesium salt (1:2); (534-17-8) **86**, 181

Cesium fluoride; (13400-13-0) **86**, 161

Chloroacetonitrile (107-14-2) **86**, 1

4-Chlorobenzaldehyde; (104-88-1) **85**, 179
Chlorobenzene; (108-90-7) **86**, 105
4-Chlorophenylboronic acid: Boronic acid, B-(4-chlorophenyl)-; (1679-18-1) **86**, 360
2-Chloropyridine; (109-09-1) **85**, 88
5-Chlorosalicylaldehyde: Benzaldehyde, 5-chloro-2-hydroxy-; (635-93-8) **86**, 172
2-Chloro-5-(trifluoromethyl)pyridine; (52334-81-3) **86**, 18
m-Chloroperbenzoic acid; Peroxybenzoic acid, *m*-chloro- (8); Benzocarboperoxoic acid,
    3-chloro- (9); (937-14-4) **86**, 308
Chlorotrimethylsilane: Silane, Chlorotrimethyl-; (75-77-4) **86**, 252
Cinnamyl alcohol: 3-Phenyl-2-propen-1-ol; (104-54-1) **85**, 96
Cinnamyl-H-phosphinic acid: [(2E)-3-phenyl-2-propenyl]-Phosphinic acid; (911128-46-
    6) **85**, 96
Condensation **85**, 27, 34, 179, 248, 267; **86**, 11, 18, 92, 121, 212, 252, 262
Copper(I) bromide; (7787-70-4) **85**, 196
Copper chloride: Cuprous chloride; (7758-89-6) **85**, 209
Copper Cyanide; (544-92-3) **85**, 131
Copper iodide; (1335-23-5) **86**, 181
Copper(I) iodide: Cuprous iodide; (7681-65-4) **86**, 225
Coupling **85**, 158, 196; **86**, 105, 225, 274
Cuprous chloride; (7758-89-6) **85**, 209
(S)-2-Cyano-pyrrolidine-1-carboxylic acid benzyl ester: (63808-36-6) **85**, 72
Cyanuric chloride: 2,4,6-Trichloro-1,3,5-triazine; (108-77-0) **85**, 72; **86**, 141
Cyclen: 1,4,7,10-Tetraazacyclododecane; (294-90-6) **85**, 10
Cyclization **86**, 18, 92, 172, 181, 194, 212, 236, 252, 262, 333
Cycloaddition, **85**, 72, 131, 138, 179
Cycloheptane-1,3-dione (1194-18-9) **85**, 138
Cyclohexanemethanol; (100-49-2) **86**, 58
Cyclohexanone; (108-94-1) **86**, 47
Cyclohexene oxide; (286-20-4) **85**, 106
Cyclohexylmethyl N-hydroxycarbamate **86**, 58
Cyclohexylmethyl *N*-hydroxycarbamate: Carbamic acid, hydroxy-, cyclohexylmethyl
    ester; (869111-30-8) **86**, 58
Cyclohexylmethyl N-tosyloxycarbamate **86**, 58
Cyclohexylmethyl *N*-tosyloxycarbamate: Benzenesulfonic acid, 4-methyl-,
    [(cyclohexylmethoxy)carbonyl]azanyl ester; (869111-41-1) **86**, 58
Cyclopropanation **85**, 172
Cyclopropanecarboxylic acid, 2-bromo-2-methyl-, ethyl ester; (89892-99-9) **85**, 172
Cyclopropanecarboxylic acid, 2-methylene-, ethyl ester; (18941-94-1) **85**, 172

(R,R)-deguPHOS: Pyrrolidine, 3,4-bis(diphenylphosphino)-1-(phenylmethyl)-, *(3R,4R)*-;
    (99135-95-2) **86**, 360
Dehydration, **85**, 34, 72
Dendrimer **86**, 151
Diallyl pimelate: Pimelic acid, diallyl ester; (91906-66-0) **86**, 194

3,3'-Diaminodipropylamine; (56-18-8) **86**, 141
(E)-2,3-Dibromobut-2-enoic acid: (2-Butenoic acid, 2,3-dibromo-, (2E)- (9); (24557-17-
    3) **85**, 231

2,5-Dibromo-1,1-dimethyl-3,4-diphenylsilole: Silacyclopenta-2,4-diene, 2,5-dibromo-1,1-dimethyl-3,4-diphenyl-; (686290-22-2) **85**, 53

1-(2,2-Dibromoethenyl)-2-nitrobenzene **86**, 36

Di(μ-bromo)bis(η-allyl)nickel(II): [allylnickel bromide dimer]; (12012-90-7) **85**, 248

2-(2,2-Dibromo-vinyl)-phenylamine **86**, 36

Di-tert-butyldiaziridinone: 3-Diaziridinone, 1,2-bis(1,1-dimethylethyl)-; (19656-74-7) **86**, 315

Di-*tert*-butyl dicarbonate: Dicarbonic acid, C,C'-bis(1,1-dimethylethyl) ester; (24424-99-5) **85**, 72, 219; **86**, 113, 374

Di(tert-butyl) (2S)-4,6-dioxo-1,2-piperidinedicarboxylate; (653589-10-7) **85**, 147

Di(tert-butyl) (2S,4S)-4-hydroxy-6-oxo-1,2-piperidinedicarboxylate; (653589-16-3) **85**, 147

trans-1,3-Di-*tert*-butyl-4-phenyl-5-vinyl-imidazolidin-2-one: 2-Imidazolidinone, 1,3-bis(1,1-dimethylethyl)-4-ethenyl-5-phenyl-, (4R,5R)-rel-; (927902-91-8) **86**, 315

2-(Di-*tert*-butylphosphino)biphenyl: Phosphine, [1,1'-biphenyl]-2-ylbis(1,1-dimethylethyl)-; (224311-51-7) **86**, 274

Di-tert-butylurea: Urea, N,N'-bis(1,1-dimethylethyl)-; (5336-24-3) **86**, 315

Dichloroacetyl chloride (79-36-7) **85**, 138

1,3-Dichloro-1,3-bis(dimethylamino)propenium chloride; (34057-61-9) **86**, 298

Dichlorobis(triphenylphosphine)palladium(II): Bis(triphenylphosphine)palladium(II) Dichloride; (13965-03-2) **86**, 225

(S)-(−)-5,5'-Dichloro-6,6'-dimethoxy-2,2'-bis(diphenylphosphino)-1,1'-biphenyl: Phosphine, [(1S)-5,5'-dichloro-6,6'-dimethoxy[1,1'-biphenyl]-2,2'-diyl]bis[diphenyl-; (185913-98-8) **86**, 47

Dichlorodimethylsilane; (75-78-5) **85**, 53

Dichloromethylen-dimethyliminium chloride; (33842-02-3) **86**, 298

Dicyclohexylmethylamine: Cyclohexanamine, *N*-cyclohexyl-*N*-methyl-; (7560-83-0) **85**, 118

Diene formation, **85**, 1

2-[(Diethylamino)methyl]benzene thiolato-copper(I) **85**, 209

(±)-Diethyl (*E,E,E*)-cyclopropane-1,2-acrylate, **85**, 15

Diethyl trans-1,2-cyclopropanedicarboxylate; (3999-55-1) **85**, 15

Diethyl(2-[(trimethylsilanyl)sulfanyl]benzyl)amine **85**, 209

3,5-Diiodosalicylaldehyde: Benzaldehyde, 2-hydroxy-3,5-diiodo-; (2631-77-8) **86**, 121

(S)-(−)-2-(N-3,5-Diiodosalicyliden)amino-3,3-dimethyl-1-butanol [(S)-1] **86**, 121

(S)-(−)-2-(N-3,5-Diiodosalicyliden)amino-3,3-dimethyl-1-butanol; (477339-39-2) **86**, 121

Diisopropylamine: 2-Propanamine, *N*-(1-methylethyl)-; (108-18-4) **86**, 47, 262

Diisopropylethylamine: 2-Propanamine, *N*-ethyl-*N*-(1-methylethyl)-; (7087-68-5) **85**, 64, 158, 278; **86**, 81

Dimedone: 1,3-cyclohexanedione, 5,5-dimethyl-; (126-81-8) **86**, 252

(4-(Dimethoxymethyl)phenoxy)(tert-butyl)dimethylsilane **86**, 130

[(E)-3,3-Dimethoxy-2-methyl-1-propenyl]benzene: Benzene, [(1E)-3,3-dimethoxy-2-methyl-1-propen-1-yl]-: (137032-32-7) **86**, 81

(2R,3S,4E)-N,3-Dimethoxy-N,2,4-trimethyl-5-phenyl-4-pentenamide

(3,5-Dimethoxy-1-phenyl-cyclohexa-2,5-dienyl)-acetonitrile **86**, 1

Dimethyl acetamide; (127-19-5) **86**, 298

Dimethylamine; (124-40-3) **86**, 298

N-(3-Dimethylaminopropyl)-*N*'-ethylcarbodiimide hydrochoride (EDC·HCl); (25952-53-8) **85**, 147

4-Dimethylaminopyridine: 4-Pyridinamine, *N,N*-dimethyl-; (1122-58-3) **85**, 64; **86**, 81

396

Formaldehyde solution; (50-00-0) **85**, 287
Formalin; (50-00-0) **86**, 344
Formic acid; (64-18-6) **86**, 344
4-Fluorobenzaldehyde; (459-57-4) **86**, 92
[2-(4-Fluorophenyl-1H-indol-4-yl]-1-pyrrolidinylmethanone **86**, 92
2-[trans-2-(4-Fluorophenyl)vinyl]-3-nitrobenzoic Acid (917614-64-3) **86**, 92
{2-[trans-2-(4-Fluorophenyl)vinyl]-3-nitrophenyl}-1-pyrrolidinylmethanone (917614-83-
6) **86**, 92

G1-[N(CH₂CH₂CH₂NHBoc)₂]₆-Cl₃ **86**, 151
G1-[N(CH₂CH₂CH₂NHBoc)₂]₆-Piperidine₃ **86**, 151

Halogenation, **85**, 53, 231
Heterocycles **85**, 10, 27, 34, 53, 64, 72, 88; **86**, 18, 59, 70, 81, 92, 105, 130, 141, 151,
181, 236, 262, 315, 333
Heterocyclic carbene, **85**, 34
Hydrazine hydrate; (10217-52-4) **86**, 105
Hydrogen peroxide; (7722-84-1) **85**, 158, 295
4-Hydroxybenzaldehyde: Benzaldehyde, 4-hydroxy-; (123-08-0) **86**, 130
3-Hydroxybutan-2-one; (513-86-0) **85**, 34
Hydroxylamine hydrochloride; (5470-11-1) **86**, 18, 58, 130
Hypophosphorous acid: Phosphinic acid; (6303-21-5) **85**, 96

Imidazole; (288-32-4) **86**, 58
Imides **86**, 70, 81
Imines **86**, 11, 121, 212
Iminodiacetic acid; (142-73-4) **86**, 344
Indium bromide: (13465-09-3) **85**, 118
Indium(III) tris(trifluoromethanesulfonate): Methanesulfonic acid, trifluoro-, indium(3+)
salt; (128008-30-0) **86**, 325
Iodine (7553-56-2) **85**, 219, 248; **86**, 1, 308
Iodomethane: Methane, iodo-; (74-88-4) **86**, 194
2-Isobutylthiazole; (18640-74-9) **86**, 105
iPrMgCl·LiCl: isopropylmagnesium chloride lithium chloride complex; (807329-97-1)
86, 374
(4*S*)-4-Isopropyl-5,5-diphenyloxazolidin-2-one (DIOZ): 2-Oxazolidinone, 4-(1-
methylethyl)-5,5-diphenyl-, (4*S*)-; (184346-45-0) **85**, 295
(4S)-4-Isopropyl-5,5-diphenyl-3-(3-phenyl-propionyl)oxazolidin-2-one: 2-Oxazolidinone,
4-(1-methylethyl)-3-(1-oxo-3-phenylpropyl)-5,5-diphenyl-, (4S)-; (213887-81-1)
85, 295
(4*S*)-Isopropyl-2-oxazolidinone: (4*S*)-4-(1-Methylethyl)-2-oxazolidinone; (17016-83-0)
85, 158
O-Isopropyl S-3-oxobutan-2-yl dithiocarbonate: carbonodithioic acid, O-(1-methylethyl)
S-(1-methyl-2-oxopropyl) ester; (958649-73-5) **86**, 333
(S)-4-Isopropyl-N-propanoyl-1,3-thiazolidine-2-thione **86**, 70

(*S*)-4-Isopropyl-*N*-propanoyl-1,3-thiazolidine-2-thione: 2-Thiazolidinethione, 4-(1-methylethyl)-3-(1-oxopropyl)-, (4*S*)-; (102831-92-5) **86**, 70

(4S)-Isopropyl-3-propionyl-2-oxazolidinone: (4S)-4-(1-Methylethyl)-3-(1-oxopropyl)-2-oxazolidinone; (77877-19-1) **85**, 158

(S)-4-Isopropyl-1,3-thiazolidine-2-thione **86**, 70

(*S*)-4-Isopropyl-1,3-thiazolidine-2-thione: 2-Thiazolidinethione, 4-(1-methylethyl)-, (4*S*); (76186-04-4) **86**, 70

(L)-*tert*-Leucine: L-Valine, 3-methyl-; (20859-02-3) **86**, 181

(L)-(+)-*tert*-leucinol: 1-Butanol, 2-amino-3,3-dimethyl-, (2*S*)-; (112245-13-3) **86**, 121

Lithium (7439-93-2) **85**, 53; **86**, 1

Lithium aluminum hydride; (16853-85-3) **85**, 158

Lithium chloride; (7447-41-8) **86**, 47

Lithium hydroxide monohydrate; (1310-66-3) **85**, 295

Lithium triethylborohydride; (22560-16-3) **85**, 64

(*R*)-(-)-Mandelic acid; (611-71-2) **85**, 106

(*S*)-(+)-Mandelic acid; (17199-29-0) **85**, 106

(R)-Mandelic acid salt of (1S,2S)-trans-2-(N-benzyl)amino-1-cyclohexanol; (882409-00-9) **85**, 106

(S)-Mandelic acid salt of (1R,2R)-trans-2-(N-benzyl)amino-1-cyclohexanol; (882409-01-0) **85**, 106

Manganese(IV) oxide; (1313-13-9)

Mesitylamine: Benzenamine, 2,4,6-trimethyl-; (88-05-1) **85**, 34

3-(Mesitylamino)butan-2-one: 2-Butanone, 3-[(2,4,6-trimethylphenyl)-amino]-; (898552-96-0) **85**, 34

Mesitylene (108-67-8) **85**, 196

N-Mesityl-N-(3-oxobutan-2-yl)formamide: Formamide, N-(1-methyl-2-oxopropyl)-N-(2,4,6-trimethylphenyl)-; (898553-01-0) **85**, 34

Metallation, **85**, 1, 45, 209; **86**, 374

Methansulfonyl chloride; (124-63-0) **85**, 219; **86**, 181

(4*S*)-3-[(2*R*,3*S*,4*E*)-3-Methoxy-2,4-dimethyl-1-oxo-5-phenyl-4-pentenyl]-4-(1-methylethyl)-2-thiazolidinethione; (332902-42-8) **86**, 81, 181

Methyl acetoacetate: Butanoic acid, 3-oxo-, methyl ester; (105-45-3) **86**, 161

Methyl 2-(2-acetylphenyl)acetate: Benzeneacetic acid, 2-acetyl-, methyl ester; (16535-88-9) **86**, 161

(R)-Methyl 2-allylpyrrolidine-2-carboxylate hydrochloride (112348-46-6) **86**, 262

2-Methyl-3-butyn-2-ol; (115-19-5) **85**, 118

(E)-3,4-Methylenedioxy-β-nitrostyrene; (22568-48-5) **85**, 179

Methylhydrazine; (60-34-4) **85**, 179

N-Methyliminodiacetic acid; (4408-64-4) **86**, 344

Methyl 2-methyl-3-nitrobenzoate; (59382-59-1) **86**, 92

4-Methyl-2'-nitrobiphenyl; (70680-21-6) **85**, 196

trans-p-Methyl-β-nitrostyrene: Benzene, 1-methyl-4-[(1*E*)-2-nitroethenyl]-; (5153-68-4) **85**, 179

*(1S,2S)-2-Methyl-3-oxo-1-phenylpropylcarbamate***86**, 11

(R)-3-Methyl-3-phenylpentene: [(1R)-1-ethyl-1-methyl-2-propenyl]-benzene]; (768392-48-9) **85**, 248

(*S*)-(−)-4-Methyl-1-phenyl-2-pentyn-1,4-diol: (321855-44-1) **85**, 118
(*E*)-2-Methyl-3-phenylpropenal; (15174-47-7) **86**, 81
Methyl propionylacetate: Pentanoic acid, 3-oxo-, methyl ester; (30414-53-0) **85**, 27
1-Methyl-2-pyrrolidone; (872-50-4) **85**, 196
1-Methyl-2-pyrrolidinone (872-50-4) **85**, 238
Methyltriphenylphosphonium bromide; (1779-49-3) **85**, 248
Microwave **86**, 18, 252

Naphthalene; (91-20-3) **85**, 53
Nitriles **86**, 1,28
2-Nitrobenzaldehyde; (552-89-6) **85**, 27; **86**, 36
2-Nitrobenzoic acid; (552-16-9) **85**, 196

Organocatalysis **86**, 11
Oxalyl chloride: HIGHLY TOXIC; Ethanedioyl dichloride; (79-37-8) **85**, 189
Oxidation, **85**, 15, 189, 267, 278; **86**, 1, 28, 121, 308. 315
Oximes **86**, 18
Oxone® monopersulfate; (37222-66-5) **85**, 278

Palladium **86**,105
Palladium(II) acetate; (3375-31-3) **85**, 96; **86**, 92, 105, 344
Palladium acetylacetonate; (140024-61-4) **85**, 196
Palladium tris(dibenzylideneacetone); (48243-18-1) **86**, 274
1,10-Phenanthroline; (66-71-7) **85**, 196 ; **86**, 92
(1,10-Phenanthroline)bis(triphenylphosphine)copper(I) nitrate; (33989-10-5) **85**, 196
Phenylacetylene: Benzene, ethynyl-; (536-74-3) **85**, 53, 118, 131; **86**, 325
(*S*)-Phenylalaninol: (3182-95-4) **85**, 267
N-Phenylbenzenecarboxamide (benzanilide); (93-98-1) **85**, 88
(*E*)-1-Phenyl-1,3-butadiene: Benzene, (1*E*)-1,3-butadienyl-: (16939-57-4) **86**, 315
2-Phenyl-1-butene; (2039-93-2) **85**, 248
5-Phenyl-1,3-dimethoxybenzene **86**, 1
(*S*)-Phenylglycine: Benzeneacetic acid, α-amino-, (α*S*)-; (2935-35-5) **85**, 219
5-Phenyl-2-isobutylthiazole; (600732-10-3) **86**, 105
3-Phenylpropanoyl chloride: Benzenepropanoyl chloride; (645-45-4) **85**, 295
5-Phenyl-1H-pyrazol-3-amine: 1*H*-pyrazol-3-amine, 5-phenyl-; (1571-10-7) **86**, 252
N-[1-Phenyl-3-(trimethylsilyl)-2-propyn-1-ylidene]-benzeneamine; (77123-64-9) **85**, 88
Phosphorus trichloride (7719-12-2) **85**, 238
Pimelic acid: Heptanedioic acid; (111-16-0) **86**, 194
Piperidinium acetate; (4540-33-4) **86**, 28
Potassium *tert*-butoxide: 2-Propanol, 2-methyl-, potassium salt (1:1); (865-47-4) **86**, 315
Potassium carbonate; (584-08-7) **85**, 287; **86**, 58
Potassium hydroxide; (1310-58-3) **85**, 196
Potassium *O*-isopropylxanthate: Carbonodithioic acid, O-(1-methylethyl) ester,
      potassium salt (1:1); (140-92-1) **86**, 333
Potassium trimethylsilanolate; (10519-96-7) **86**, 274
Propanol, 2-amino-, 3-phenyl, (*S*); (3182-95-4) **85**, 267
Propionyl chloride : Propanoyl chloride; (79-03-8) **85**, 158; **86**, 70

(*S*)-Proline; (147-85-3) **86**, 262

2-Propenoic acid, 3,3'-(1,2-cyclopropanediyl)bis-, diethyl ester, [1α(E),2β(E)]-; (58273-88-4) **85**, 15

1*H*-Pyrazole, 5-(1,3-benzodioxol-5-yl)-3-(4-chlorophenyl)-1-methyl-; (908329-89-5) **85**, 179

(4S)-5-Pyrimidinecarboxylic acid, 1,2,3,4-tetrahydro-6-methyl-2-oxo-4-phenyl-1-(phenylmethyl)-, methyl ester; (865086-56-2) **86**, 236

1-Pyrrolidinecarboxylic acid, 2-(aminocarbonyl)-, phenylmethyl ester, (2*S*)-; (34079-31-7) **85**, 72

1,2-Pyrrolidinedicarboxylic acid, 1-(phenylmethyl) ester, (2*S*)-; (1148-11-4) **85**, 72

(S)-5-Pyrrolidin-2-yl-1H-tetrazole: 2H-Tetrazole, 5-[(2S)-2-pyrrolidinyl]-; (33878-70-5) **85**, 72

Quinoline; (91-22-5) **85**, 196

Rearrangements **86**, 18, 113, 172

Reduction, **85**, 15, 27, 53, 64, 72, 138, 147, 158, 219, 248; **86**, 1, 28, 36, 92, 181, 344

Resolution, **85**, 106

Rhodium (II) tetrakis(triphenylacetate);(142214-04-8) **86**, 58

Rhodium **86**, 59

Rhodium acetate dimer: Rhodium, tetrakis[μ-(acetato-κO:κO')]di-, (Rh-Rh); (15956-28-2) **85**, 172; **86**, 58

[RhCl(cod)]₂; (chloro(1,5-cyclooctadiene)rhodium(I) dimer) (12092-47-6) **86**, 360

Ring Expansion, **85**, 138

Saponification, **85**, 27

Semicarbazide hydrochloride: Hydrazinecarboxamide, hydrochloride (1:1); (563-41-7) **86**, 194

Silver(I) fluoride; (7775-41-9) **86**, 225

Silver nitrate; (7761-88-8) **86**, 225

Sodium; (7440-23-5) **86**, 262

Sodium acetate; (127-09-3) **86**, 105, 194

Sodium acetate trihydrate; (6131-90-4) **85**, 10

Sodium amide; (7782-92-5) **86**, 298

Sodium azide; (26628-22-8) **85**, 72, 278; **86**, 113

Sodium borohydride: Borate(1-), tetrahydro-, sodium (1:1); (16940-66-2) **85**, 147; **86**, 181

Sodium carbonate; (497-19-8) **86**, 36

Sodium cyanide; (143-33-9) **85**, 219

Sodium hydride; (7646-69-7) **85**, 45, 172; **86**, 18, 194

Sodium tetrafluoroborate; (13755-29-8) **85**, 248

Sodium tetrakis[(3,5-trifluoromethyl)phenyl]borate; (79060-88-1) **85**, 248

Sodium thiosulfate (7772-98-7) **86**, 1

4-Spirocyclohexyloxazolidinone **86**, 58

4-Spirocyclohexyloxazolidinone: 3-Oxa-1-azaspiro[4.5]decan-2-one: (81467-34-7) **86**,58

Substitution **86**, 18, 141, 151, 181, 333

Sulfonation **86**, 11

Sulfur; (7704-34-9) **85**, 209
Super-Hydride®: Lithium triethylborohydride; (22560-16-3) **85**, 64

1-(Trimethylsilyl)acetylene; (1066-54-2) **85**, 88
Trimethylsilyl chloride: Silane, chlorotrimethyl-; (75-77-4) **85**, 209
1-Trimethylsilyloxybicyclo[3.2.0]heptan-6-one (125302-44-5) **85**, 138
1-Trimethylsilyloxycyclopentene (19980-43-9) **85**, 138
1-Trimethylsilyloxy-7,7-dichlorobicyclo[3.2.0]heptan-6-one (66324-01-4) **85**, 138
2-(Trimethylsilyl)phenyl trifluoromethanesulfonate: Methanesulfonic acid, 1,1,1-
Triphenylacetic acid: Benzeneacetic acid, α,α-diphenyl-; (595-91-5) **86**, 58
Triphenylphospine oxide; (791-28-6) **85**, 248; **86**, 274
Triphosgene: Methanol, 1,1,1-trichloro-, 1,1'-carbonate; (32315-10-9) **86**, 315
Tris(dibenzylideneacetone) dipalladium(0): Palladium, tris[μ-[(1,2-η:4,5-η)-(1*E*,4*E*)-1,5-
    diphenyl-1,4-pentadien-3-one]]di-; (51364-51-3) **86**, 194
Tris(dibenzylideneacetone)-dipalladium(0) chloroform adduct, (52522-40-4) **86**, 47
1,3,5-[Tris-piperazine]-triazine: 1,3,5-Triazine, 2,4,6-tri-1-piperazinyl-; (19142-26-8) **86**,
    141, 151

(*S*)-Valinol: 1-Butanol, 2-amino-3-methyl-, (2*S*)-; (2026-48-4) **86**, 70
Vanadyl acetylacetonate; (3153-26-2) **86**, 121
Vandium **86**, 121
4-Vinylbenzophenone: Methanone, (4-ethenylphenyl)phenyl-; (3139-85-3) **86**, 274
3-Vinylquinoline: Quinoline, 3-ethenyl-; (67752-31-2) **86**, 274

Zinc (II) chloride; (7646-85-7) **85**, 27, 53
Zinc triflate; (54010-75-2) **86**, 113